Removal of Cyanobacteria and Cyanotoxins in Waters

Removal of Cyanobacteria and Cyanotoxins in Waters

Editors

Albert Serrà
Elvira Gómez
Laëtitia V.S. Philippe

MDPI • Basel • Beijing • Wuhan • Barcelona • Belgrade • Manchester • Tokyo • Cluj • Tianjin

Editors

Albert Serrà
Department of Materials Science and Physical Chemistry
University of Barcelona
Barcelona (Catalonia)
Spain

Elvira Gómez
Department of Materials Science and Physical Chemistry
University of Barcelona
Barcelona
Spain

Laëtitia V.S. Philippe
Laboratory of Mechanics of Materials and Nanostructures
Swiss Federal Laboratories of Materials Science and Technology
Thun
Switzerland

Editorial Office
MDPI
St. Alban-Anlage 66
4052 Basel, Switzerland

This is a reprint of articles from the Special Issue published online in the open access journal *Toxins* (ISSN 2072-6651) (available at: www.mdpi.com/journal/toxins/special_issues/Removal_Cyanobacteria_Cyanotoxins).

For citation purposes, cite each article independently as indicated on the article page online and as indicated below:

LastName, A.A.; LastName, B.B.; LastName, C.C. Article Title. *Journal Name* **Year**, *Volume Number*, Page Range.

ISBN 978-3-0365-2108-4 (Hbk)
ISBN 978-3-0365-2107-7 (PDF)

Contents

About the Editors vii

Albert Serrà, Laetitia Philippe and Elvira Gómez
Removal of Cyanobacteria and Cyanotoxins in Waters
Reprinted from: *Toxins* **2021**, *13*, 636, doi:10.3390/toxins13090636 1

Renan Silva Arruda, Natália Pessoa Noyma, Leonardo de Magalhães, Marcella Coelho Berjante Mesquita, Éryka Costa de Almeida, Ernani Pinto, Miquel Lürling and Marcelo Manzi Marinho
'Floc and Sink' Technique Removes Cyanobacteria and Microcystins from Tropical Reservoir Water
Reprinted from: *Toxins* **2021**, *13*, 405, doi:10.3390/toxins13060405 5

Maíra Mucci, Iame A. Guedes, Elisabeth J. Faassen and Miquel Lürling
Chitosan as a Coagulant to Remove Cyanobacteria Can Cause Microcystin Release
Reprinted from: *Toxins* **2020**, *12*, 711, doi:10.3390/toxins12110711 23

Miquel Lürling, Maíra Mucci and Guido Waajen
Removal of Positively Buoyant *Planktothrix rubescens* in Lake Restoration
Reprinted from: *Toxins* **2020**, *12*, 700, doi:10.3390/toxins12110700 43

Kanarat Pinkanjananavee, Swee J. Teh, Tomofumi Kurobe, Chelsea H. Lam, Franklin Tran and Thomas M. Young
Potential Impacts on Treated Water Quality of Recycling Dewatered Sludge Supernatant during Harmful Cyanobacterial Blooms
Reprinted from: *Toxins* **2021**, *13*, 99, doi:10.3390/toxins13020099 59

Farhad Jalili, Hana Trigui, Juan Francisco Guerra Maldonado, Sarah Dorner, Arash Zamyadi, B. Jesse Shapiro, Yves Terrat, Nathalie Fortin, Sébastien Sauvé and Michèle Prévost
Can Cyanobacterial Diversity in the Source Predict the Diversity in Sludge and the Risk of Toxin Release in a Drinking Water Treatment Plant?
Reprinted from: *Toxins* **2021**, *13*, 25, doi:10.3390/toxins13010025 73

Matheus Almeida Ferreira, Cristina Celia Silveira Brandão and Yovanka Pérez Ginoris
Oxidation of Cylindrospermopsin by Fenton Process: A Bench-Scale Study of the Effects of Dose and Ratio of H_2O_2 and Fe(II) and Kinetics
Reprinted from: *Toxins* **2021**, *13*, 604, doi:10.3390/toxins13090604 93

Majdi Benamara, Elvira Gómez, Ramzi Dhahri and Albert Serrà
Enhanced Photocatalytic Removal of Cyanotoxins by Al-Doped ZnO Nanoparticles with Visible-LED Irradiation
Reprinted from: *Toxins* **2021**, *13*, 66, doi:10.3390/toxins13010066 107

Sabrina Sorlini, Carlo Collivignarelli, Marco Carnevale Miino, Francesca Maria Caccamo and Maria Cristina Collivignarelli
Kinetics of Microcystin-LR Removal in a Real Lake Water by UV/H_2O_2 Treatment and Analysis of Specific Energy Consumption
Reprinted from: *Toxins* **2020**, *12*, 810, doi:10.3390/toxins12120810 121

Mark W. Lusty and Christopher J. Gobler
The Efficacy of Hydrogen Peroxide in Mitigating Cyanobacterial Blooms and Altering Microbial Communities across Four Lakes in NY, USA
Reprinted from: *Toxins* **2020**, *12*, 428, doi:10.3390/toxins12070428 . . . . **139**

Saber Moradinejad, Hana Trigui, Juan Francisco Guerra Maldonado, Jesse Shapiro, Yves Terrat, Arash Zamyadi, Sarah Dorner and Michèle Prévost
Diversity Assessment of Toxic Cyanobacterial Blooms during Oxidation
Reprinted from: *Toxins* **2020**, *12*, 728, doi:10.3390/toxins12110728 . . . . **159**

Katherine E. Greenstein, Arash Zamyadi, Caitlin M. Glover, Craig Adams, Erik Rosenfeldt and Eric C. Wert
Delayed Release of Intracellular Microcystin Following Partial Oxidation of Cultured and Naturally Occurring Cyanobacteria
Reprinted from: *Toxins* **2020**, *12*, 335, doi:10.3390/toxins12050335 . . . . **181**

Soukaina El Amrani Zerrifi, Fatima El Khalloufi, Richard Mugani, Redouane El Mahdi, Ayoub Kasrati, Bouchra Soulaimani, Lillian Barros, Isabel C. F. R. Ferreira, Joana S. Amaral, Tiane Cristine Finimundy, Abdelaziz Abbad, Brahim Oudra, Alexandre Campos and Vitor Vasconcelos
Seaweed Essential Oils as a New Source of Bioactive Compounds for Cyanobacteria Growth Control: Innovative Ecological Biocontrol Approach
Reprinted from: *Toxins* **2020**, *12*, 527, doi:10.3390/toxins12080527 . . . . **197**

Maranda Esterhuizen and Stephan Pflugmacher
Large-Scale Green Liver System for Sustainable Purification of Aquacultural Wastewater: Construction and Case Study in a Semiarid Area of Brazil (Itacuruba, Pernambuco) Using the Naturally Occurring Cyanotoxin Microcystin as Efficiency Indicator
Reprinted from: *Toxins* **2020**, *12*, 688, doi:10.3390/toxins12110688 . . . . **217**

About the Editors

Albert Serrà

Dr. Serrà is a lecturer in Materials Science and Physical Chemistry at the Institute of Nanoscience and Nanotechnology at the University of Barcelona (IN^{2}UB). He received a Ph.D. degree (cum laude and honors) in the field of electrodeposition from the University of Barcelona (2016). He is interested on the development of new micro- and nanomaterials for water remediation, hydrogen production, and biomedicine. His professional career has been developed primarily at the University of Barcelona, the University of Newcastle, and the Swiss Federal Laboratories of Materials Science and Technology. He received the Jóvenes Talentos award from the electrochemistry group CIC energiGUNE, given to the best researcher under 35 years of age in the field of electrochemistry. He also received a MASCA postdoctoral fellowship and has published around 47 peer-reviewed papers (h-index = 18).

Elvira Gómez

Elvira Gómez is a full Professor of the Department of Materials Science and Physical Chemistry and of the Institute of Nanoscience and Nanotechnology of the University of Barcelona. She is the group leader of the thin layer electroplating and nanostructures laboratory. She received her doctorate from the University of Barcelona (1983). Her main research focuses on the electrodeposition field, and throughout her careeer she has explored a wide spectrum of solutions, from aqueous to ionic liquids. She uses electrochemical means to create nanostructured materials, compact layers and well-ordered microdevices with exquisitely defined geometry, controlled surface chemistry and adjustable physical, mechanical and/or magnetic properties. Her research is mainly applied to problems related to mechanical devices, renewable energy, magnetism, biomedicine and water decontamination, without forgetting the basic study of the first stages of deposition processes. She has published 173 peer-reviewed articles (index h = 37) and book chapters.

Laëtitia V.S. Philippe

Dr. Philippe is currently responsible for the R&D process sector at ROLEX. Laetitia Philippe was group leader in Electrochemistry in the Laboratory of Mechanics of Materials and Nanostructures at Swiss Federal Laboratories of Materials Science and Technology (EMPA) until 2020. She received a PhD degree in Physical chemistry at the University of Manchester (ex UMIST), UK in 2002 and she undertook a postdoctoral position at TU-Delft afterwards. She has been a lecturer at EPFL since 2012. In her research, she uses electrochemical means to create well-ordered nanostructured materials, compact layers and microdevices, featuring exquisitely defined geometry, controlled surface chemistry, and tunable physical/mechanical properties. She is the author or coauthor of around 100 publications, has organized several international conferences, and is co-founder of one start-up.

toxins

MDPI

Editorial

Removal of Cyanobacteria and Cyanotoxins in Waters

Albert Serrà [1,2,*], Laetitia Philippe [3] and Elvira Gómez [1,2,*]

1 Thin Films and Nanostructures Electrodeposition Group (GE-CPN), Department of Materials Science and Physical Chemistry, University of Barcelona, Martí i Franquès 1, E-08028 Barcelona, Catalonia, Spain
2 Institute of Nanoscience and Nanotechnology (IN2UB), Universitat de Barcelona, E-08028 Barcelona, Catalonia, Spain
3 Manufacture des Montres ROLEX SA, Research & Development, CH-2501 Bienne, Switzerland; laetitia.philippe@gmail.com
* Correspondence: a.serra@ub.edu (A.S.); e.gomez@ub.edu (E.G.)

Harmful cyanobacterial algal blooms and cyanotoxins currently pose a major threat to global society, one that exceeds local and national interests due to their extremely destructive effects on the environment and human health. In the near future, the formation of harmful cyanobacterial algal blooms and, in turn, cyanotoxins is expected to become widespread, driven by eutrophication and anthropogenic causes such as water pollution and promoted by escalating global temperatures. Such trends, studied since the late 1990s, have attracted increased interest due to (i) the high environmental impact of cyanobacterial blooms and cyanotoxins worldwide, (ii) the ineffective removal of those pollutants by conventional water treatment processes, (iii) the transformational capacity to completely destroy those organic toxins via alternative treatments such as advanced oxidation processes, and (iv) engineering challenges when transitioning toward the treatment of large volumes of water. The global context of this threat thus urges the innovation of simple, sustainable, low-cost strategies and technologies for water decontamination that can be readily implemented worldwide [1].

In this Special Issue, titled "Removal of Cyanobacteria and Cyanotoxins in Waters," we have attempted to provide readers with a comprehensive overview of different strategies and alternative treatments currently being studied for the effective removal of cyanobacteria and cyanotoxins from water. The following presents a brief synopsis of the 13 research papers that constitute this Special Issue.

Coagulation is a key process employed by conventional water treatment technologies. Worldwide, the most widely applied water treatment technology at water treatment plants involves a combination of coagulation, flocculation, sedimentation, filtration, and disinfection to treat water. Herein, five papers focus either directly or indirectly on coagulation and its alternatives or on the potential impacts of conventional water treatment technologies on drinking water containing cyanobacteria and/or cyanotoxins [2]. Arruda et al. [2] have demonstrated that the so-called "floc-and-skin" technique, based on the combination of a ballast (i.e., natural soil or modified clay) and a coagulant, can effectively remove cyanobacterial biomass comprised of *Dolichospermum circinalis* and *Microcystis aeruginosa*, depending on the ballast's capacity to adsorb microcystin. At the laboratory scale, the authors studied the effect of the floc-and-skin technique on the release of microcystins from cyanobacterial biomass in real water from the Funil Reservoir, a eutrophic system in southern Brazil. Their results demonstrate that the technique is more effective than using a sole coagulant (i.e., polyaluminum chloride) to remove cyanobacteria and extracellular microcystins from water [2]. Next, Mucci et al. [3] studied the effect of using chitosan as a coagulant to remove cyanobacteria from the water column. Although chitosan's coagulation efficiency has been extensively studied, little is known about its effect on the viability of cyanobacteria cells and the release of cyanotoxins. The authors' study confirmed that chitosan was able to damage *Microcystis aeruginosa* cells by inducing lysis and, consequently, the release of cyanotoxins,

check for updates

Citation: Serrà, A.; Philippe, L.; Gómez, E. Removal of Cyanobacteria and Cyanotoxins in Waters. *Toxins* **2021**, *13*, 636. https://doi.org/10.3390/toxins13090636

Received: 2 September 2021
Accepted: 8 September 2021
Published: 9 September 2021

which thus inhibit chitosan's use in lake restoration without directly testing its effects on the natural target biota because, as the authors show, the side effects depend on the strain [3]. Lürling et al. [4] also examined the efficiency of deploying the floc-and-skin technique to control the growth of cyanobacterial blooms using a combination of a coagulant (i.e., polyaluminum chloride) and a ballast (i.e., lanthanum modified bentonite) from Lake De Kuil and Lake Rauwbraken in the Netherlands. Above all, the authors demonstrate the necessity of testing the technique's process in the water to be treated, for the technique, when used with similar species of *Planktothrix rubescens,* yielded different results depending on the lake. After one day, the filaments of a *Planktothrix rubescens* biomass from Lake De Kuil resurfaced but remained precipitated in Lake Rauwbraken, possibly due to water matrix effects. By contrast, Pinkanjananavee et al. [5] studied the effectiveness of using conventional treatment processes consisting of coagulation, flocculation, sedimentation, filtration, sludge dewatering, and disinfection to remove *Microcystis aeruginosa* and microcystin-LR prior to disinfection. Noting that several conventional drinking water treatment plants recycle dewatered supernatant from sludge-dewatering operations, the authors aimed to determine the impact of recycling dewatered sludge supernatant on the quality of treated water in laboratory-scale experiments. Ultimately, their study demonstrated that recycling dewatered sludge supernatant can indeed affect the quality of treated water. The fifth paper, by Jalili et al. [6], describes the effects of various cyanobacteria in a treatment plant for drinking water before, during, and after the occurrence of cyanobacterial blooms by monitoring raw water, sludge in the holding tank, and the sludge supernatant. In view of their results, the authors conclude that predicting the cyanobacterial community in the sludge supernatant continues to be a challenge.

Three other papers address different advanced oxidation processes used to remove cyanotoxins [7–9]. Advanced oxidation processes involve a set of oxidative water treatments, including treatment with UV–O_3, UV–H_2O_2, the Fenton process, the photo-Fenton process, and nonthermal plasmas as well as sonolysis, photocatalysis, and radiolysis, all based on the production of powerful, highly reactive oxidants. Among those oxidants, hydroxyl radicals are the most commonly used and are the strongest known oxidants after fluorine, and, similar to other highly reactive oxidants, can nonselectively destroy the majority of organic matter and can mineralize organic pollutants in water. Against that background, Ferreira et al. [7] investigated the applicability of the Fenton process to mineralize cylindrosmpermopsin by analyzing how the dosage of Fenton reagents, namely H_2O_2 and Fe(II), and the H_2O_2-to-Fe(II) molar ratio affected the removal of cylindrosmpermopsin in ultrapure water (i.e., unrealistic conditions that do not consider water matrix effects or real pH) [7]. Although the Fenton process is a promising advanced oxidation technique that can be easily implemented at full scale worldwide due to its simplicity and highly cost-effective technology, the water matrix effect and large-scale volumes need to be investigated to evaluate the process for application in treating cyanotoxins because the presence of scavengers and competing species identified in real water matrices may affect degradation kinetics. Next, Benamara et al. [8] report that Al-doped ZnO nanoparticles are effective photocatalysts for degrading and mineralizing two typical cyanotoxins, microcystin-LR and anatoxin-A, under visible-light irradiation with light emitting diodes (LEDs). Central to that process is using easily fabricated Al-doped ZnO nanoparticles to act as photocatalysts for water decontamination. Their paper, describing the visible-light-driven removal of cyanotoxins in the presence of ZnO-based photocatalysts, shows that high removal efficiencies can be achieved under simple conditions. The authors also found that the system's outstanding performance derives from the excellent photocatalytic performance and the high chemical and photochemical stability of Al-doped ZnO nanoparticles under visible-light irradiation with LEDs [8]. In another study, Sorlini et al. [9] investigated the viability of using UV–H_2O_2 to treat microcystin-LR, particularly by analyzing specific energy consumption in the waters of Lake Iseo in the Province of Brescia, Lombardy, Italy. Those authors aimed to preliminarily study the effectiveness of UV–H_2O_2 on real water with respect to water matrix effects in order to clarify how the type of oxidant influences kinetics by analyzing the effects of the initial microcystin-LR concentration and the H_2O_2 dosage. The authors also examined the total specific energy consumption of UV–H_2O_2 compared to UV treatment to determine the optimal operational conditions [9]. Although the authors suggest the potential of using those three advanced oxidation processes to remove various cyanotoxins, unrealistic laboratory settings and conditions that allow the treatment of small volumes of water with highly energy-intensive lamps underscore some of the challenges of applying those technologies in the near future.

Because different chemical oxidants can be used to treat water, the effects of some of those oxidants are discussed in three additional papers. In one of those papers, because H_2O_2 is a highly investigated Fenton reagent used in different water treatment strategies, Lusty et al. [10] investigated the effects of H_2O_2 on cyanobacteria and microbial communities in order to analyze its use to mitigate potentially toxic freshwater cyanobacterial blooms. The study revealed that H_2O_2 can effectively reduce but not fully eliminate cyanobacteria from eutrophic bodies of water; thus, the authors conclude that H_2O_2 is not the best candidate for use in high biomass ecosystems containing harmful cyanobacterial algal blooms. By comparison, Moradinejad et al. [11] investigated variation and shifts in the structure of cyanobacterial communities during chemical oxidation (i.e., with Cl_2, $KMnO_4$, O_3, and H_2O_2) using the metagenomic shotgun approach, which allows the consideration of the diversity of cyanobacterial communities in pre-oxidant selection in the on-site management of cyanobacterial blooms. According to their results, only pre-oxidation with H_2O_2 exhibited a clear decrease in the abundance of cyanobacterial biomasses, and the authors add that the selection of the pre-oxidant translates into considerable reductions in cost. Beyond that, the paper by Greenstein et al. [12] provides guidance on how five oxidants (i.e., Cl_2, NH_2Cl, ClO_2, $KMnO_4$, and O_3) affect the delayed release of intracellular microcystins after the partial oxidation of cyanobacteria as a critical parameter to consider, especially for drinking water treatments, because the concertation of cyanotoxins in water can increase after several hours of treatment.

From another angle, El Amrani Zerrifi et al. [13] investigated the potential of employing natural compounds extracted from seaweeds to control harmful algae in aquatic ecosystems. The authors report the results of a series of anti-cyanobacterial assays with different extracts from seaweeds from Morocco, which demonstrate the anti-cyanobacterial properties of *Cystoseira tamariscifolia* and its potential use in developing environmentally friendly procedures to control the growth of toxic cyanobacteria.

Last, the paper by Esterhuizen [14] presents a systematic approach to planning, constructing, monitoring, and optimizing a phytoremediation system for the treatment of wastewater from aquaculture. In their study, the authors constructed a large-scale Green Liver System requiring minimal maintenance and low construction costs based on the use of macrophytes. Because aquacultural wastewaters are eutrophic, cyanobacterial blooms flourish in them and can often result in the presence of cyanobacterial toxins. However, the authors' study demonstrated that various microcystins were effectively removed in their bioremediation process.

Author Contributions: Writing—original draft preparation, A.S., L.P. and E.G.; writing—review and editing, A.S., L.P. and E.G. All authors have read and agreed to the published version of the manuscript.

Funding: This research received no external funding.

Institutional Review Board Statement: Not applicable.

Informed Consent Statement: Not applicable.

Acknowledgments: The gest editors of this Special Issue, A.S., L.P. and E.G. are grateful to all of the authors for their contributions and particularly to the expert peer reviewers for their rigorous work.

Conflicts of Interest: The authors declare no conflict of interest.

References

1. Serrà, A.; Philippe, L.; Perreault, F.; Garcia-Segura, S. Photocatalytic treatment of natural waters. Reality or hype? The case of cyanotoxins remediation. *Water Res.* **2021**, *188*, 116543. [CrossRef] [PubMed]
2. Arruda, R.S.; De Magalh, L.; Coelho, M.; Mesquita, B.; De Almeida, É.C.; Pinto, E.; Lürling, M.; Marinho, M.M. 'Floc and Sink' Technique Removes Cyanobacteria and Microcystins from Tropical Reservoir Water. *Toxins* **2021**, *13*, 405. [CrossRef] [PubMed]
3. Mucci, M.; Guedes, I.A.; Faassen, E.J.; Lürling, M. Chitosan as a Coagulant to Remove Cyanobacteria Can Cause Microcystin Release. *Toxins* **2020**, *12*, 711. [CrossRef] [PubMed]
4. Lürling, M.; Mucci, M.; Waajen, G. Removal of Positively Buoyant Planktothrix rubescens in Lake Restoration. *Toxins* **2020**, *12*, 700. [CrossRef] [PubMed]
5. Pinkanjananavee, K.; Teh, S.J.; Kurobe, T.; Lam, C.H.; Tran, F.; Young, T.M. Potential Impacts on Treated Water Quality of Recycling Dewatered Sludge Supernatant during Harmful Cyanobacterial Blooms. *Toxins* **2021**, *13*, 99. [CrossRef] [PubMed]
6. Jalili, F.; Trigui, H.; Guerra Maldonado, J.F.; Dorner, S.; Zamyadi, A.; Shapiro, B.J.; Terrat, Y.; Fortin, N.; Sauvé, S.; Prévost, M. Can Cyanobacterial Diversity in the Source Predict the Diversity in Sludge and the Risk of Toxin Release in a Drinking Water Treatment Plant? *Toxins* **2021**, *13*, 25. [CrossRef] [PubMed]
7. Almeida-Ferreira, M.; Silveira, C.C.; Pérez, Y. Oxidation of Cylindrosmpermopsin by Fenton Process: A Bench-Scale Study of the Effects of Dose and Ratio of H_2O_2 and Fe(II) and Kinetics. *Toxins* **2021**, *13*, 604. [CrossRef]
8. Benamara, M.; Gómez, E.; Dhahri, R.; Serrà, A. Enhanced Photocatalytic Removal of Cyanotoxins by Al-Doped ZnO Nanoparticles with Visible-LED Irradiation. *Toxins* **2021**, *13*, 66. [CrossRef] [PubMed]
9. Sorlini, S.; Collivignarelli, C.; Carnevale Miino, M.; Caccamo, F.M.; Collivignarelli, M.C. Kinetics of Microcystin-LR Removal in a Real Lake Water by UV/H_2O_2 Treatment and Analysis of Specific Energy Consumption. *Toxins* **2020**, *12*, 810. [CrossRef] [PubMed]
10. Lusty, M.W.; Gobler, C.J. The Efficacy of Hydrogen Peroxide in Mitigating Cyanobacterial Blooms and Altering Microbial Communities across Four Lakes in NY, USA. *Toxins* **2020**, *12*, 428. [CrossRef] [PubMed]
11. Moradinejad, S.; Trigui, H.; Guerra Maldonado, J.F.; Shapiro, J.; Terrat, Y.; Zamyadi, A.; Dorner, S.; Prévost, M. Diversity Assessment of Toxic Cyanobacterial Blooms during Oxidation. *Toxins* **2020**, *12*, 728. [CrossRef] [PubMed]
12. Greenstein, K.E.; Zamyadi, A.; Glover, C.M.; Adams, C.; Rosenfeldt, E.; Wert, E.C. Delayed Release of Intracellular Microcystin following Partial Oxidation of Cultured and Naturally Occurring Cyanobacteria. *Toxins* **2020**, *12*, 335. [CrossRef] [PubMed]
13. El Amrani Zerrifi, S.; El Khalloufi, F.; Mugani, R.; El Mahdi, R.; Kasrati, A.; Soulaimani, B.; Barros, L.; Ferreira, I.C.F.R.; Amaral, J.S.; Finimundy, T.C.; et al. Seaweed Essential Oils as a New Source of Bioactive Compounds for Cyanobacteria Growth Control: Innovative Ecological Biocontrol Approach. *Toxins* **2020**, *12*, 527. [CrossRef] [PubMed]
14. Esterhuizen, M.; Pflugmacher, S. Large-Scale Green Liver System for Sustainable Purification of Aquacultural Wastewater: Construction and Case Study in a Semiarid Area of Brazil (Itacuruba, Pernambuco) Using the Naturally Occurring Cyanotoxin Microcystin as Efficiency Indicator. *Toxins* **2020**, *12*, 688. [CrossRef] [PubMed]

toxins

MDPI

Article

'Floc and Sink' Technique Removes Cyanobacteria and Microcystins from Tropical Reservoir Water

Renan Silva Arruda [1,*], Natália Pessoa Noyma [1], Leonardo de Magalhães [1], Marcella Coelho Berjante Mesquita [1], Éryka Costa de Almeida [2], Ernani Pinto [2], Miquel Lürling [3,4] and Marcelo Manzi Marinho [1]

1 Laboratory of Ecology and Physiology of Phytoplankton, Department of Plant Biology, University of Rio de Janeiro State, Rua São Francisco Xavier 524—PHLC Sala 511a, Rio de Janeiro 20550-900, Brazil; np.noyma@gmail.com (N.P.N.); demagalhaesleonardo@gmail.com (L.d.M.); macobeme@gmail.com (M.C.B.M.); manzi.uerj@gmail.com (M.M.M.)
2 Department of Clinical and Toxicological Analyses, School of Pharmaceutical Sciences, University of São Paulo, São Paulo 05508-900, Brazil; erykaca@usp.br (É.C.d.A.); ernani@usp.br (E.P.)
3 Aquatic Ecology & Water Quality Management Group, Department of Environmental Sciences, Wageningen University, P.O. Box 47, 6700 AA Wageningen, The Netherlands; miquel.lurling@wur.nl
4 Department of Aquatic Ecology, Netherlands Institute of Ecology (NIOO-KNAW), P.O. Box 50, 6700 AB Wageningen, The Netherlands
* Correspondence: renan.arruda@ymail.com

Abstract: Combining coagulants with ballast (natural soil or modified clay) to remove cyanobacteria from the water column is a promising tool to mitigate nuisance blooms. Nevertheless, the possible effects of this technique on different toxin-producing cyanobacteria species have not been thoroughly investigated. This laboratory study evaluated the potential effects of the "Floc and Sink" technique on releasing microcystins (MC) from the precipitated biomass. A combined treatment of polyaluminium chloride (PAC) with lanthanum modified bentonite (LMB) and/or local red soil (LRS) was applied to the bloom material (mainly *Dolichospermum circinalis* and *Microcystis aeruginosa*) of a tropical reservoir. Intra and extracellular MC and biomass removal were evaluated. PAC alone was not efficient to remove the biomass, while PAC + LMB + LRS was the most efficient and removed 4.3–7.5 times more biomass than other treatments. Intracellular MC concentrations ranged between 12 and 2.180 $\mu g\ L^{-1}$ independent from the biomass. PAC treatment increased extracellular MC concentrations from 3.5 to 6 times. However, when combined with ballast, extracellular MC was up to 4.2 times lower in the top of the test tubes. Nevertheless, PAC + LRS and PAC + LMB + LRS treatments showed extracellular MC concentration eight times higher than controls in the bottom. Our results showed that Floc and Sink appears to be more promising in removing cyanobacteria and extracellular MC from the water column than a sole coagulant (PAC).

Keywords: toxic bloom; cyanobacteria mitigation; geo-engineering; *Dolichospermum*; *Microcystis*; eutrophication control

Key Contribution: The Floc and Sink technique can effectively settle the biomass comprised of *Dolichospermum circinalis* and *Microcystis aeruginosa* and showed potential microcystin adsorption depending on the ballast used.

Citation: Arruda, R.S.; Noyma, N.P.; de Magalhães, L.; Mesquita, M.C.B.; de Almeida, É.C.; Pinto, E.; Lürling, M.; Marinho, M.M. 'Floc and Sink' Technique Removes Cyanobacteria and Microcystins from Tropical Reservoir Water. *Toxins* **2021**, *13*, 405. https://doi.org/10.3390/toxins13060405

Received: 12 April 2021
Accepted: 31 May 2021
Published: 8 June 2021

1. Introduction

Cyanobacteria perform many vital functions to the health of ecosystems, especially as photosynthetic organisms (e.g., oxygen production and nitrogen-fixing); however, they have become an increasing issue worldwide, mainly due to the massive proliferation of toxin-producing species caused by anthropogenic eutrophication [1–3]. Cyanobacterial blooms represent a threat to human health and aquatic biota, mainly due to the potential contamination by toxins [1–3].

Cyanotoxins comprise many compounds with wide variations in the chemical structure, causing effects in exposed invertebrates and vertebrates. In freshwaters, the most common cyanotoxin is the hepatotoxic MC, which presents more than 300 known variants [4]. MCs are cyclic heptapeptides with a general structure (Figure S1) composed of D-alanine, a variable amino acid (X1), D-MeAsp (D-erythro-β-methylaspartic acid), another variable amino acid (X2), Adda ((2S,3S,4E,6E,8S,9S)-3-amino-9-methoxy-2,6,8-trimethyl-10-phenyldeca-4,6-dienoic acid), D-glutamic acid, and Mdha (*N*-methyldehydroalanine). MCs have been named according to the standard one-letter amino acid code applied to the variable amino acids X1 and X2 (present in the positions 2 and 4, respectively). For example, the one-letter codes A, R, L, F, Y, and W present in some MCs variants (e.g., MC-RR, MC-LR, MC-LA, MC-LF, MC-LW, MC-YR, Figure S2, Table S1) represent the amino acids alanine (A), arginine (R), leucine (L), phenylalanine (F), tyrosine (Y), and tryptophan (W). Other structural modifications are added to the name as a suffix, such as [D-Asp3] which represents a missing methyl group in the position 3 of MCs structure (e.g., [D-Asp3]MC-RR and [D-Asp3]MC-LR, Figure S2) [3]. Studies have reported that MCs were shown to be tumor promoters, immunotoxicants, and endocrine disruptors [5–7]. This toxin can inhibit protein phosphatases 1 and 2A, promote DNA fragmentation, necrosis, apoptosis, and intrahepatic bleeding, leading to death [1,8]. MC-producing cyanobacteria blooms have been described in 80 countries, and the genera *Microcystis* and *Planktothrix* are considered the most potent sources because they are often associated with high toxin levels [9]. However, the filamentous genus *Dolichospermum* (formerly *Anabaena)* contains species capable of producing MCs [9] and are considered as potential MC producers in risk assessments [3], mainly strains from high latitudes [10–12]. MC-producing strains have also been recorded at moderate latitude [13] and low latitude [14,15].

Although the most coherent way to mitigate eutrophication is to reduce the external nutrient input [2,16], it is also an onerous approach in developing countries, where the sewage system and treatment are inefficient, requiring expensive investments [17]. Furthermore, there is a need to control the internal nutrient load to accelerate the system's recovery [18].

The coagulation and precipitation of cyanobacteria biomass and phosphate (P) is a promising tool to manage eutrophication and its nuisance [19]. This technique combines a ballast and a coagulant (Floc and Sink) that moves intact cells and P out of the water column, bound to ballast, toward the sediment, and is proven safe [19–23]. However, the coagulation step is a critical part of this technique because the coagulant can cause physiological or chemical stress to cell membranes, releasing intracellular toxins and P into water [20,21,24]. Polyaluminium chloride (PAC) is widely used as an inorganic flocculant in drinking water supply plants [25,26], wastewater treatment [27], and in-lake restoration measures to mitigate cyanobacterial blooms [23,28–30]. Its mechanism of coagulation includes charge neutralization, sweep coagulation, and bridge aggregation [31]. Algal blooms coagulation can suffer interference from the release of algogenic organic matter (AOM) into the water, extracellularly (EOM), and intracellularly (IOM) when cell lysis occurs [32]. One mechanism AOM interferes in the coagulation efficiency is by forming complexes with cations [33]. More specifically, studies have shown that proteins from *Microcystis aeruginosa* interfere with coagulation by forming chelate complexes with the coagulant [34,35]. PAC also interacts with humic acid, reducing its effectiveness [36,37]. The reduction of humic acid is important in lake restoration measurements because of it interferes in phosphorus adsorption by lanthanum modified clay [38,39]. Although all these cited interferences can affect the flocculation efficiency, each water body is unique and has physical, chemical and biological characteristics. In each system, algal removal can be achieved by conducting experimental trials aiming to find the right flocculant and the best dosage [28,40,41]. In our study, PAC was chosen because it previously showed high efficiency to flocculate cyanobacterial blooms at the Funil reservoir [23,42]. In addition to PAC, LMB and LRS as a ballast present similar efficiency to settle the cyanobacteria with similar dosages [21,23,28]. Both compounds are solid-phase P sorbents (SPB), but also the removal of extracellular harmful algal toxins from the water has already been recorded for them [21,43]. Floc and Sink proved to be efficient in laboratory tests, cleaning the water column without indicating cell lysis [19]. Moreover, water from a tropical eutrophic reservoir could be cleared from cyanobacteria [23,42]. At the time of those studies, the samples comprised predominantly small spherical/flat colonial cyanobacteria (e.g., *Microcystis aeruginosa*, *Microcystis brasiliensis*, *Microcystis panniformis*). The majority of Floc and Sink studies have been performed with *Microcystis* [41,42,44]. Some studies, however, indicated that the efficiency of the Floc and Sink technique in cell removal varied among cyanobacteria species [21]. A recent study also mentioned that "the applicability of the technique to genera of cyanobacteria that have not yet been studied is unknown" [44]. One of the genera of which no information exists about how well it can be removed from water by the Floc and Sink technique is the filamentous genus *Dolichospermum* which recently became dominant in the phytoplankton community of a tropical eutrophic reservoir in the southeast region of Brazil (Funil Reservoir).

Since controlling eutrophication and mitigating cyanobacteria nuisance have been considered a crucial challenge to agencies and companies responsible for producing and distributing potable water [21,23], insight into the removal efficacy of filamentous *Dolichospermum* species is needed. The experiments executed here tested the hypothesis that the Floc and Sink technique efficiently removes the cyanobacterial biomass composed mainly of *Dolichospermum circinalis* with undergrowth of *Microcystis aeruginosa* in a deep tropical reservoir without MC releasing.

2. Results

2.1. Coagulant Range

In the experiment testing different doses of PAC, the cyanobacteria suspension presented no positive buoyancy, and after the incubation (2 h), most of the biomass was in the bottom of the tube, as shown in the control (Figure 1). The chlorophyll-*a* concentration in the top 5 mL of the tubes containing PAC varied from 1.6 to 1.9 times less than in the bottom 5 mL of the control tube (Figure 1). The best dosage was 3 mg L^{-1}, yielding the high biomass concentration in the bottom of the tube. The pH decreased gradually with the increasing dose of PAC concentrations, from 1 mg L^{-1} (Figure 1). Moreover, the Photosystem II efficiency (Φ_{PSII}) reduced progressively in the top and bottom 5 mL in the PAC range tubes from 1 mg L^{-1}.

2.2. Floc and Sink Assays

The cyanobacterial suspension in this experiment presented similar distribution in the water column. As indicated in the controls, the chlorophyll-*a* concentrations in the top of the cylinders were similar to the bottom, even after incubation (Figure 2). The addition of only PAC doubled the concentration of cyanobacterial biomass in both areas of the cylinders (Figure 2). This effect was strongly modified when PAC was combined with a ballast. Virtually all biomass was precipitated to the bottom of the cylinders with the addition of LMB and/or LRS (Figure 2). The chlorophyll-*a* concentration in the top of the cylinders in these treatments with ballast was 20 to 60 times lower than in the controls (Figure 2A), reaching a removal efficiency of 97 ± 1.8%, whereas in the bottom of the cylinders, there was 4.3 to 7.5 times more chlorophyll-*a* than in the control (Figure 2B). One-way ANOVA indicated a significant difference in chlorophyll-*a* concentrations among treatments for the top ($F_{4,14}$ = 28.275; $p < 0.001$) and the bottom ($F_{4,14}$ = 22.910; $p < 0.001$) of the cylinders. In the bottom of the cylinders, three homogeneous groups for chlorophyll-*a* concentrations were found: (i) the lowest concentration was observed in the control; (ii) similar to each other, but higher values than control were found for PAC, PAC + LMB, and PAC + LRS; (iii) the highest concentrations were detected in the PAC + LMB + LRS treatment (Figure 2B).

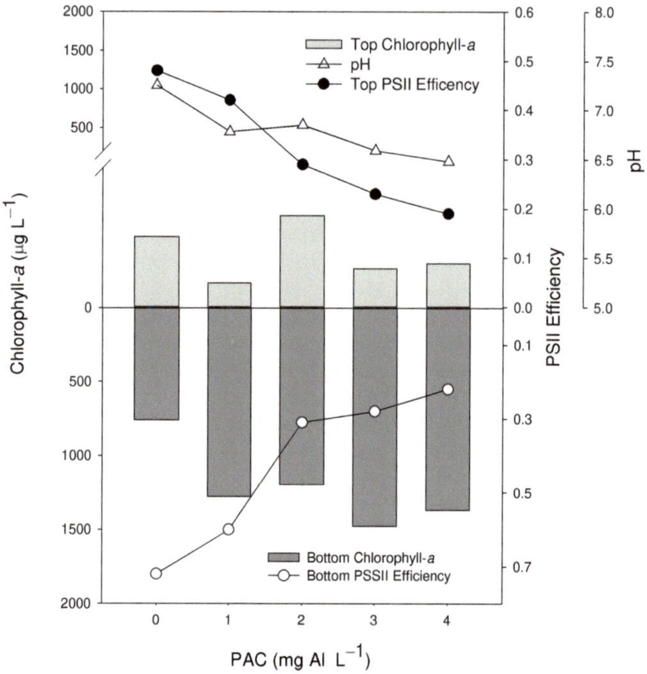

Figure 1. Chlorophyll-*a* concentrations (µg L^{-1}) in the top 5 mL (top light gray bars) and bottom 5 mL (lower dark gray bars), Photosystem II efficiency (Φ_{PSII}) (circles), and pH values (triangles) of 60 mL cyanobacteria suspensions incubated for 2 h in the presence of different concentrations (1, 2, 3, and 4 mg Al L^{-1}) of the coagulant PAC (polyaluminium chloride). The control is represented by 0 mg Al L^{-1}.

The Φ_{PSII} did not show a difference among treatments in the top of the cylinders ($F_{4,10} = 3.259$; $p > 0.001$), but Φ_{PSII} of PAC + LMB + LRS treatment was significantly lower than the control and other treatments in the bottom of the cylinders ($p = 0.002$) (Figure 2A,B). Differences were observed in pH between the control and all treatments ($F_{4,10} = 79.083$; $p < 0.001$) (Figure 2).

2.3. *Effects on the Microcystins Concentrations*

Total intracellular MC concentration in the top of the cylinders varied from 12 to 2049 µg L^{-1} (Figure 3A) and in the bottom, from 176 to 2180 µg L^{-1} (Figure 3B). Total intracellular MC values in the top of the cylinders (Figure 3A) were significantly higher in the control and the PAC treatment alone than in the other treatments ($F_{4,12} = 59.980$; $p < 0.001$). Total intracellular MC concentration in the top of the treatments combining PAC with a ballast was 40 times (PAC + LMB) to 148 times (PAC + LRS) lower than in the control. We also observed a significant difference between total intracellular MC concentrations in the bottom of the cylinders ($F_{4,13} = 23.326$; $p < 0.001$). The PAC + LMB + LRS treatment had the highest intracellular MC concentration in the bottom and was different from the control and PAC treatment (Figure 3B). No difference was recorded in the bottom of the tubes among treatments that combined PAC and a ballast (Figure 3B). Total intracellular MC concentration in treatments with a ballast was 7 to 12 times higher than in the control.

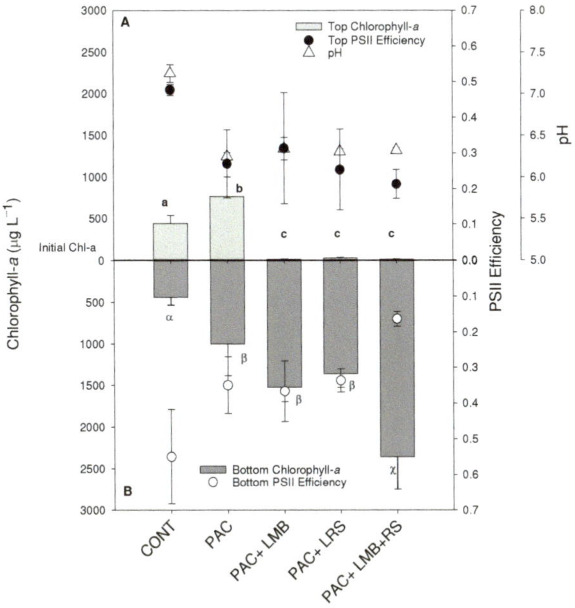

Figure 2. Chlorophyll-*a* concentrations (µg L^{-1}) in the top 15 mL ((**A**) top light gray bars) and bottom 15 mL ((**B**) lower dark gray bars), Photosystem II efficiency (Φ_{PSII}) in the top ((**A**) filled circles), and bottom ((**B**) open circles) and pH values ((**A**) triangles) of 1 L cyanobacteria suspensions from the Funil Reservoir incubated for 2 h in the absence (control) or presence of the coagulant (polyaluminium chloride, PAC 3 mg Al L^{-1}) and coagulant combined with ballast (lanthanum modified bentonite, LMB 0.2 mg L^{-1}, and local red soil, LRS 0.2 mg L^{-1}) separately or in binary mixtures (lanthanum modified bentonite, LMB 0.1 mg L^{-1}, and local red soil, LRS 0.1 mg L^{-1}). The dotted line indicates the initial chlorophyll-*a* concentration in the cylinders, error bars represent one standard deviation ($n = 3$), and similar letters indicate homogeneous groups according to the Holm–Sidak post-hoc test ($p < 0.05$).

Six MC variants were detected in the intracellular samples (WR, YR, FR, LR, and [D-Asp^3] RR), while only three of those were found in extracellular samples (YR, LR, and RR). RR was the most abundant variant in both the intra- and extracellular samples, followed by LR and YR (Figures 3 and 4). In samples with chlorophyll-*a* <11 µg L^{-1}, the variants WR, FR, and [D-Asp^3] RR were not detected. The concentration of the intracellular MC variants varied between treatments in the top and the bottom of the cylinders. The RR variant was the most abundant, representing an average of 69% in the top and 73% at the bottom of the cylinders, whereas the YR was less abundant, representing an average of 6% and 3% in the tube's top and bottom, respectively.

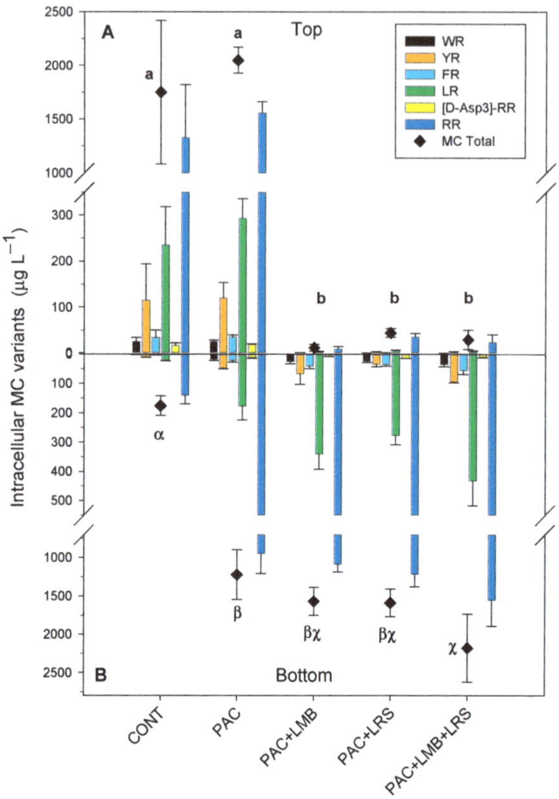

Figure 3. Intracellular concentrations (µg L^{-1}) of MCs and their variants in the top of the cylinder (**A**) and bottom (**B**) in 1 L of cyanobacteria suspensions from the Funil Reservoir incubated for 2 h in the absence (control) or presence of the coagulant (polyaluminium chloride, PAC 3 mg Al L^{-1}) and coagulant combined with ballast (lanthanum modified bentonite, LMB 0.2 mg L^{-1}, and local red soil, LRS 0.2 mg L^{-1}) separately or in binary mixtures (lanthanum modified bentonite, LMB 0.1 mg L^{-1}, and local red soil, LRS 0.1 mg L^{-1}). The black diamonds represent the total intracellular MC in the treatments. Similar letters indicate homogeneous groups in the total MC according to the Holm–Sidak post-hoc test ($p < 0.05$).

Total extracellular MC concentrations in the top of the cylinders were considerably lower than intracellular concentrations and varied between 0.33 and 3.14 µg L^{-1} (Figure 4A). Although PAC increased 2.3 times the extracellular MC concentration, when combined with LMB, it reduced 75% of the MC concentration. PAC + LMB + LRS and PAC + LRS reduced 64% and 50%, respectively. One-way ANOVA indicated significant differences in total extracellular MC concentrations in the top of the cylinders ($F_{4,10} = 8.174$; $p = 0.003$). The PAC treatment was significantly different from the PAC + LMB and PAC + LRS and similar to PAC + LMB + LRS (Figure 4A). Nevertheless, no difference was recorded among treatments with ballast and control at the top (Figure 4A). The treatments combining PAC and ballast had 1 to 4.2 times less extracellular MC in the top than treatment with PAC only. Extracellular MC concentrations were also different among treatments in the bottom of the cylinders ($F_{4,6} = 9.433$; p = 0.009). The Holm–Sidak post-hoc test separated the extracellular MC concentrations in the bottom of the cylinders into two groups: (i) PAC, PAC + LRS, and PAC + LMB + LRS and (ii) control (Figure 4B). The highest concentration of extracellular MCs (5.67 µg L^{-1}) in the bottom of the cylinders was observed in the PAC + LMB + LRS treatment (Figure 4B). The ANCOVA analysis showed no influence of biomass on the extracellular MC concentration in the top ($F_{1,9} = 2.127$; $p = 0.179$) and in the bottom ($F_{1,8} = 1.534$; $p = 0.251$).

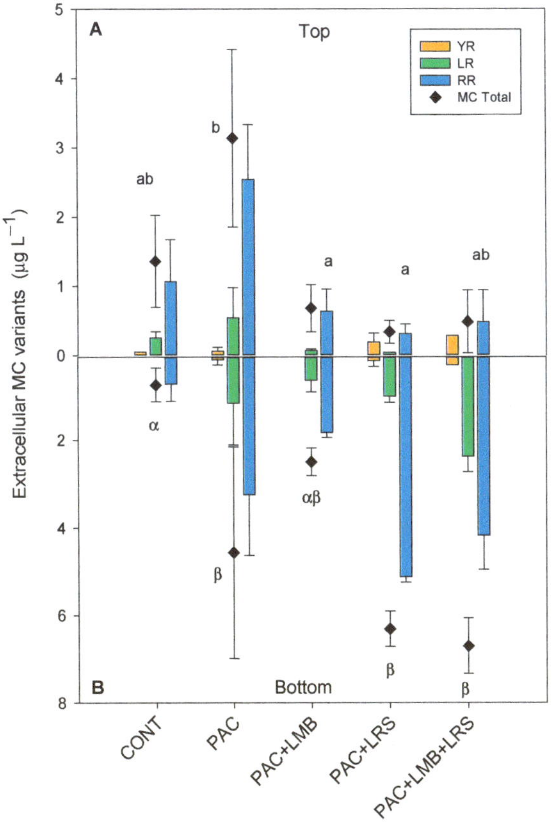

Figure 4. Extracellular concentrations (µg L^{-1}) MCs and their variants in the top of the cylinder (**A**) and bottom (**B**) in 1 L of cyanobacteria suspensions from the Funil Reservoir incubated for 1 h in the absence (control) or presence of the coagulant (polyaluminium chloride, PAC 3 mg Al L^{-1}) and coagulant combined with ballast (lanthanum modified bentonite, LMB 0.2 mg L^{-1}, and local red soil, LRS 0.2 mg L^{-1}) separately or in binary mixtures (lanthanum modified bentonite, LMB 0.1 mg L^{-1}, and local red soil, LRS 0.1 mg L^{-1}). The black diamonds represent the total intracellular MC in the treatments. Similar letters indicate homogeneous groups in the total MC according to the Holm–Sidak post-hoc test ($p < 0.05$).

3. Discussion

This study tested the hypothesis that the Floc and Sink technique could efficiently remove the cyanobacterial biomass comprised of filamentous *Dolichospermum* and small colonial *Microcystis* species without releasing MCs in water from a tropical reservoir. Our results indicated that biomass composed predominantly of *Dolichospermum circinalis* and *Microcystis aeruginosa* could be efficiently removed from the water column using a mixture of low-dose coagulants and ballasts. PAC in combination with LMB or LRS had similar efficacy, and this result is in agreement with previous studies carried out in tropical systems [21–24]. At odds with what was observed by Miranda et al. [21] and Lucena-Silva et al. [22], we recorded the release of MCs with a low dose of PAC (3 mg Al L^{-1}) and in all combinations of PAC and ballast. The highest extracellular MC concentration was recorded in PAC + LMB + LRS, the most efficient treatment in removing biomass. Nevertheless, the decrease in MC concentration in the top of the cylinders treated with PAC + LMB and PAC + LRS and the lower concentration of MC compared to solely PAC suggests the clays' potential capacity as ballast to remove dissolved MC from the water column [45–47].

3.1. Coagulant Range

Algal morphology and other characteristics (e.g., motility, surface charge, and extracellular organic matter) influence the coagulation and sedimentation processes [48]. In laboratory conditions, Miranda et al. [21] showed that similar concentrations of PAC resulted in distinct responses for blooms of the filamentous *Raphidiopsis raciborskii* (formerly *Cylindrospermopsis raciborskii*) and the colonial *Microcystis aeruginosa*. The filamentous species (*R. raciborskii*) formed flocs and subsequently sank, whereas the colonial species accumulated at the surface. Experiments with water samples dominated by the filamentous species *Planktothrix agardhii* and *R. raciborskii* also presented a tendency to sink in a wide range of PAC concentrations (2–32 mg Al L^{-1}) [22]. In our experiment, the biomass concentration was higher at the bottom of the tube than at the top, possibly reflecting the coagulant effect on the dominant filamentous species, *Dolichospermum circinalis*. Based on these studies executed in freshwater, we could suggest a pattern where the filamentous species response to PAC is to sink, and colonial spherical tends to float. This hypothesis is also supported by differences between our results and the other study performed in the same system [23], in which the dominant species were the colonial spherical *Microcystis brasiliensis* (formerly *Sphaerocavum brasiliense*) and *Microcystis panniformis*. With a similar design, the experiment tested a range from 1 to 32 mg Al^{-1} of PAC, and in all concentrations, the biomass presented positive buoyance, accumulating in the top of the tubes [19].

We recorded a gradual decline in pH during the experiment, starting from 7 in control and 6.8 in the dose of 1 mg Al L^{-1} to 6.5 in the highest PAC dose of 4 mg L^{-1}, values considered safe to the cyanobacteria's physiological status [18]. Because hydrogen ions are released during hydrolysis of aluminum-based coagulants, they may cause a decline in the pH, promoting cell lysis [23,49]. There are reports of this phenomenon in the literature from doses $\geq$8 mg Al^{-1} [21,23], almost three times more than the 3 mg Al L^{-1} dose chosen as the optimal dose in this study. The Φ_{PSII} was not affected up to 1 mg of AL L^{-1} and decrease sharply from this concentration, being at odds with other studies where the decrease was recorded from 8 mg Al L^{-1} [41,50,51]. Miranda et al. [21] observed a reduction not only in the Φ_{PSII} but also in the pH in water dominated by *Microcystis* when using higher doses of PAC (16 and 32 mg Al L^{-1}) in the PAC range test. These results suggest that there is no direct action of the PAC on the Φ_{PSII}, which may vary according to the organism's physiological state or vary between species.

3.2. Floc and Sink of Cyanobacteria

As we expected, the Floc and Sink experiment showed that *D. circinalis* and *M. aeruginosa* could be precipitated using a combination of a low dose of PAC and LMB and LRS as a ballast. These results agree with other studies conducted in tropical freshwater systems [19,52,53]. Several coagulants and clays (as ballast) have been tested, aiming to find a safe, efficient, and cheaper treatment [21,23,42]. In tropical aquatic ecosystems, both combinations (PAC + LMB and PAC + LRS) have shown similar efficacy in removing cyanobacterial biomass in laboratory tests [21,23]. The LMB is known as the commercial name Phoslock [54], is more expensive than LRS, a material easily found in banks of natural aquatic systems in the southeast of Brazil [51]. However, LMB has a great advantage over LRS: it immobilizes more P per unit product than LRS [51]. We also decided to test the efficacy of these two materials together (LMB + LRS) combined with PAC to reduce the cost of the potential application. All tested treatments had similar total ballast dosages. However, the treatment with PAC + LMB + LRS removed 38% more biomass than the treatment with just a single ballast (PAC + LMB or LRS).

The lanthanum of LMB has a strong affinity for P, forming an electrostatic interaction with flocks of biomass previously formed by charge neutralization of PAC [19,21,23,28].

This process turns these flocks heavy, sinking the cyanobacteria biomass. Natural soils present less affinity for P but can perform other mechanisms inducing sedimentation [23]. Previous tests in the Funil Reservoir using PAC + LRS and PAC + LMB showed excellent flocculation and sedimentation of biomass formed by *D. circinalis* and *M. aeruginosa* and was achieved without effects on the zeta potential [23]. On that occasion, the authors suggested inter-particle bridging and sweeping as a mechanism of flocculation followed by sedimentation. In other laboratory studies, the combined use of a coagulating metal salt and colloidal local soil solution has been shown to increase the electrostatic interaction, bridging, and enmeshment, which enhanced the effective collision between algal cells and clay particles and inducing sedimentation [44]. Considering this information, we suggest that the combined use of PAC + LMB + LRS were able to increase removal efficiency due to firstly, the action of the different chemical flocculation mechanisms provided by two different ballast, and secondly, the presence of inorganic particles with different granulometry, increasing the action of the physical mechanism in biomass flocks of different sizes, which led to improved sweep flocculation. It is noteworthy that the LRS is composed of coarse sand (2–0.0 mm), fine sand (0.20–0.05 mm), silt (0.05–0.002 mm), and clay (<0.002 mm) [23].

Surprisingly, a reduction in Φ_{PSII} was observed in the bottom of the cylinders when LMB and LRS were used together. This parameter expresses the Photosystem II efficiency, giving a more realistic evaluation of the physiological status of the cells under the tested conditions [23,55,56]. However, we cannot assume that a reduction of 30% in Φ_{PSII} reflects cell lysis [57]. A reduction in ΦPSII was already recorded using PAC (2 mg Al L^{-1}) + $\leq$100 mg LMB L^{-1} in tests with the filamentous species *Planktothrix rubescens*, but there was no evidence of cell damage [19,50,58]. Miranda et al. [21] used 4 mg of Al L^{-1} in the filamentous species *R. raciborskii* and 8 mg of Al L^{-1} in the small colonial *Microcystis* in the Floc and Sink experiment and observed a marginal effect on the pH and Φ_{PSII}, indicating no damaging of the cells during the incubation time. Thus, based on our data of pH and $\Phi_{PSII,}$ we cannot affirm the occurrence of the cell lysis, although it seems the stress in Φ_{PSII} may have stimulated MC liberation.

3.3. Effects of Floc and Sink Technique on the MCs Concentration

The Floc and Sink technique was proposed as a tool to manage eutrophication and cyanobacterial blooms [19,50,58]. One of the advantages of this technique is the possible coagulation and sinking of cells without lysis [19,50,58] and a later degradation of cyanobacteria and their toxins in the sediment [59,60]. In this way, we expected that the pH and Φ_{PSII} would remain steady or with low variations, not causing cell damage, and that the intra and extracellular MC amounts would remain unchanged after treatment.

In this study, six variants of MC were identified in the intracellular form. The less abundant variants could not be quantified in the low biomass. Only the three most abundant intracellular MC variants were also identified in the extracellular MC samples. Although in our experiments, the dosage of PAC (3 mg Al L^{-1}) was below a critical dosage described in the literature, and the pH remained stable and under a safe level [19,21–23,49], we recorded variations in intra- and extracellular MC values among the control and the treatments. Firstly, we cannot generalize the physiological responses; different cyanobacteria species will respond differently to the same treatment, and different strains of the same species may also respond differently [57], so this precludes direct comparison with other studies. Second, we also cannot consider that a variation in the intracellular MC portion is just a consequence of the release. We have biomass comprised of a natural phytoplankton community, predominantly composed of two different species with morphological differences and probably various strains producing MC or not. In that view, different species or strains may possess different buoyancy, as exemplified in the controls by the discrepancy between the biomass (chlorophyll-*a*) and the measured intracellular MC concentration. Whereas chlorophyll-*a* concentrations in both the top and the bottom of the tubes were similar (see Figure 2), intracellular MC concentrations in the top were 10 times higher than in the bottom (see Figure 3A,B). In addition, we have the influence of the treatment under all these variables mentioned above. An evaluation of the coagulation properties of ten microalgae and cyanobacteria species showed that not just the type of coagulant and dosage influence the effectiveness of the coagulation process but also the species [61]. Based on this, we presume that species characteristics (e.g., morphology, mucilage, surface charge, etc.) could affect the coagulation process. Thus, it was not expected that the distribution of species and, consequently, the intracellular MC of flocculated biomass in the cylinders occurs homogeneously. In this way, with the methodology adopted in this study, we did not consider any variation in the intracellular MC portion as an irrefutable indicator of toxin release.

Among the coagulants commonly used to remove cyanobacterial biomass, PAC is safer than other aluminum-based salt coagulants because a lower dose is needed to obtain efficient results. Consequently, there is less pH reduction, preventing cell lysis during the settling process and the liberation of intracellular toxins [62,63]. However, we observed an increase of 2.3 and 4.6 times in the extracellular MC in the bottom and top of the cylinders, respectively, for PAC treatment compared to control. Unlike our findings, Miranda et al. [21] applied a dose of 4 mg Al L^{-1} in water samples dominated by *R. raciborskii* and *Microcystis* spp. and had not observed any significant alteration in saxitoxins or MC concentration. Lucena et al. [22] reported that PAC had not affected the extracellular MC fraction, indicating no cell lysis in experiments using water samples dominated by *P. agardhii* and *R. raciborskii.* An explanation for the divergences between our results and those in the published literature would be the initial pH; here it was ~7 while the similar tests were around 8–10 [21,22]. Another explanation is that the tests already carried out were comprised of biomass of different cyanobacteria species. There are no data about the coagulation process with *Dolichospermum* species. Another possibility is the physiological stage of cells in our samples. The release of toxins appears to occur mainly, but not exclusively, during cellular senescence [3]. From a physiological perspective, the PAC solution can be more toxic when the cells are in the senescence phase [3].

In the Floc and Sink experiment, we sampled both the top and bottom of the cylinders. We observed that the extracellular MC concentration was higher in the bottom of the cylinders, giving the impression that it was higher in more dense biomasses. Although the covariance analysis (ANCOVA) has shown that there is no effect of biomass on extracellular MC, we cannot ignore that the cell lysis may not happen immediately after the application of the treatment, occurring after cells sedimentation [24] and promoting more extracellular MC concentration in more dense biomasses. It may also happen during the filtration process due to initial slight cell damage promoted by PAC added to the mechanical action of the sampling process. Furthermore, MC release may not occur due to cell lysis specifically but because of the chemical stress response. Coagulation tests with polyaluminium ferric chloride and the filamentous species *R. raciborskii* reported increasing extracellular saxitoxins even without any indicator of cell lysis [21]. The authors attributed this result to the fact that filamentous species are more susceptible to external stress than small colonial species. In this study, we did not use any direct test of cell lysis. However, considering the incubation time and our results, we can say that lysis did not occur immediately after applying of the technique but after the sinking of the cyanobacteria.

Additionally, MCs' bioavailability in the water column is influenced by suspended particles' adsorption, especially clays [45,64]. The MC desorption process was already described, and the stability of this bound MC depends on the clays' composition [65,66]. Hence, MC molecules, which are unstably connected to ballast particles after sedimentation, were released by adverse reactions, increasing extracellular MC levels in the bottom of the cylinders. Nonetheless, more research is needed to disentangle possible coagulant MC-liberating and ballast MC-adsorbing/desorbing effects. Some studies have indicated that a Floc and Sink treatment can reduce biomass and MCs strongly [57], while LMB itself could lower extracellular MC concentration by 61–86% [46].

There are some possibilities for the non-homogeneity in dissolved MC in the water column, first when the tests are carried out with a natural phytoplankton population and large water volumes. Another reason is the potential property of clays in reducing the concentration of dissolved MC. LMB and LRS adsorption may have indirectly influenced dissolved MC concentration. The PAC + LMB treatment showed a significant reduction in extracellular MC (73%) compared to only PAC treatment on the top of the cylinders. If PAC promoted MC release, as observed by comparing to control, the reduction recorded in PAC + LMB treatment could be due to LMB sorption. These findings agree with the recently reported capacity of LMB to lower extracellular MC concentration [45,64]. A similar activity could be observed in PAC + LRS treatment. The extracellular MC decreased by 88% in the top of the cylinders. Laboratory studies showed that sediment particles applied in the water column effectively reduced MC concentration to less than the detection limit [45,64]. Noteworthy, no significant difference was observed in the bottom of the cylinders in all treatments.

Unexpectedly, no difference was recorded in MC-extra in the top of the cylinders between PAC + LMB + LRS and PAC treatment. A possible explanation for the divergence between this treatment and others with ballast is that the ballast compound showed no linear adsorption capacity. Investigating the ability to absorb MC by LMB, the authors observed that at the doses of 50, 100, and 150 ppm, the decrease in MC concentrations was 61.2%, 86.0%, and 75.4%, respectively, relative to the control [50,60]. Many studies have focused on the adsorptive capacity of MC in sediment [50,60], but little is known about the potential of MC adsorption by lanthanum modified bentonite (LMB). These facts would justify the differences observed in extracellular MC concentration between treatments that contained ballast since we think that MCs' release was promoted by adding PAC in all ballast treatments.

We emphasize that in this study, at the top of the cylinders in all treatments combining coagulant and ballast, the extracellular MC concentration was below 1 $\mu g\ L^{-1}$, the maximum acceptable values adopted by WHO [3] and the Brazilian Government. Although we observed an increase in extracellular MC concentration after application of the Floc and Sink technique, especially in the bottom of the cylinders of PAC, PAC + LRS, and PAC + LMB + LRS treatments, these extracellular MC amounts are negligible compared with the intracellular concentrations that range from 1.2 to 2.2 $\mu g\ L^{-1}$ in the same treatments. For instance, in the most efficient treatment in removing biomass, PAC + LMB + LRS, it is less than 0.3%. Moreover, some authors consider that lysis of cyanobacteria after sedimentation of biomass is beneficial in reducing the possibility of recolonization of the water column by perturbations or bioturbation process, whereas near the sediment, liberated toxins can be adsorbed and degraded by decomposing bacteria [50,60]. These results show that Floc and Sink is a promising tool to manage toxic blooms. However, we suggest a detailed investigation of PAC effects on different species of toxin-producing cyanobacteria, and additionally, an investigation of the potential capacity for MCs adsorption in LMB and clays is needed.

4. Conclusions

Our results showed that combining a low dose of coagulant with a ballast compound effectively removed biomass composed predominantly of *Dolichospermum circinalis* and *Microcystis aeruginosa* from the water column. PAC + LMB + LRS was the most efficient combination to settle the cyanobacteria biomass in the bottom of the tubes. PAC promoted the release of MC but combined with ballast, and the concentrations remained similar to control in the top. PAC + LMB was efficient to remove biomass and maintained concentrations of extracellular MC similar to the control. Although we observed an increase in extracellular concentrations of MC in the bottom of the tube after the application of PAC + LRS and LMB + LRS, the concentrations were extremely low, which does not preclude technical use.

5. Materials and Methods

5.1. Sampling

The Funil Reservoir is a eutrophic system located in southern Brazil (22°30′ S and 44°45′ W) at 440 m of altitude, in a warm-rainy tropical climate area (Cwa in the Köppen system). Since the 1980s, toxigenic cyanobacterial blooms with a dominance of *Microcystis aeruginosa* have been registered in this reservoir, especially during warm periods [67,68].

For the experiments, water samples were collected in the mild-cold season and concentrated with a plankton net (50 µm mesh) to create a cyanobacterial suspension, yielding an initial concentration of ~155 µg L^{-1} of chlorophyll-*a*. The total cyanobacteria biomass in the sample was composed of 72% *Dolichospermum circinalis* (Rabenhorst) Wacklin, Hoffmann and Komárek, 23% *Microcystis aeruginosa* (Kützing) Kützing, 3% *Dolichospermum spiroides* (Klebhan) Wacklin, L. Hoffmann and Komárek, and 1% *Microcystis panniformis* Komárek et al. At the time of sampling, the pH was 7.38, alkalinity was 499 µEq L^{-1}, turbidity was 16 NTU, and the water temperature was 21 °C.

5.2. Chemicals and Materials

The coagulant PAC (polyaluminium chloride; $Al_n(OH)_mCl3_{n-m}$, r ~1.37 kg L^{-1}, 9.5% Al, 21.0% Cl) was obtained from Purewater Efluentes (São Paulo, Brazil). Local red soil (LRS) was collected from the banks of the Funil Reservoir as described by Noyma et al. (2016), and the lanthanum modified bentonite Phoslock® (LMB) was obtained from HydroScience (Porto Alegre, Brazil). LRS and LMB were used as ballast.

5.3. Floc and Sink Assays

5.3.1. Experiment 1—Coagulant Range

A range of PAC concentrations were tested (0, 1, 2, 3, and 4 mg Al L^{-1}). This experiment has no replicates because it followed a regression design to evaluate the most appropriate and effective coagulant dosage. The assay was set up in glass tubes (10 × 200 mm) containing 60 mL of cyanobacteria suspensions with an initial concentration of ~155 µg L^{-1} of chlorophyll-*a*. PAC was added at the surface and mixed with a metal rod for 30 s; the tubes were then incubated for two hours. After the incubation time, the tubes were visually inspected for flocs formation, and 5 mL aliquots were taken from the top and bottom of the tubes to determine the precipitation of cyanobacteria biomass. The chlorophyll-*a* concentration and the Photosystem II efficiency (Φ_{PSII}) [55] were measured using a PHYTO-PAM phytoplankton analyzer (Heinz Walz GmbH, Effeltrich, Germany). The pH was also measured in the tubes.

5.3.2. Experiment 2—Floc and Sink Assays

Based on the first experiment, the coagulant (PAC) dose of 3 mg of Al L^{-1} was chosen as an effective dose. This 3 mg Al L^{-1} of PAC was also combined with LMB (0.2 g L^{-1}), LRS (0.2 g L^{-1}), and LMB + LRS (0.1 g^{-1} of each). The dosage of ballast was based on previous tests [21,23]. The experiment was run in triplicate in acrylic cylinders containing 1 L of cyanobacteria suspensions from the Funil Reservoir. We tested four treatments: (1) only PAC, (2) PAC + LMB, (3) PAC + LRS, and (4) PAC + LMB + LRS, while the fifth series remained untreated (Controls). The PAC was added first, followed by the immediate addition of a slurry of LMB and/or LRS in the top of the cylinders. Subsequently, the suspensions were mixed with a glass stirring rod, and after two hours, 15 mL samples were collected from the top and bottom of the cylinders. Then, 8 mL of sample were filtered through 1.2 µm glass fiber filters (85/70 BF, Macherey-Nagel) to quantify MCs. The filters were used to quantify intracellular MC, and the filtrate was used to quantify extracellular MC. Filter and filtrate samples were immediately frozen and kept at −20 °C until the analysis. Then, 5 mL of the samples were used to quantify chlorophyll-*a* and Φ_{PSII} by PHYTO-PAM phytoplankton analyzer (Heinz Walz GmbH, Effeltrich, Germany).

5.4. Sample Analysis

Toxins Analysis

Before the extractions, filters, and filtrates were freeze-dried (Sartorius GmbH, Germany), for the analysis of the intracellular toxins, 2.5 mL of 75% (*v*/*v*) methanol (MeOH) (Merck®, São Paulo, Brazil) was added to the filters in an 8 mL glass tube. After vortexing the samples for 15 s, they were placed in a water bath (Buchi Heating Bath b-491, São Paulo, Brazil) for 30 min at 60 °C. The suspensions were transferred to new glass tubes. This extraction step was repeated twice, but this time with 2.0 mL of 75% (*v*/*v*) methanol. The new glass tubes containing 6.5 mL of extract were placed in the Speedvac (Savant SPD121P, Thermo Scientific, Waltham, MA, USA). After drying, they were resuspended with 2 mL of MeOH 100% (*v*/*v*), vortexed for 15 s and filtered through 0.45 µm PVDF membrane syringe filters (Analítica, São Paulo, Brazil) into amber glass vials for LC-MS/MS analysis. For the extracellular toxins analysis, the freeze-dried samples were resuspended with 2 mL of MeOH 100% (*v*/*v*), vortexed for 15 s, and filtered through 0.45 µm PVDF membrane syringe filters (Analítica, Brazil) into amber glass vials. If needed, the samples with high MC concentrations were diluted in methanol 75% *v*/*v* before re-analysis. The lyophilized samples were resuspended with 2 mL of MeOH (100%), vortexed for 15 s, and filtered through 0.45 µm PVDF membrane syringe filters (Analítica, Brazil) into amber vials for the analysis of dissolved toxins. Samples were then stored in a freezer (−20 °C) until LC-MS/MS analysis. See the Supplementary Materials for additional details on MC recovery (Table S2).

Toxin determination in all samples was performed by liquid chromatography-tandem mass spectrometry (LC-MS/MS) using a series 200 HPLC system (Perkin Elmer, Waltham, MA, USA) coupled to electrospray ionization (ESI) mass spectrometer. Chromatographic separations were carried out on a Luna C18 column (150 × 2 mm; 5 µm particles, Phenomenex, Torrance, CA, USA). The mobile phases consisted of 5 mM ammonium formate and 53 mM formic acid in water (mobile phase A) and 90% (*v*/*v*) acetonitrile (mobile phase B). Gradient elution was performed at a flow rate of 300 $\mu L\ min^{-1}$ and followed a linear increase from 10 to 90% B within 15 min, then it was held at 90% B for 2 min, and returned to the initial condition (10% B) within 12 min. MS/MS experiments were performed using an API 365 triple quadrupole (QqQ) mass spectrometer (AB Sciex, Concord, ON, Canada) equipped with a turbo ion spray source. The instrument was operated using the selected reaction monitoring (SRM) mode, with specific *m/z* transitions selected for the highest sensitivity and selectivity. Single and double-charged ions were monitored in positive ion mode. Characteristic precursor ions for SRM were *m/z* 519 (RR), 910 (LA), 986 (LF), 995 (LR), 1025 (LW), 1045 (YR), 512 ([D-Asp^3] RR), and 981 ([D-Asp^3] LR). As only MCs were detected in the studied samples, MCs' quantification in SRM mode was based on the characteristic Adda fragment at *m/z* 135. Calibration standards of non-demethylated MCs were obtained from Abraxis (Eurofins®, Nantes, France) and prepared in methanol 75%. Samples were quantified using a calibration curve and subsequently corrected for recovery. Quantification of the demethylated MC structures was performed by relative quantification using the corresponding non-demethylated MC as analytical standards.

5.5. Statistical Analysis

A one-way ANOVA analysis was performed to evaluate the differences in chlorophyll-*a* and MC concentration between treatments in the tool pack SigmaPlot (12.5 version, Rio de Janeiro, Brazil). A pairwise multiple comparison post-hoc test (Holm–Sidak) was applied to distinguish means that were significantly different ($p < 0.05$). The ANCOVA analysis was also performed in IBM SPSS Statistics® (2.0 version, Rio de Janeiro, Brazil) to evaluate the linear relationship between biomass and MC concentration.

Supplementary Materials: The following are available online at https://www.mdpi.com/article/10.3390/toxins13060405/s1, Figure S1: Microcystins (MCs) general structure, cyclo-(*D*-Ala^1-$X_1{}^2$-*D*-MeAsp3-$X_2{}^4$-$Adda^5$-*D*-Glu^6-$Mdha^7$). *D*-Ala: D-alanine, X_1 and X_2: variable amino acids, *D*-MeAsp:

D-erythro-β-methylaspartic acid, Adda: (2S,3S,4E,6E,8S,9S)-3-amino-9-methoxy-2,6,8-trimethyl-10-phenyldeca-4,6-dienoic acid, D-Glu: D-glutamic acid, Mdha: N-methyldehydroalanine. R_1 and R_2: H or CH_3., Figure S2: Chemical structure for the MC variants MC-RR, [D-Asp3]MC-RR, MC-YR, MC-LR, [D-Asp3]MC-LR, MC-LA, MC-LF, and MC-LW. The one-letter codes A, R, L, F, Y, and W represent the amino acids alanine, arginine, leucine, phenylalanine, tyrosine, and tryptophan, respectively. The suffix [D-Asp3] represents a missing methyl group in position 3 of MCs structure. Table S1: Chemical and physical properties of the MC variants MC-RR, [D-Asp3]MC-RR, MC-YR, MC-LR, [D-Asp3]MC-LR, MC-LA, MC-LF, and MC-LW., Table S2: Recovery in % for MC in two different matrices (tap water and cultures of Chlorella vulgaris) for three concentrations.

Author Contributions: Conceptualization, R.S.A., N.P.N., M.L. and M.M.M.; Data curation, R.S.A.; Formal analysis, R.S.A. and N.P.N.; Methodology, R.S.A.; Resources, E.P., M.L. and M.M.M.; Supervision, M.M.M.; Writing—original draft, R.S.A.; Writing—review & editing, R.S.A., N.P.N., L.d.M., M.C.B.M., É.C.d.A., E.P., M.L. and M.M.M. All authors have read and agreed to the published version of the manuscript."

Funding: R.S.A.'s PhD scholarship was funded by Coordenação de Aperfeiçoamento de Pessoal de Nível Superior (CAPES). This study was financed in part by the Coordenação de Aperfeiçoamento de Pessoal de Nivel Superior, Brazil (CAPES)- Finance Code 001; Conselho Nacional de Desenvolvimento Científico e Tecnológico, Brazil (CNPq), Grant 403515/2016-5; Fundação de Amparo à Pesquisa do Estado do Rio de Janeiro, Brasil (FAPERJ), Grant 303572/2017-5.

Institutional Review Board Statement: Not applicable.

Informed Consent Statement: Not applicable.

Acknowledgments: We would like to cordially thank the members of Laboratory of Ecology and Physiology of Phytoplankton (UERJ) and Laboratory of Toxins and Natural Products from Algae and Cyanobacteria (LTPNA - USP) for technical support.

Conflicts of Interest: The authors declare no conflict of interest.

References

1. World Health Organization. *Toxic Cyanobacteria in Water: A Guide to Their Public Health Consequences, Monitoring, and Management*; Chorus, I., Bertram, J., Eds.; F & FN Spon: London, UK; New York, NY, USA, 1999.
2. Huisman, J.; Codd, G.A.; Paerl, H.W.; Ibelings, B.W.; Verspagen, J.M.H.; Visser, P.M. Cyanobacterial Blooms. *Nat. Rev. Microbiol.* **2018**, *16*, 471–483. [CrossRef]
3. World Health Organization. *Toxic Cyanobacteria in Water: A Guide to Their Public Health Consequences, Monitoring and Management*, 2nd ed.; Chorus, I., Welker, M., Eds.; CRC Press: Boca Raton, FL, USA, 2021; ISBN 9781003081449.
4. Jones, M.R.; Pinto, E.; Torres, M.A.; Dörr, F.; Mazur-Marzec, H.; Szubert, K.; Tartaglione, L.; Dell'Aversano, C.; Miles, C.O.; Beach, D.G.; et al. Comprehensive Database of Secondary Metabolites from Cyanobacteria. *Water Res.* **2020**, *196*, 15. [CrossRef]
5. Lankoff, A.; Carmichael, W.W.; Grasman, K.A.; Yuan, M. The Uptake Kinetics and Immunotoxic Effects of Microcystin-LR in Human and Chicken Peripheral Blood Lymphocytes in Vitro. *Toxicology* **2004**, *204*, 23–40. [CrossRef]
6. Nishiwaki-Matsushima, R.; Ohta, T.; Nishiwaki, S.; Suganuma, M.; Kohyama, K.; Ishikawa, T.; Carmichael, W.W.; Fujiki, H. Liver Tumor Promotion by the Cyanobacterial Cyclic Peptide Toxin Microcystin-LR. *J. Cancer Res. Clin. Oncol.* **1992**, *118*, 420–424. [CrossRef] [PubMed]
7. Rojas, M.; Nuñez, M.T.; Zambrano, F. Inhibitory Effect of a Toxic Peptide Isolated from a Waterbloom of Microcystis sp. (Cyanobacteria) on Iron Uptake by Rabbit Reticulocytes. *Toxicon* **1990**, *28*, 1325–1332. [CrossRef]
8. Dziga, D.; Maksylewicz, A.; Maroszek, M.; Budzyńska, A.; Napiorkowska-Krzebietke, A.; Toporowska, M.; Grabowska, M.; Kozak, A.; Rosińska, J.; Meriluoto, J. The Biodegradation of Microcystins in Temperate Freshwater Bodies with Previous Cyanobacterial History. *Ecotoxicol. Environ. Saf.* **2017**, *145*, 420–430. [CrossRef]
9. Li, X.; Dreher, T.W.; Li, R. An Overview of Diversity, Occurrence, Genetics and Toxin Production of Bloom-Forming Dolichospermum (Anabaena) Species. *Harmful Algae* **2016**, *54*, 54–68. [CrossRef] [PubMed]
10. Harada, K.I.; Ogawa, K.; Kimura, Y.; Murata, H.; Suzuki, M.; Thorn, P.M.; Evans, W.R.; Carmichael, W.W. Microcystins from Anabaena Flos-Aquae NRC 525-17. *Chem. Res. Toxicol.* **1991**, *4*. [CrossRef]
11. Rantala, A.; Rajaniemi-Wacklin, P.; Lyra, C.; Lepistö, L.; Rintala, J.; Mankiewicz-Boczek, J.; Sivonen, K. Detection of Microcystin-Producing Cyanobacteria in Finnish Lakes with Genus-Specific Microcystin Synthetase Gene E (McyE) PCR and Associations with Environmental Factors. *Appl. Environ. Microbiol.* **2006**, *72*, 6101. [CrossRef]
12. Kobos, J.; Błaszczyk, A.; Hohlfeld, N.; Toruńska-Sitarz, A.; Krakowiak, A.; Hebel, A.; Sutryk, K.; Grabowska, M.; Toporowska, M.; Kokociński, M.; et al. Cyanobacteria and Cyanotoxins in Polish Freshwater Bodies. *Oceanol. Hydrobiol. Stud.* **2013**, *42*, 358–378. [CrossRef]

13. Dreher, T.W.; Collart, L.P.; Mueller, R.S.; Halsey, K.H.; Bildfell, R.J.; Schreder, P.; Sobhakumari, A.; Ferry, R. Anabaena/Dolichospermum as the Source of Lethal Microcystin Levels Responsible for a Large Cattle Toxicosis Event. *Toxicon X* **2019**, *1*, 100003. [CrossRef]
14. Sá, L.L.C.D.; Vieira, J.M.D.S.; Mendes, R.D.A.; Pinheiro, S.C.C.; Vale, E.R.; Alves, F.A.D.S.; Jesus, I.M.D.; Santos, E.C.D.O.; Costa, V.B.D. Ocorrência de Uma Floração de Cianobactérias Tóxicas Na Margem Direita Do Rio Tapajós, No Município de Santarém (Pará, Brasil). *Rev. Pan-Amaz. Saúde* **2010**, *1*, 1. [CrossRef]
15. Sant'Anna, C.L.; Azevedo, M.T.D.P. Contribution to the Knowledge of Potentially Toxic Cyanobacteria from Brazil. *Nova Hedwig.* **2000**, *71*, 359–385. [CrossRef]
16. Paerl, H.W.; Gardner, W.S.; Havens, K.E.; Joyner, A.R.; McCarthy, M.J.; Newell, S.E.; Qin, B.; Scott, J.T. Mitigating Cyanobacterial Harmful Algal Blooms in Aquatic Ecosystems Impacted by Climate Change and Anthropogenic Nutrients. *Harmful Algae* **2016**, *54*, 213–222. [CrossRef] [PubMed]
17. van Loosdrecht, M.C.M.; Brdjanovic, D. Anticipating the next Century of Wastewater Treatment. *Science* **2014**, *344*, 1452–1453. [CrossRef] [PubMed]
18. Cooke, G.D.; Welch, E.B.; Peterson, S.; Nichols, S.A. *Restoration and Management of Lakes and Reservoirs*; CRC Press: Boca Raton, FL, USA, 2016.
19. Lürling, M.; Kang, L.; Mucci, M.; van Oosterhout, F.; Noyma, N.P.; Miranda, M.; Huszar, V.L.M.; Waajen, G.; Marinho, M.M. Coagulation and Precipitation of Cyanobacterial Blooms. *Ecol. Eng.* **2020**, *158*, 106032. [CrossRef]
20. Liu, H.; Du, Y.; Wang, X.; Sun, L. Chitosan Kills Bacteria through Cell Membrane Damage. *Int. J. Food Microbiol.* **2004**, *95*, 147–155. [CrossRef]
21. Miranda, M.; Noyma, N.; Pacheco, F.S.; de Magalhães, L.; Pinto, E.; Santos, S.; Soares, M.F.A.; Huszar, V.L.; Lürling, M.; Marinho, M.M. The Efficiency of Combined Coagulant and Ballast to Remove Harmful Cyanobacterial Blooms in a Tropical Shallow System. *Harmful Algae* **2017**, *65*, 27–39. [CrossRef]
22. de Lucena-Silva, D.; Molozzi, J.; dos Santos Severiano, J.; Becker, V.; de Lucena Barbosa, J.E. Removal Efficiency of Phosphorus, Cyanobacteria and Cyanotoxins by the "Flock & Sink" Mitigation Technique in Semi-Arid Eutrophic Waters. *Water Res.* **2019**, *159*, 262–273. [CrossRef]
23. Noyma, N.P.; de Magalhães, L.; Furtado, L.L.; Mucci, M.; van Oosterhout, F.; Huszar, V.L.M.; Marinho, M.M.; Lürling, M. Controlling Cyanobacterial Blooms through Effective Flocculation and Sedimentation with Combined Use of Flocculants and Phosphorus Adsorbing Natural Soil and Modified Clay. *Water Res.* **2016**, *97*, 26–38. [CrossRef]
24. Sun, F.; Pei, H.Y.; Hu, W.R.; Ma, C.X. The Lysis of Microcystis Aeruginosa in AlCl 3 Coagulation and Sedimentation Processes. *Chem. Eng. J.* **2012**, *193*, 196–202. [CrossRef]
25. Zouboulis, A.; Traskas, G.; Samaras, P. Comparison of Efficiency between Poly-Aluminium Chloride and Aluminium Sulphate Coagulants during Full-Scale Experiments in a Drinking Water Treatment Plant. *Sep. Sci. Technol.* **2008**, *43*, 1507–1519. [CrossRef]
26. Zarchi, I.; Friedler, E.; Rebhun, M. Polyaluminium Chloride as an Alternative to Alum for the Direct Filtration of Drinking Water. *Environ. Technol.* **2013**, *34*, 1199–1209. [CrossRef]
27. Delgado, S.; Diaz, F.; Garcia, D.; Otero, N. Behaviour of Inorganic Coagulants in Secondary Effluents from a Conventional Wastewater Treatment Plant. *Filtr. Sep.* **2003**, *40*, 42–46. [CrossRef]
28. Lürling, M.; Oosterhout, F. van Controlling Eutrophication by Combined Bloom Precipitation and Sediment Phosphorus Inactivation. *Water Res.* **2013**, *47*, 6527–6537. [CrossRef]
29. Araújo, F.; dos Santos, H.R.; Becker, V.; Attayde, J.L. The Use of Polyaluminium Chloride as a Restoration Measure to Improve Water Quality in Tropical Shallow Lakes. *Acta Limnol. Bras.* **2018**, *30*, e109. [CrossRef]
30. Kasprzak, P.; Gonsiorczyk, T.; Grossart, H.P.; Hupfer, M.; Koschel, R.; Petzoldt, T.; Wauer, G. Restoration of a Eutrophic Hard-Water Lake by Applying an Optimised Dosage of Poly-Aluminium Chloride (PAC). *Limnologica* **2018**, *70*, 33–48. [CrossRef]
31. Wei, N.; Zhang, Z.; Liu, D.; Wu, Y.; Wang, J.; Wang, Q. Coagulation Behavior of Polyaluminum Chloride: Effects of PH and Coagulant Dosage. *Chin. J. Chem. Eng.* **2015**, *23*, 1041–1046. [CrossRef]
32. Her, N.; Amy, G.; Park, H.R.; Song, M. Characterizing Algogenic Organic Matter (AOM) and Evaluating Associated NF Membrane Fouling. *Water Res.* **2004**, *38*, 1427–1438. [CrossRef]
33. Bernhardt, H.; Hoyer, O.; Lusse, B.; Schell, H. Investigations on the Influence of Algal-Derived Organic Substances on Flocculation and Filtration. In Proceedings of the National Conference on Drinking water: Treatment of Drinking Water for Organic Contaminants, Edmonton, AB, Canada, 7–8 April 1987.
34. Takaara, T.; Sano, D.; Konno, H.; Omura, T. Cellular Proteins of Microcystis Aeruginosa Inhibiting Coagulation with Polyaluminum Chloride. *Water Res.* **2007**, *41*, 1653–1658. [CrossRef]
35. Takaara, T.; Sano, D.; Konno, H.; Omura, T. Affinity Isolation of Algal Organic Matters Able to Form Complex with Aluminium Coagulant. *Water Supply* **2004**, *4*, 95–102. [CrossRef]
36. Zhang, P.; Ren, B.Z.; Wang, F. Humic Acid Removal from Water by Poly Aluminium Chloride (PAC). *Appl. Mech. Mater.* **2013**, *253–255*, 892–896. [CrossRef]
37. Sudoh, R.; Islam, M.S.; Sazawa, K.; Okazaki, T.; Hata, N.; Taguchi, S.; Kuramitz, H. Removal of Dissolved Humic Acid from Water by Coagulation Method Using Polyaluminum Chloride (PAC) with Calcium Carbonate as Neutralizer and Coagulant Aid. *J. Environ. Chem. Eng.* **2015**, *3*, 770–774. [CrossRef]

38. Lürling, M.; Waajen, G.; van Oosterhout, F. Humic Substances Interfere with Phosphate Removal by Lanthanum Modified Clay in Controlling Eutrophication. *Water Res.* **2014**, *54*, 78–88. [CrossRef] [PubMed]
39. Reitzel, K.; Balslev, K.A.; Jensen, H.S. The Influence of Lake Water Alkalinity and Humic Substances on Particle Dispersion and Lanthanum Desorption from a Lanthanum Modified Bentonite. *Water Res.* **2017**, *125*, 191–200. [CrossRef] [PubMed]
40. Mucci, M.; Noyma, N.P.; de Magalhães, L.; Miranda, M.; van Oosterhout, F.; Guedes, I.A.; Huszar, V.L.M.; Marinho, M.M.; Lürling, M. Chitosan as Coagulant on Cyanobacteria in Lake Restoration Management May Cause Rapid Cell Lysis. *Water Res.* **2017**, *118*, 121–130. [CrossRef] [PubMed]
41. de Magalhães, L.; Noyma, N.P.; Furtado, L.L.; Mucci, M.; van Oosterhout, F.; Huszar, V.L.M.; Marinho, M.M.; Lürling, M. Efficacy of Coagulants and Ballast Compounds in Removal of Cyanobacteria (Microcystis) from Water of the Tropical Lagoon Jacarepaguá (Rio de Janeiro, Brazil). *Estuaries Coasts* **2017**, *40*, 121–133. [CrossRef]
42. Noyma, N.P.; de Magalhães, L.; Miranda, M.; Mucci, M.; van Oosterhout, F.; Huszar, V.L.M.; Marinho, M.M.; Lima, E.R.A.; Lurling, M. Coagulant plus Ballast Technique Provides a Rapid Mitigation of Cyanobacterial Nuisance. *PLoS ONE* **2017**, *12*, e0178976. [CrossRef]
43. Pierce, R.H.; Henry, M.S.; Higham, C.J.; Blum, P.; Sengco, M.R.; Anderson, D.M. Removal of Harmful Algal Cells (Karenia Brevis) and Toxins from Seawater Culture by Clay Flocculation. *Harmful Algae* **2004**, *3*, 141–148. [CrossRef]
44. Thongdam, S.; Kuster, A.C.; Huser, B.J.; Kuster, A.T. Low Dose Coagulant and Local Soil Ballast Effectively Remove Cyanobacteria (Microcystis) from Tropical Lake Water without Cell Damage. *Water* **2021**, *13*, 111. [CrossRef]
45. Chen, W.; Song, L.; Peng, L.; Wan, N.; Zhang, X.; Gan, N. Reduction in Microcystin Concentrations in Large and Shallow Lakes: Water and Sediment-Interface Contributions. *Water Res.* **2008**, *42*, 763–773. [CrossRef]
46. Dail, H.; Iv, L.; Lefler, F.; Berthold, D.E. Sorption of Dissolved Microcystin Using Lanthanum-Modified Bentonite Clay. *J. Aquat. Plant Manag.* **2020**, *58*, 72–75.
47. Prochazka, E.; Hawker, D.; Hwang, G.S.; Shaw, G.; Stewart, I.; Wickramasinghe, W. The Removal of Microcystins in Drinking Water by Clay Minerals. In Proceedings of the 14th International Conference on Harmful Algae, Hersonissos, Greece, 1–5 November 2010.
48. Li, H.; Pei, H.; Xu, H.; Jin, Y.; Sun, J. Behavior of Cylindrospermopsis Raciborskii during Coagulation and Sludge Storage-Higher Potential Risk of Toxin Release than Microcystis Aeruginosa? *J. Hazard. Mater.* **2018**, *347*, 307–316. [CrossRef] [PubMed]
49. Han, J.; Jeon, B.S.; Park, H.D. Cyanobacteria Cell Damage and Cyanotoxin Release in Response to Alum Treatment. *Water Sci. Technol. Water Supply* **2012**, *12*, 549–555. [CrossRef]
50. Pan, G.; Zou, H.; Chen, H.; Yuan, X. Removal of Harmful Cyanobacterial Blooms in Taihu Lake Using Local Soils. III. Factors Affecting the Removal Efficiency and an in Situ Field Experiment Using Chitosan-Modified Local Soils. *Environ. Pollut.* **2006**, *141*, 206–212. [CrossRef] [PubMed]
51. Mucci, M.; Maliaka, V.; Noyma, N.P.; Marinho, M.M.; Lürling, M. Assessment of Possible Solid-Phase Phosphate Sorbents to Mitigate Eutrophication: Influence of PH and Anoxia. *Sci. Total. Environ.* **2018**, *619*, 1431–1440. [CrossRef]
52. Lürling, M.; Faassen, E.J. Dog Poisonings Associated with a Microcystis Aeruginosa Bloom in the Netherlands. *Toxins* **2013**, *5*, 556–567. [CrossRef]
53. Pan, G.; Dai, L.; Li, L.; He, L.; Li, H.; Bi, L.; Gulati, R.D. Reducing the Recruitment of Sedimented Algae and Nutrient Release into the Overlying Water Using Modified Soil/Sand Flocculation-Capping in Eutrophic Lakes. *Environ. Sci. Technol.* **2012**, *46*, 5077–5084. [CrossRef] [PubMed]
54. Valsami-Jones, E.; International Water Association. *Phosphorus in Environmental Technology: Principles and Applications*; IWA Publishing: London, UK, 2004; ISBN 1843390019.
55. Genty, B.; Briantais, J.M.; Baker, N.R. The Relationship between the Quantum Yield of Photosynthetic Electron Transport and Quenching of Chlorophyll Fluorescence. *Biochim. Biophys. Acta Gen. Subj.* **1989**, *990*, 87–92. [CrossRef]
56. Weenink, E.F.J.; Luimstra, V.M.; Schuurmans, J.M.; van Herk, M.J.; Visser, P.M.; Matthijs, H.C.P. Combatting Cyanobacteria with Hydrogen Peroxide: A Laboratory Study on the Consequences for Phytoplankton Community and Diversity. *Front. Microbiol.* **2015**, *6*, 714. [CrossRef] [PubMed]
57. Lürling, M.; Mucci, M.; Waajen, G. Removal of Positively Buoyant Planktothrix Rubescens in Lake Restoration. *Toxins* **2020**, *12*, 700. [CrossRef]
58. Faassen, E.J.; Lürling, M. Occurrence of the Microcystins MC-LW and MC-LF in Dutch Surface Waters and Their Contribution to Total Microcystin Toxicity. *Mar. Drugs* **2013**, *11*, 2643–2654. [CrossRef]
59. Grützmacher, G.; Wessel, G.; Klitzke, S.; Chorus, I. Microcystin Elimination during Sediment Contact. *Environ. Sci. Technol.* **2010**, *44*, 657–662. [CrossRef]
60. Holst, T.; Jørgensen, N.O.G.; Jørgensen, C.; Johansen, A. Degradation of Microcystin in Sediments at Oxic and Anoxic, Denitrifying Conditions. *Water Res.* **2003**, *37*, 4748–4760. [CrossRef]
61. Lama, S.; Muylaert, K.; Karki, T.B.; Foubert, I.; Henderson, R.K.; Vandamme, D. Flocculation Properties of Several Microalgae and a Cyanobacterium Species during Ferric Chloride, Chitosan and Alkaline Flocculation. *Bioresour. Technol.* **2016**, *220*, 464–470. [CrossRef] [PubMed]
62. de Julio, M.; Fioravante, D.A.; de Julio, T.S.; Oroski, F.I.; Graham, N.J.D. A Methodology for Optimising the Removal of Cyanobacteria Cells from a Brazilian Eutrophic Water. *Braz. J. Chem. Eng.* **2010**, *27*, 113–126. [CrossRef]

63. Gebbie, P. Using Polyaluminium Coagulants in Water Treatment. In Proceedings of the 64th Annual Water industry Engineers and Operators' Conference, Bendigo, Australia, 5–6 September 2001.
64. Chen, X.; Yang, X.; Yang, L.; Xiao, B.; Wu, X.; Wang, J.; Wan, H. An Effective Pathway for the Removal of Microcystin LR via Anoxic Biodegradation in Lake Sediments. *Water Res.* **2010**, *44*, 1884–1892. [CrossRef]
65. Miller, M.J.; Hutson, J.; Fallowfield, H.J. The Adsorption of Cyanobacterial Hepatoxins as a Function of Soil Properties. *J. Water Health* **2005**, *3*, 339–347. [CrossRef]
66. Miller, M.J.; Critchley, M.M.; Hutson, J.; Fallowfield, H.J. The Adsorption of Cyanobacterial Hepatotoxins from Water onto Soil during Batch Experiments. *Water Res.* **2001**, *35*, 1461–1468. [CrossRef]
67. Rangel, L.M.; Silva, L.H.S.; Rosa, P.; Roland, F.; Huszar, V.L.M. Phytoplankton Biomass Is Mainly Controlled by Hydrology and Phosphorus Concentrations in Tropical Hydroelectric Reservoirs. *Hydrobiologia* **2012**, *693*, 13–28. [CrossRef]
68. Soares, M.C.S.; Maria, M.I.; Marinho, M.M.; Azevedo, S.M.F.O.; Branco, C.W.C.; Huszar, V.L.M. Changes in Species Composition during Annual Cyanobacterial Dominance in a Tropical Reservoir: Physical Factors, Nutrients and Grazing Effects. *Aquat. Microb. Ecol.* **2009**, *57*, 137–149. [CrossRef]

toxins MDPI

Article

Chitosan as a Coagulant to Remove Cyanobacteria Can Cause Microcystin Release

Maíra Mucci [1,*], Iame A. Guedes [2], Elisabeth J. Faassen [1,3] and Miquel Lürling [1]

1 Aquatic Ecology and Water Quality Management Group, Department of Environmental Sciences, Wageningen University, Droevendaalsesteeg 3a, 6708 PB Wageningen, The Netherlands; els.faassen@wur.nl (E.J.F.); miquel.lurling@wur.nl (M.L.)
2 Laboratory of Microbiology, Wageningen University, Stippeneng 4, 6708 WE Wageningen, The Netherlands; iame.alvesguedes@wur.nl
3 Wageningen Food Safety Research, Wageningen Research, Akkermaalsbos 2, 6708 WB Wageningen, The Netherlands
* Correspondence: maira.mucci@wur.nl

Received: 16 October 2020; Accepted: 6 November 2020; Published: 10 November 2020

check for updates

Abstract: Chitosan has been tested as a coagulant to remove cyanobacterial nuisance. While its coagulation efficiency is well studied, little is known about its effect on the viability of the cyanobacterial cells. This study aimed to test eight strains of the most frequent bloom-forming cyanobacterium, *Microcystis aeruginosa*, exposed to a realistic concentration range of chitosan used in lake restoration management (0 to 8 mg chitosan L^{-1}). We found that after 1 h of contact with chitosan, in seven of the eight strains tested, photosystem II efficiency was decreased, and after 24 h, all the strains tested were affected. EC_{50} values varied from 0.47 to >8 mg chitosan L^{-1} between the strains, which might be related to the amount of extracellular polymeric substances. Nucleic acid staining (Sytox-Green®) illustrated the loss of membrane integrity in all the strains tested, and subsequent leakage of pigments was observed, as well as the release of intracellular microcystin. Our results indicate that strain variability hampers generalization about species response to chitosan exposure. Hence, when used as a coagulant to manage cyanobacterial nuisance, chitosan should be first tested on the natural site-specific biota on cyanobacteria removal efficiency, as well as on cell integrity aspects.

Keywords: lake restoration; cyanobacteria bloom control; membrane integrity; *Microcystis aeruginosa*; microcystin

Key Contribution: Chitosan is used as a coagulant to remove cyanobacterial nuisance, but it damages membranes of *Microcystis aeruginosa* in a strain-specific manner, leading to the release of intracellular compounds, including microcystins.

1. Introduction

Cyanobacteria play an essential role in oxygen production, being responsible for half of the ocean's primary production [1]. However, cyanobacterial species may form intense blooms under certain conditions, which have severe impacts on water bodies, such as increased water turbidity, nocturnal depletion of oxygen, fish kills, and malodour [2,3]. In addition, cyanobacteria can produce toxins that are harmful to aquatic and terrestrial organisms, including humans and dogs, impeding water bodies use for recreational activities, drinking water production, fishing, and agricultural use and, consequently, causing severe economic losses [4–8]. The main cause of cyanobacterial blooms is the excess of nutrient supply to waterbodies (eutrophication) [9]. Thus, to manage the problem, nutrients must be limited. The classical and most straightforward approach is to reduce the external nutrient input [10–12]; however, adequate catchment control is not always feasible for economic reasons [13]. In addition, in cases where the internal loading is the primary nutrient source due to long-term diffuse load (e.g., [14,15]) the reduction in external nutrient sources will be inefficient [16,17]. Hence, to speed-up system recovery and minimize nuisance, in-lake measures have been recognized as a feasible solution [13,18].

In this context, geo-engineering materials, like the use of low doses of flocculants (e.g., PolyAluminium Chloride (PAC) or iron chloride—"Floc") followed by the addition of natural soils or modified clays (e.g., lanthanum modified bentonite or aluminium modified zeolite—"Lock/Sink") have gained attention as useful tools to mitigate the effects of eutrophication. This "Floc & Lock/Sink" technique can remove cyanobacteria from the water column while blocking P efflux from the sediment [19]. This approach has been implemented effectively using PAC or iron chloride as coagulant [14,15,20]. Recently, an organic coagulant, chitosan, has gained attention as a possible alternative for inorganic metal-based coagulants [21].

Chitosan is an organic polymer synthesized by alkaline deacetylation of chitin, a biopolymer extracted from shellfish and crustaceans [22]. Chitosan acts as a cationic polyelectrolyte when protonated in an acidic medium; thus, its free amino groups interact with the negatively charged cyanobacterial cell wall [22–24]. Due to its long polymer chain, chitosan can also attach to the cells, forming bridges that entrap the cells [22]. Chitosan is frequently viewed as an eco-friendly and non-toxic coagulant [21,24–26], and besides its coagulation property, chitosan is also known for its antimicrobial activities [27–31], and it has been even used to preserve food [32–34].

Several studies have used chitosan to remove cyanobacteria or dinoflagellates from the water column, some using chitosan-Modified Local Soils/Sand (MLS, e.g., [21,25,35–39]), others adding first only chitosan, followed by soils/clays (e.g., [40–43]) and recently a chitosan fiber has been used [44]. However, in only a few of these studies, the possible chitosan effects on the algal cells viability were investigated. From these studies, some did not find any adverse effect on the cyanobacterium *Microcystis aeruginosa* (e.g., [40–42,45]), whereas others showed a detrimental effect on the cyanobacterium *Cylindrospermopsis raciborskii* and growth inhibition in the dinoflagellate *Amphidinium carterae* [25,41]. A more recent study from our group indicated rapid cell lysis of some cyanobacterial species when incubated with chitosan, but a less severe impact on *M. aeruginosa* [46], and in this study, cyanotoxin release was not analyzed. In fact, only a few studies so far have addressed cyanotoxin release caused by chitosan; in some of these studies, toxins were released, and in others, toxins were not released [35,41,47–49].

Our present study aims to extend the knowledge of possible side effects caused by chitosan on the cyanobacterium, *Microcystis aeruginosa*. Possible materials to manage blooms must be efficient, easy to apply, cheap, and safe [50]. Therefore, an environmentally safe management strategy should be selected, and methods that cause cell damage and toxin release must be applied carefully or should be avoided [51].

We tested the response of the most frequently encountered bloom-forming cyanobacterium, *M. aeruginosa* [52–54], to a realistic concentration range of chitosan as used in lake restoration management [49]. Since intraspecific variation was observed in other species [46], we tested eight different strains of *M. aeruginosa*. The effect of chitosan was evaluated by analysing the photosystem II efficiency and filterable Chlorophyll-*a* concentration. Besides, we analysed cell membrane integrity and the extracellular microcystin concentration. We hypothesized (1) that chitosan would negatively affect all the *M. aeruginosa* strains tested only at the highest chitosan dose, (2) that sensitivity to chitosan will not differ between strains, and (3) that cell lysis followed by toxin release will be observed only at the highest chitosan dose.

2. Results

The eight *M. aeruginosa* strains tested were affected differently by chitosan. Considering the effect of chitosan on the Photosystem II (PSII) efficiency, we could divide the *M. aeruginosa* strains based on their response into two groups: (1) the strains that have a delayed response to chitosan (MiRF-1, PCC 7806 ΔmcyB, PCC 7806 and PCC 7820) and (2) the more sensitive strains with an earlier response (SAG 14.85, CYA 140, PCC 7005 and SAG 17.85) (Figure 1). After 1 h of contact with chitosan, hardly any effect on the PSII efficiency from the first group of strains was observed (Figure 1A; Table A1). After 4 h, the effects on PSII efficiency became visible at the highest concentration for the strain PCC 7806 ΔmcyB and from 1 mg L^{-1} for the strain PCC 7820 (Figure 1B), while after 24 h, these effects became more pronounced (Figure 1C), and the PSII efficiencies in strains PCC 7806 and MiRF-1 were reduced at 8 mg chitosan L^{-1} (Figure 1C).

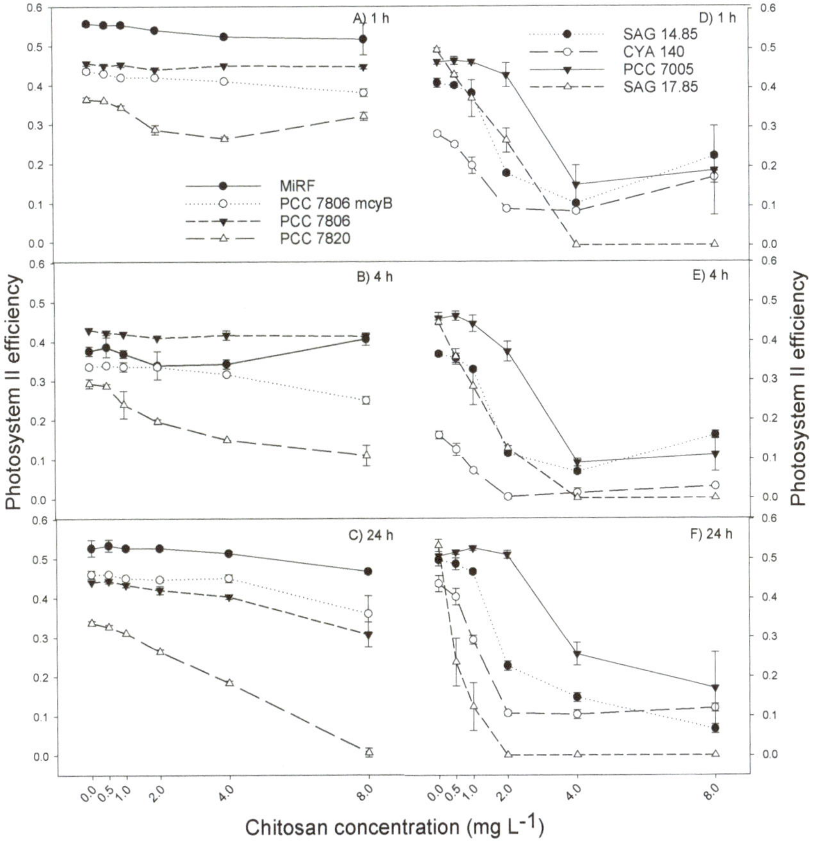

Figure 1. (**A**,**D**) Photosystem II efficiency (PSII) for all 8 strains tested after 1 h, (**B**,**E**) 4 h, and (**C**,**F**) 24 h exposure to different concentrations of chitosan (0 to 8 mg L^{-1}). Error bars indicate standard deviation (n = 3). Grey graphs on the left show the strains with slow response (MiRF-1, PCC 7806 ΔmcyB-, PCC7806, and PCC 7820), and graphs on the right show strains with an earlier response (SAG 1485, CYA 140, PCC 7005 and SAG 1785).

The response of strains from the second group was different; SAG 14.85, CYA 140, PCC 7005, and SAG 17.85 showed already after 1 h a sigmoidal decrease in PSII efficiency with higher chitosan concentrations (Figure 1D). This pattern persisted after 4 and 24 h of chitosan incubation (Figure 1E,F).

The strains PCC 7820 and SAG 17.85 showed an increase in total and extracellular Chlorophyll-*a* concentrations as a response to the chitosan treatments (Figures 2 and A1). At the end of the experiment, the total Chlorophyll-*a* concentration of both strains, when exposed to 8 mg chitosan L^{-1}, was three times higher than in the control. The other six strains used (MiRF-1, PCC 7806 ΔmcyB, PCC 7806, SAG 14.85, PCC 7005, and CYA 140) only increased total Chlorophyll-*a* compared to control after 24 h and in the highest chitosan concentration used. In all strains, pH variation between treatments remained low and below 0.5 units (Figure A2). The addition of acetic acid did not affect PSII efficiency (Figure A3).

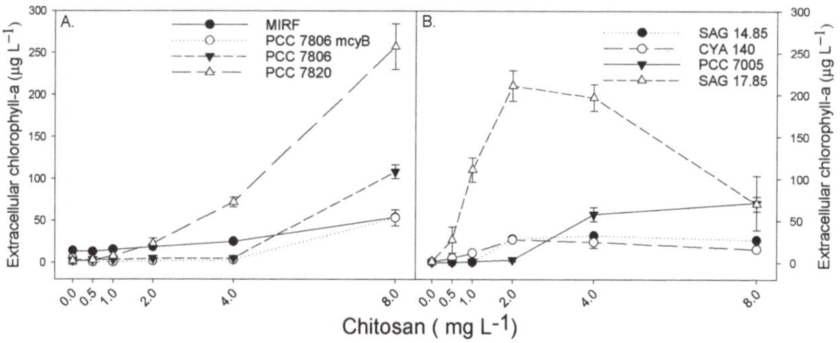

Figure 2. Extracellular Chlorophyll-*a* concentration for MiRF, PCC 7806 ΔmcyB, PCC 7806 and PCC 7820 (**A**) and for SAG 14.85, CYA 140, PCC 7005 and SAG 17.85 (**B**) after 24 h exposure to different concentrations of chitosan (0 to 8 mg L^{-1}). Error bars indicate standard deviation (n = 3).

In the first group of strains (MiRF-1, PCC 7806 ΔmcyB, PCC 7806, and PCC 7820), extracellular Chlorophyll-*a* concentrations were elevated at the highest chitosan dose (Figure 2A). On the contrary, in the second group, at lower chitosan doses, elevated extracellular Chlorophyll-*a* concentrations were observed (Figure 2B). Extracellular Chlorophyll-*a* concentrations differed considerably among strains, with the highest concentration found in strain PCC 7820 (257 µg extracellular Chlorophyll-*a* L^{-1}) and the lowest in strain CYA 140 (16 µg extracellular Chlorophyll-*a* L^{-1}), both at the 8 mg chitosan L^{-1} treatment (Figure 2).

EC_{50} values for MiRF-1, PCC 7806 ΔmcyB, and PCC 7806 could not be calculated because the values exceeded the highest dose used (8 mg L^{-1}) (Table 1). SAG 17.85 was the most sensitive strain with the lowest EC_{50} value (0.47 mg chitosan L^{-1}) followed by CYA 140 (1.06 mg chitosan L^{-1}), SAG 14.85 (1.71), PCC 7005 (3.44) and PCC 7820 (4.51, Table 1). One-way ANOVA showed a difference between the strains ($F_{4,10} = 47.74$; $p < 0.001$) and the Tukey post-hoc test divided the strains into three different groups: (1) SAG 17.85, CYA 140 and SAG 14.85 were the most sensitive, (2) followed by PCC 7005 and 3) PCC 7820 (Table A1). MiRF-1, PCC 7806 ΔmcyB, and PCC 7806 were the least sensitive.

Table 1. Mean EC_{50} values (mg L^{-1}; with standard deviation, n = 3) of chitosan for the Photosystem II efficiency in different *M. aeruginosa* strains. Letters (A, B and C) represent homogenous groups (Tukey pairwise comparisons).

***M. aeruginosa* Strain**	**EC_{50}-24 h (mg L^{-1})**
MiRF-1	>8
PCC 7806 ΔmcyB	>8
PCC 7806	>8
PCC 7820	4.51 (0.37) $p < 0.0001^{A}$
PCC 7005	3.44 (0.42) $p < 0.0001^{B}$
SAG 14.85	1.71 (0.08) $p < 0.0001^{C}$
CYA 140	1.06 (0.04) $p < 0.0001^{C}$
SAG 17.85	0.47 (0.05) $p < 0.0001^{C}$

Extracellular microcystins (MCs) concentrations were below the detection level in the filtrates from incubations of MiRF-1 exposed to 0 to 4 mg chitosan L^{-1}, while the variant MC-LR was detected at 8 mg chitosan L^{-1}, but below the level of quantification. Likewise, in strain CYA 140, no extracellular MCs were detected in incubations exposed to 0 to 1 mg chitosan L^{-1}, whereas MC-LR was detected, yet not quantifiable, at 2, 4, and 8 mg chitosan L^{-1}. On the other hand, in the strain PCC 7820, extracellular MCs increased with an increasing chitosan concentration (one-way ANOVA; $F_{5,11} = 4516.5$; $p < 0.001$), and already at 1 mg chitosan L^{-1}, the extracellular MC concentration was significantly higher than in the control (Figure 3A). In the strain PCC 7806, the extracellular MC variants LR and dmLR increased only at the highest chitosan dose (*Kruskal–Wallis* One Way Analysis of Variance on Rank; s $H_5 = 16.251$; $p = 0.006$) (Figure 3B). The MC analysis was not affected by the presence of 8 mg of chitosan L^{-1}, as demonstrated in an incubation experiment (Student's *t*-test, $p = 0.552$, Figure A4).

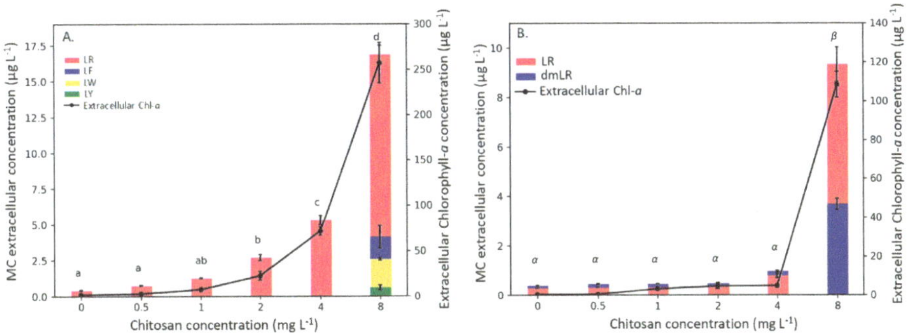

Figure 3. Extracellular MCs (bars) and extracellular Chlorophyll-*a* (line) after 24 h of chitosan exposure for PCC 7820 (**A**) and PCC 7806 (**B**). Errors bars indicate standard deviation ($n = 3$). Letters represent a statistical difference (Tukey pairwise comparisons $p < 0.05$).

For the four most sensitive strains (SAG 14.85, PCC 7005, PCC 7820, SAG 17.85), except for CYA 140, the cell membrane permeability test showed differences between cells exposed to chitosan at each of the concentrations used and the non-exposed (control) cells (Figure A5). Most of the strains had similar results as the strain PCC 7820: the non-exposed cells showed only the natural red fluorescence (Figure 4; panel control B) and no intracellular accumulation of Sytox Green (Figure 4; panel control C). However, in the treatment with chitosan, intracellular accumulation of Sytox Green was observed (Figure 4; panel 2 mg L^{-1} C), while the accumulation was even more substantial at the highest chitosan dose (8 mg L^{-1}), indicating membrane damage (Figure 4; panel 8 mg L^{-1} C). In the less sensitive strains MiRF-1, PCC 7806 ΔmcyB, and PCC7806, such intracellular accumulation of Sytox Green was only strongly observed at the highest concentration (Figure A6). For certain strains (SAG 14.85, PCC 7005, PCC 7806 and PCC 7806 ΔmcyB), it was not possible to analyze at the highest concentration the cell damage via Sytox because the cells were already destroyed.

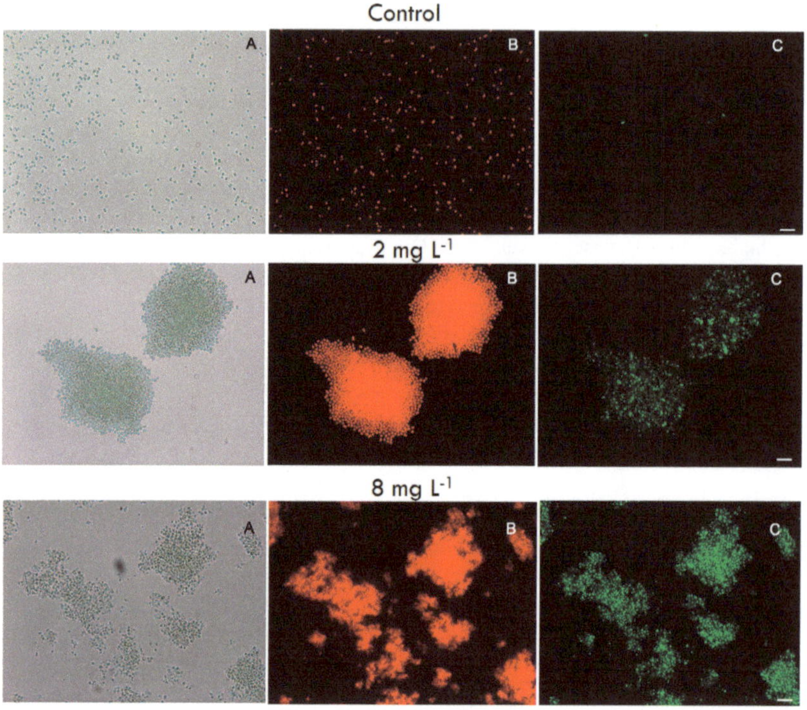

Figure 4. Fluorescence images of PCC 7820 cells in the control (upper pictures **A–C**), 2 mg chitosan L^{-1} (middle pictures **A–C**), and 8 mg chitosan L^{-1} (lower pictures **A–C**). The panels A shows bright-field images, B the cyanobacteria autofluorescence (red), and C the intracellular accumulation of Sytox® Green (green). Scale bar: 20 µm.

3. Discussion

Chitosan has recently received attention as a coagulant to remove cyanobacteria from waterbodies. It has good flocking properties depending on water chemistry [55] and, combined with a ballast, can remove cyanobacteria effectively from the water column [21,25,45]. However, effects on the viability of flocked cyanobacteria have received less attention but are of great importance as chitosan may cause cell membrane damage in bacteria [56] and may cause cyanotoxin release [41]. Our study filled in this research gap by testing the response of eight different *Microcystis* strains to chitosan, while also measuring extracellular MCs.

Our results are not in agreement with the hypothesis that chitosan would affect the *M. aeruginosa* strains only at a high dose. In five strains, an immediate negative impact was detected, while in three less sensitive strains (MiRF-1, PCC 7806, and PCC 7806 ΔmcyB), a significant reduction in PSII efficiency was observed only at the highest chitosan dose tested. PSII is one of the reaction centers responsible for transporting energized electrons to accomplish photosynthesis [57]; thus, a decrease in PSII efficiency reflects damage to the thylakoid membrane and gives insight into the physiological status of the cells. The reduction in PSII efficiency most probably reflects increased membrane permeability and cell lysis that is a result of the cationic amino groups ($C\text{-}H_3^+$) of chitosan interacting with negatively charged cyanobacterial cell membranes [56]. A significant reduction in the PSII efficiency of MiRF-1 cells was also observed in our previous work [46]. In contrast, some studies found no decrease in PSII efficiency at similar chitosan concentrations [40–42,45]. These studies had incubated lake water infested with cyanobacteria only for one hour, which might be too short to evoke a measurable effect. The strains

SAG 14.85, CYA 140, PCC 7005, and SAG 17.85 were strongly affected after 1 h incubation, but in MIRF-1, PCC 7806, PCC 7820, and PCC 7806 ΔmcyB, negative effects became apparent after 24 h incubation. Hence, the exposure time might be an important factor.

We refute the second hypothesis that the sensibility to chitosan would be similar in all strains; here, we showed that the EC_{50} varied from 0.47 to >8 mg chitosan L^{-1}. Mucci et al. [46] also found an intraspecific variability (EC_{50} of 0.41 and >8 mg chitosan L^{-1}) between two strains of *Planktothrix agardhii* equal to what we observed here for *M. aeruginosa*. All the strains tested in this study were uni-and bicellular, which implies that the among strain variability seems to be caused by strain specific characteristics rather than a colonial or unicellular appearance. In addition, the presence or absence of MCs is not related to chitosan sensitivity, as both the MC-producing wild-type PCC 7806 and its MC-lacking mutant PCC 7806 ΔmcyB were equally sensitive.

The among strain variability can be explained by differences in the composition of the outer layer and amount/composition of extracellular polysaccharides (EPS) [58]. EPS are mainly composed of polysaccharides and proteins [59]. Due to a large number of negatively charged functional groups, their efficiency in removing heavy metals and organic contaminants protects the cells [60–64]. EPS also protects *M. aeruginosa* against strong oxidizers like hydrogen peroxide [60], and consequently, follow-up studies could explore the role of EPS in among strain variability and among species variability in sensitivity to chitosan. Another factor that might play a role in chitosan's sensibility is the charge density of the membrane in each species/strain. Positively charged chitosan will have electrostatic interactions with the negatively charged cell wall of the cyanobacteria, where a higher negative charge density will lead to a stronger interaction with chitosan [65]. Stronger interactions can cause membrane destabilization and disruption of the membrane, leading to leakage of intracellular substances [29], as observed in our study.

A literature research on the use of chitosan to flocculate cyanobacterial revealed that about half of the studies did not analyse any cell health aspects (Table A2). Considering chitosan as a tool to be applied in water bodies to remove cyanobacteria, it is important not only to look at removal/coagulation efficiency but also on possible side effects on the aquatic community. From the studies that included a cell viability indicator, one third showed a negative effect of chitosan on the cells, and two-thirds did not report any adverse effect (Table A2).

In analogy with our expectation that *M. aeruginosa* would only be affected at the highest chitosan dose, we hypothesized that only at these exposures, microcystins would be released from the cells. Our results, however, are not in line with this third hypothesis. Extracellular microcystins (MCs) could be detected in the filtrates from all the strains tested, albeit not always at levels allowing quantification. Nonetheless, in strain PCC 7820 already at 2 mg chitosan L^{-1} MC release was significantly higher than in the controls. In the strain PCC 7806, extracellular MC concentration was elevated at 4 mg L^{-1} and increased at 8 mg chitosan L^{-1}. Since the MC concentrations in these treatments were high, such chitosan doses should be used with care when used to treat blooms in drinking water supplies. The chitosan capacity to remove extracellular MCs has been reported and can be substantial (e.g., [47,66]). In addition, Miranda et al. [41] using field samples dominated by *M. aeruginosa* showed that exposure for two hours to chitosan significantly lowered extracellular MC concentration. The positively charged chitosan molecules probably interact with negatively charged microcystins that have a -1 charge over a broad pH range [67]. In our study, however, a clearly different chitosan effect was observed, namely the release of intracellular MCs. It is likely that extracellular MCs were first reduced, but evidently the chitosan-induced cell leakage led to significantly enhanced extracellular MC concentrations compared to non-exposed cells. Likewise, when Miranda et al. [41] used water dominated by the sensitive cyanobacterium *Cylindrospermopsis raciborskii*, they found not only strongly reduced PSII efficiency but also enhanced extracellular saxitoxins. Hence, differences in sensitivity of *Microcystis* used, exposure duration too short to evoke cell lysis (e.g., [41,66]) or matrix effects on the MC detection could underlie the apparent differences. Pei et al. [47] used an ELISA kit to measure MCs, but possible matrix effects on the antibodies were not determined. Our study showed that chitosan did not interfere with our LC-MS/MS method for MC analysis.

Studies that combined chitosan and a ballast compound revealed a reduction in extracellular MCs. For instance, a mesocosm experiment performed by Pan et al. [49] showed that chitosan-modified soil (MLS) decrease dissolved MC. Similarly, Li and Pan [35] found a reduction in MC when MLS was applied, however, when only chitosan was added, an increase in MC was observed, indicating that the decreased MC concentrations might be related to the soil MC adsorption capacity instead of to chitosan. Miranda et al. [41] found that extracellular MC concentrations were significantly reduced in treatments where chitosan was combined with a ballast compared to in chitosan only treatments. In contrast, they found higher extracellular saxitoxin concentrations when *C. raciborskii* was exposed either to chitosan alone or chitosan combined with soils and clay. While electrostatic interactions of chitosan with MCs can be expected, this is less likely for positively charged saxitoxins [68]. The study of Miranda et al. [41] underpins that when chitosan is to be applied in drinking water reservoirs, depending on the cyanobacteria prevailing, corresponding cyanotoxin analysis is strongly advised.

Any material used to mitigate cyanobacterial nuisance that causes the release of toxins is a double-edged sword. On one side, if the nuisance is reduced, this will be viewed as positive, but if cells are killed rapidly and toxins released, the water body might not be suitable for drinking, irrigation, or recreational purposes [69]. However, when such cell death happens later and close to the sediment, released toxins can be degraded (e.g., [35]) with far less impact on ecosystem functionality. Thus, the use of ballast together with chitosan (a "floc and sink" approach or the MLS technique) seems a better strategy than using only chitosan, not only because it might prevent higher concentrations of toxins in the water column but also because a ballast prevents cell accumulation at the water surface. Recently, it has been shown that damaging the cells first with hydrogen peroxide before adding the coagulant and the ballast (kill, Floc & Sink) could be a promising approach to keep *P. rubescens* precipitated [70]; in this case, chitosan could also be an alternative if used together with a ballast.

The increase in total Chlorophyll-*a* (Figure A2) does not reflect an increase in biomass but is a result of pigments leaking out of the cells, which was confirmed by the increase in extracellular Chlorophyll-*a* (Figure 2). A rapid increase in extracellular Chlorophyll-*a* is a strong indicator of cell lysis, as is a rapid increase in extracellular MC concentration. Cell lysis implies membrane damage, which was confirmed by the membrane viability assay, wherein for all the strains tested, a green fluorescence was observed at the highest chitosan concentration used (for example Figures 4, A5 and A6). Sytox Green has a high affinity with nucleic acids, however, it is not able to penetrate living cells. Yet, when the membrane integrity is compromised, the stain can colour the genetic material with a bright green colour [71], as observed here in the chitosan treatments. The absence of green colour in the control, but the presence of natural red fluorescence of cyanobacteria, indicates no membrane damage in the controls (Figure 4). In the strains in which it was possible to quantify MCs, a significant positive linear relation between MC concentration and extracellular Chlorophyll-*a* was observed ($r^2 = 0.98$ $p < 0.0001$ for PCC7820 and $r^2 = 0.99$ $p < 0.0001$ for PCC 7806) (Figure 2). Hence, when dissolved toxins analysis is not possible, extracellular Chlorophyll-*a* might be used as a surrogate to give insight into possible toxin release.

The strain CYA 140 was the second most sensitive strain (Table 1), yet the extracellular Chlorophyll-*a* in the chitosan treatments was not as high as for other strains such as PCC 7820 and PCC 7806, which showed higher EC_{50}. The absence of high concentrations of extracellular Chlorophyll-*a* for CYA 140 could be related to quick degradation of Chlorophyll-*a*. It is well known that dissolved chlorophyll-a is extremely unstable, and it will be degraded when exposed to light. It could also be related to a slower membrane damage, so the cells are not physiologically well, thus EC_{50} is low but intercellular contents are released slower. Clearly, more research is needed to decipher the cause of the observed differences between strains.

Geo-engineering materials used to manage eutrophication and control cyanobacterial blooms must be efficient, easy to apply, cheap and safe, which means it is important to be aware of all the consequences that any material can cause in the ecosystem [50]. Here, Chitosan was able to damage *M. aeruginosa* cells causing cell lysis and consequently microcystin and pigment release. These effects were, however, strain dependent. Evidently, these trials need a follow up with natural seston dominated by *M. aeruginosa*, which in the field is usually found in its typical colonial form contrasting the unicellular morphology in laboratory cultures [72]. Considering the high diversity of cyanobacteria when chitosan is considered to be used in lake restoration, the best approach to understand its effects is to test it directly on the natural biota being targeted. Such tests should include not only coagulation efficiency but also cell viability. We highlighted the importance of controlled experiments to understand the implications and efficiency of materials used to mitigate cyanobacterial nuisance. Such tests are the first step, and to predict real effects a tiered approach from laboratory to field tests is needed.

4. Materials and Methods

4.1. *Microcystis aeruginosa* Cultures

The eight different strains used in the experiments were obtained from different culture collections (Table 2) and were cultivated on modified WC medium [73] under controlled conditions at 22 °C with a 16:8 h light–dark cycle and 45 µmol quanta m^{-2} s^{-1} light intensity. Before the experiment, the cultures were refreshed twice (around two weeks interval), always in the exponential phase.

Table 2. *Microcystis aeruginosa* strains used in the experiments and the microcystin (MC) variants that have been found in them.

Strain ID	Acquired from	Microcystins (MCs) Produced
MiRF-1	Laboratory of Ecophysiology and Toxicology of Cyanobacteria (Brazil)	dm-MC-LR, MC-LR, MC-LY, MC- LW, MC-LF [74]
PCC 7806 ΔmcyB	Pasteur Culture Collection (France)	None [75]
PCC 7806	Pasteur Culture Collection (France)	dm-MC-LR, MC-LR [76]
PCC 7005	Pasteur Culture Collection (France)	None detected (this study)
PCC 7820	Pasteur Culture Collection (France)	dm-MC-LR, MC-LR, MC-LY, MC-LW, MC-LF [77]
SAG 14.85	Sammlung von Algenkulturen der Universität Göttingen (Germany)	dm-MC-LR, MC-LR (unpublished data)
SAG 17.85	Sammlung von Algenkulturen der Universität Göttingen (Germany)	dm-MC-LR, MC-LR, MC-YR [76]
CYA 140	Norwegian Institute for Water Research (Norway)	dm-MC-LR, MC-LR [76]

4.2. Chitosan

Chitosan was obtained from Polymar Ciência e Nutrição S/A, and the deacetylation degree was 86.3% (Batch-010913, Fortaleza, CE, Brazil), and there is no information on the molecular weight. The chitosan (made of shrimp shells) was acidified with 96% acetic acid solution (Merck, analytical grade, VWR International B.V., Amsterdam, The Netherlands), yielding a final concentration of 0.1% acetic acid.

4.3. Experiment Design

Aliquots of *M. aeruginosa* were transferred to 100 mL Erlenmeyer containing 50 mL of modified WC medium, yielding a final concentration of 100 µg Chlorophyll-*a* L^{-1}. Six concentrations of Chitosan were used (0, 0.5, 1, 2, 4 and 8 mg L^{-1}) based on the concentrations frequently used to flocculate cyanobacteria in lake restoration [26,35,37,40,45]. The experiment was done in triplicate. An extra control was added in which only acetic acid was added in the same dose as in chitosan treatment to check if the acetic acid in which chitosan was dissolved had any influence on *M. aeruginosa* cells. After the addition of chitosan or acid acetic, the flasks were mixed and placed in the laboratory at 22 °C in 16:8 h light–dark cycle at 45 µmol quanta m^{-2} s^{-1}. After 1, 4, and 24 h, subsamples were taken to measure the total Chlorophyll-*a* concentration and Photosystem efficiency II (PSII) through PHYTOPAM phytoplankton analyser (Heinz WalzGmbH, Effeltrich, Germany). Additionally, at the end of the experiment, 3 mL samples from each flask were filtered through a filter unit (Aqua 30/0.45CA, Whatman®, VWR International B.V., Amsterdam, The Netherlands) and measured again in the PHYTOPAM to quantify Chlorophyll-*a* released from the cells. After 24 h, pH was measured in each flask, and 8 mL samples were filtered through glass fiber filters (GF/C, Whatman®, VWR International B.V., Amsterdam, The Netherlands) and placed in glass tubes for dissolved microcystin (MC) analysis. The samples were dried in a Speedvac concentrator (Savant™ SPD121P, Thermo Fisher Scientific, Asheville, NC, USA) and were reconstituted in 900 µL methanol (J.T. Baker®, 97%, VWR International B.V., Amsterdam, The Netherlands). After that, the reconstituted samples were transferred to a 1.5 mL tube with a cellulose-acetate filter and centrifuged for 5 min at 16,000× *g*. The filtrates were transferred to amber glass vials and analysed for eight MC variants (MC–dmRR, RR, YR, dmLR, LR, LY, LW, and LF) using LC-MS/MS according to Lürling and Faassen [6]. The MC analysis was performed for the strains MiRF-1, PCC 7806, PCC 7820, and CYA 140 (Table 2).

4.4. Cell Membrane Permeability

An aliquot from each replica was taken, joined, and centrifuged at 5000× *g* for 10 min to evaluate chitosan's effect on membrane integrity, immediately after 24 h of exposure. The pellet was stained with Sytox® Green (Thermo Fisher Scientific, Waltham, MA, USA) at a final concentration of 1nM for 30 min in the dark. The samples were observed under a fluorescence microscope (ZEISS, Axioimager D2, Jena, Germany) using the filter long pass for Fluorescein (450–490 for excitation and 515 nm for emission). Sytox® Green binds to nucleic acid, but it cannot penetrate the cell membrane. However, a damaged membrane allows the stain to infiltrate, resulting in a green fluorescence colour when analysed in a fluorescence microscope.

4.5. Matrix Effect on MC Analysis

The possible effect of chitosan on the toxins analysis was evaluated by incubating pure microcystin mix standards (all eight variants: MC–dmRR, RR, YR, dmLR, LR, LY, LW, and LF) for 24 h in a solution with 8 mg Chitosan L^{-1} dissolved in WC medium. The control series contained only WC medium and pure microcystin mix standards. The test was performed using three replicas, and MC analysis was executed as mentioned before.

4.6. Data Analysis

The PSII for each strain at each time point was compared between different chitosan concentrations using one-way ANOVA or Kruskal–Wallis One Way Analysis of Variance on Ranks when the normality test (Shapiro–Wilk) or Equal Variance test (Brown–Forsythe) failed. For each strain, the chitosan concentrations that caused a 50% reduction in their PSII efficiency compared to the control (EC_{50}) were determined by non-linear regression using four logistic parameter curves in the software Sigma Plot 13.0. EC_{50} values were statistically compared between strains using one-way ANOVA. Extracellular MC and filterable Chlorophyll-*a* concentration were compared between different chitosan concentrations using the one-way ANOVA or Kruskal–Wallis One Way Analysis of Variance on Ranks when the normality test (Shapiro–Wilk) failed. The effect of chitosan on MC standards was tested by Student's *t*-test.

Author Contributions: Conceptualization, M.M., I.A.G., E.J.F. and M.L.; methodology, M.M., I.A.G., E.J.F. and M.L.; validation, M.M., I.A.G., E.J.F. and M.L.; formal analysis, M.M., I.A.G., E.J.F. and M.L.; investigation, M.M. and I.A.G. writing—original draft preparation, M.M.; writing—review and editing, I.A.G., E.J.F. and M.L.; visualization, M.M. All authors have read and agreed to the published version of the manuscript.

Funding: M.M. was funded by SWB/CNPq (201328/2014-3).

Acknowledgments: Alba Lorente is cordially thanked for the assistance with Figure 3.

Conflicts of Interest: The authors declare no conflict of interest. The funders had no role in the design of the study; in the collection, analyses, or interpretation of data; in the writing of the manuscript, or in the decision to publish the results.

Appendix A

Table A1. F- and *p*-values of one-way ANOVAs and H- and *p*-values of Kruskal–Wallis One Way Analysis of Variance on Ranks when normality tests failed (Shapiro–Wilk) for Photosystem II efficiencies in eight different *M. aeruginosa* strain exposed for 1, 4 and 24 h to six different concentrations chitosan (0 to 8 mg L^{-1}).

M. aeruginosa Strain	Exposure Duration 1 h	4 h	24 h
MiRF-1	$H_5 = 9.93$; $p = 0.077$	$F_{5,12} = 4.65$; $p = 0.014$ *	$F_{5,12} = 13.87$; $p < 0.001$ *
PCC 7806 ΔmcyB	$H_5 = 16.70$; $p = 0.005$ *	$F_{5,12} = 66.34$; $p < 0.001$ *	$H_5 = 12.70$; $p = 0.026$ *
PCC 7806	$H_5 = 12.24$; $p = 0.032$ *	$H_5 = 11.28$; $p = 0.046$ *	$H_5 = 15.81$; $p = 0.007$ *
PCC 7820	$F_{5,12} = 89.62$; $p < 0.001$ *	$F_{5,12} = 47.06$; $p < 0.001$ *	$F_{5,12} = 1079.0$ $p < 0.001$ *
SAG 14.85	$H_5 = 16.317$; $p = 0.006$ *	$F_{5,12} = 876.16$; $p < 0.001$ *	$H_5 = 16.1$; $p = 0.007$ *
CYA 140	$F_{5,12} = 149.94$; $p < 0.001$ *	$F_{5,12} = 144.96$; $p < 0.001$ *	$F_{5,12} = 338.1$; $p < 0.001$ *
PCC 7005	$H_5 = 14.13$; $p = 0.015$ *	$F_{5,12} = 187.49$; $p < 0.001$ *	$H_5 = 15.16$; $p = 0.01$ *
SAG 17.85	$H_5 = 16.74$; $p = 0.005$ *	$H_5 = 16.74$; $p = 0.005$ *	$H_5 = 16.56$; $p = 0.005$ *

* represents the significant statistical differences.

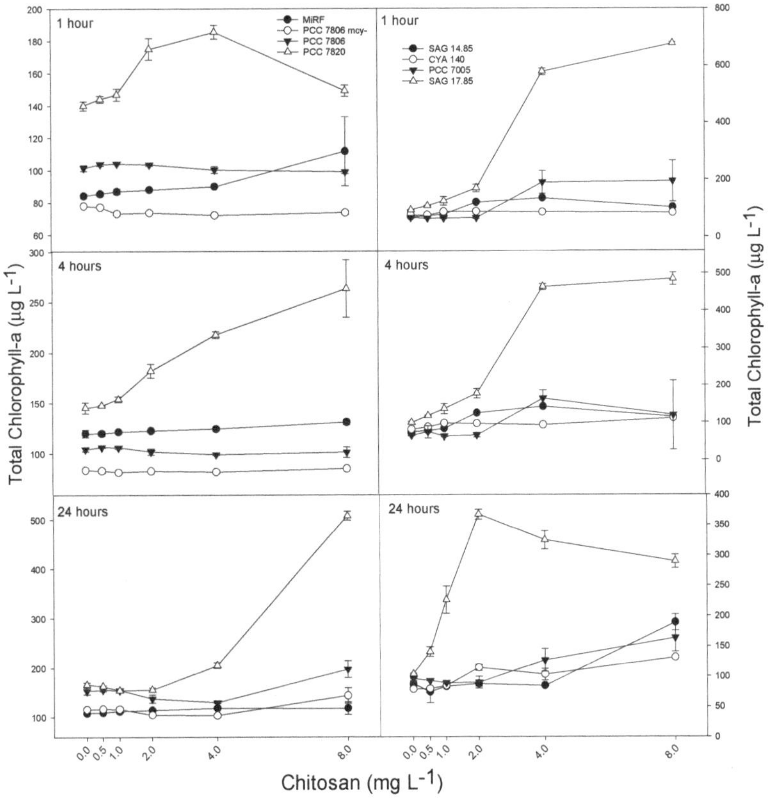

Figure A1. Total Chlorophyll-*a* concentration for all 8 strains tested after 1 h, 4 h and 24 h exposure to different concentrations of chitosan (0 to 8 mg L^{-1}). Error bars indicate standard deviation ($n = 3$).

Graphs on the left show strains with a slow response (MiRF-1, PCC 7806 ΔmcyB, PCC7806, and PCC 7820) and graphs on the right show strains with an earlier response (SAG 14.85, CYA 140, PCC 7005 and SAG 17.85).

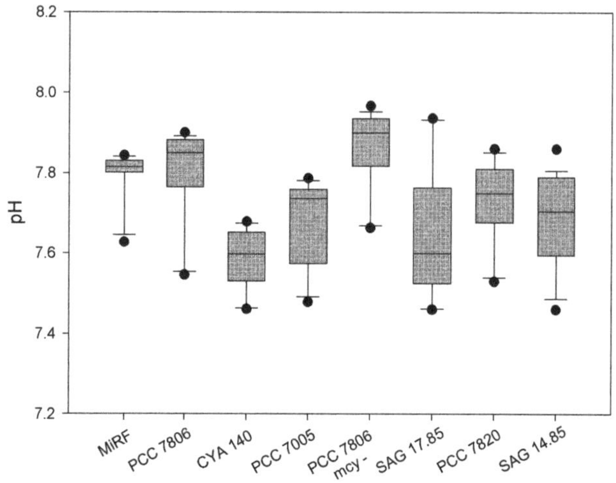

Figure A2. Box plot showing median pH between the treatment, 10th, 25th, 75th and 90th percentiles with error bars (n = 18) for each strain.

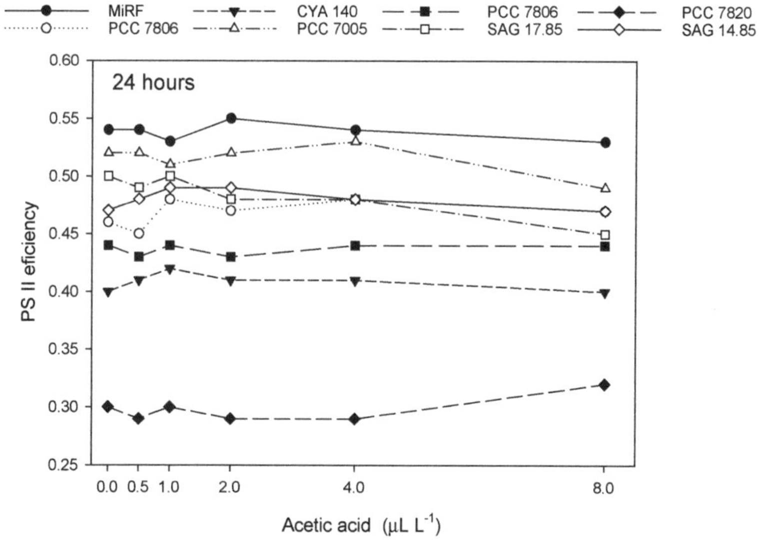

Figure A3. PSII efficiency of all the eight strains tested after 24 h exposure to different acetic acid concentration, the same used for the chitosan treatments.

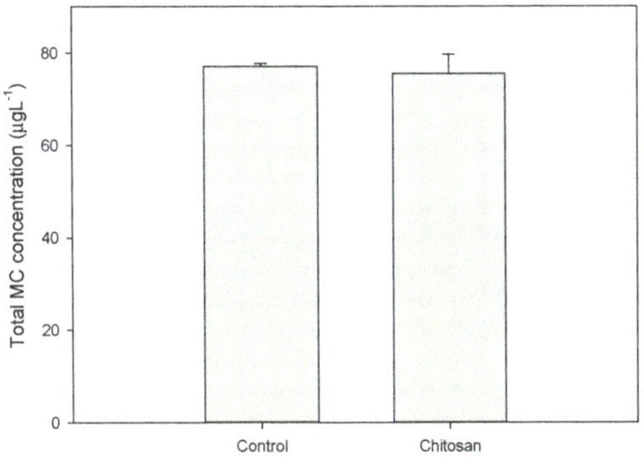

Figure A4. Total MC incubated for 24 h with only WC medium (Control) and with WC medium plus 8 mg chitosan L^{-1} (Chitosan).

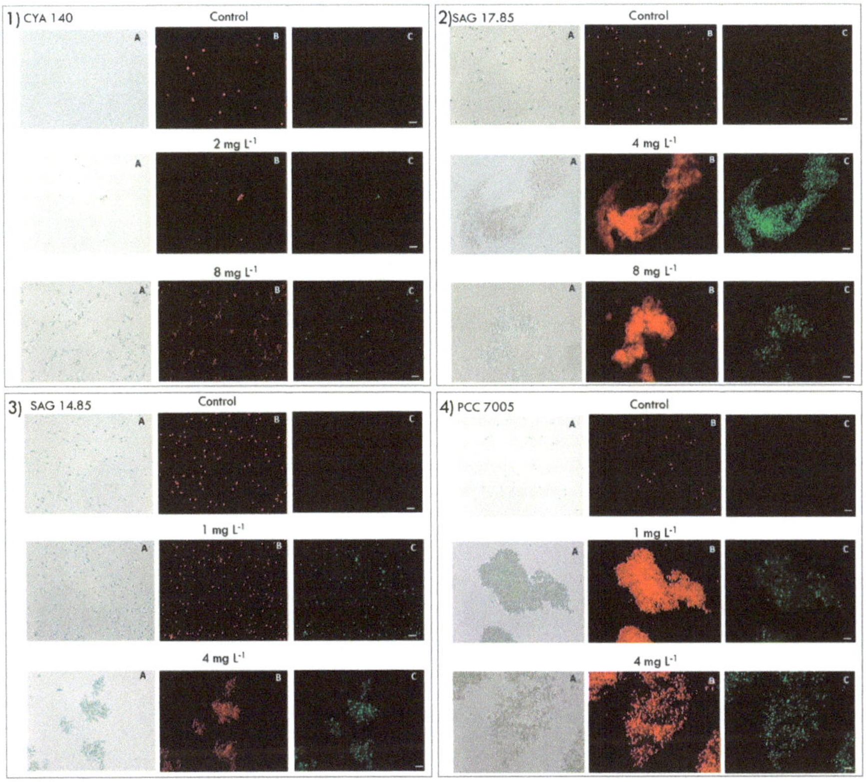

Figure A5. Four panels showing the fluorescence images of four strains (1: CYA 140, 2: SAG 17.85, 3: SAG 14.85 and 4: PCC 7005. For each panel the cells in the control are the upper pictures (**A–C**), 2, 1 or 4 mg chitosan L^{-1} are the middle pictures (**A–C**), and 8 or 4 mg chitosan L^{-1} are the lower pictures (**A–C**). For all the panels, picture A shows bright-field images, B the cyanobacteria autofluorescence (red), and C the intracellular accumulation of Sytox® Green (green). Scale bar: 20 µm.

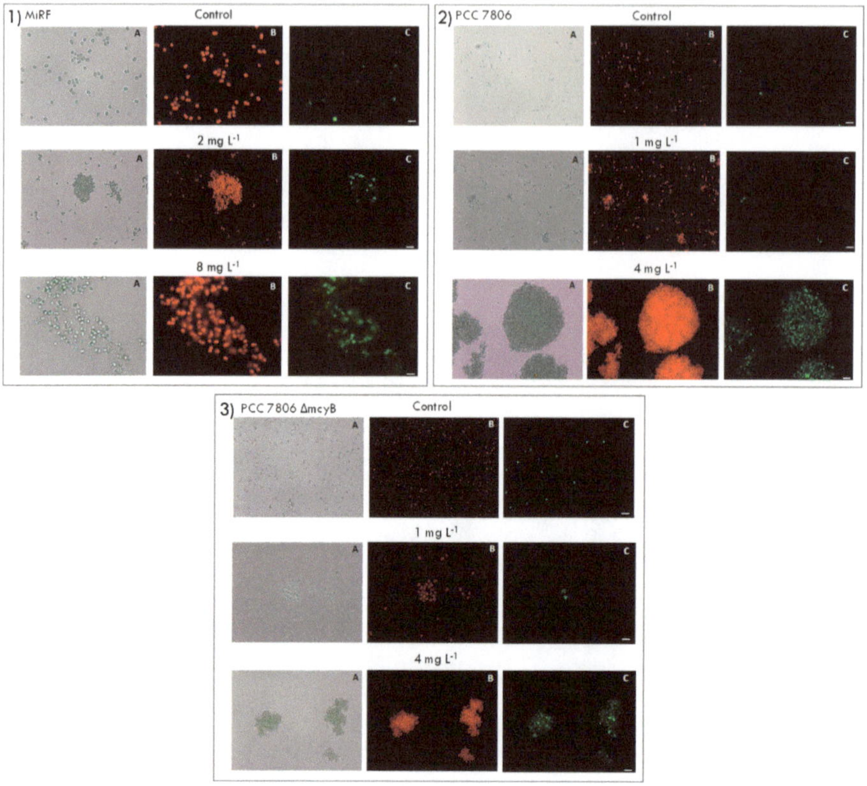

Figure A6. Three panels showing the fluorescence images of three strains (1: MiRF, 2: PCC 7806, and 3: PCC 7806 ΔmcyB. For each panel, the cells in the control are the upper pictures (**A–C**), 2, 1 or 4 mg chitosan L^{-1} are the middle pictures (**A–C**), and 8 or 4 mg chitosan L^{-1} are the lower pictures (**A–C**). For all the panels, picture A shows bright-field images, B the cyanobacteria autofluorescence (red), and C the intracellular accumulation of Sytox® Green (green). Scale bar: 20 μm.

Table A2. Scientific papers that studied chitosan coagulation efficiency or/and the effect on cyanobacteria/dinoflagellates.

Literature Available	Endpoints	Effect
[46]	PSII efficiency and Membrane permeability (MP)	–
[47]	K^+ release and dissolved toxins	–
[78]	Phycocyanin and allophycocyanin release	–
[48]	K^+/M^{2+} release and dissolved toxins	–
This study	PSII efficiency, dissolved toxins, MP and FChl-*a*	–
[40]	PSII efficiency	0
[79]	PSII efficiency	0
[35]	Dissolved toxins	0
[21]	Cell viability and recovery	0
[55]	PSII efficiency	0
[66]	K^+ release	0
[80]	Growth	0
[45]	PSII efficiency	0

Table A2. *Cont.*

Literature Available	Endpoints	Effect
[42]	PSII efficiency	0
[41]	PSII efficiency and dissolved toxins	0 and –
[43]	Dissolved toxins	0 and –
[81]	NR	NR
[82]	NR	NR
[83]	NR	NR
[84]	NR	NR
[35]	NR	NR
[85]	NR	NR
[86]	NR	NR
[38]	NR	NR
[87]	NR	NR
[88]	NR	NR
[37]	NR	NR
[36]	NR	NR
[25]	NR	NR
[26]	NR	NR
[89]	NR	NR

– means negative effect on the biota target, 0 means no negative effect observed and NR means no effect reported.

References

1. Gadd, G.M.; Raven, J.A. Geomicrobiology of Eukaryotic Microorganisms. *Geomicrobiol. J.* **2010**, *27*, 491–519. [CrossRef]
2. Paerl, H.W.; Huisman, J. Climate change: A catalyst for global expansion of harmful cyanobacterial blooms. *Environ. Microbiol. Rep.* **2009**, *1*, 27–37. [CrossRef] [PubMed]
3. Smith, V.H.; Schindler, D.W. Eutrophication science: Where do we go from here? *Trends Ecol. Evol.* **2009**, *24*, 201–207. [CrossRef] [PubMed]
4. Azevedo, S.M.F.; Carmichael, W.W.; Jochimsen, E.M.; Rinehart, K.L.; Lau, S.; Shaw, G.R.; Eaglesham, G.K. Human intoxication by microcystins during renal dialysis treatment in Caruaru—Brazil. *Toxicology* **2002**, *181–182*, 441–446. [CrossRef]
5. Carmichael, W.W.; Azevedo, S.M.F.O.; An, J.S.; Molica, R.J.R.; Jochimsen, E.M.; Lau, S.; Rinehart, K.L.; Shaw, G.R.; Eaglesham, G.K. Human Fatalities from Cyanobacteria: Chemical and Biological Evidence for Cyanotoxins. *Environ. Health Perspect.* **2001**, *109*, 663. [CrossRef] [PubMed]
6. Lürling, M.; Faassen, E.J. Dog poisonings associated with a *Microcystis aeruginosa* bloom in The Netherlands. *Toxins* **2013**, *5*, 556–567. [CrossRef] [PubMed]
7. Dodds, W.K.; Bouska, W.W.; Eitzmann, J.L.; Pilger, T.J.; Pitts, K.L.; Riley, A.J.; Schloesser, J.T.; Thornbrugh, D.J. Eutrophication of U.S. Freshwaters: Analysis of Potential Economic Damages. *Environ. Sci. Technol.* **2009**, *43*, 12–19. [CrossRef]
8. Hamilton, D.P.; Wood, S.A.; Dietrich, D.R.; Puddick, J. Costs of harmful blooms of freshwater cyanobacteria. In *Cyanobacteria*; John Wiley & Sons, Ltd.: Chichester, UK, 2013; pp. 245–256, ISBN 9781118402238.
9. Huisman, J.; Codd, G.A.; Paerl, H.W.; Ibelings, B.W.; Verspagen, J.M.H.; Visser, P.M. Cyanobacterial blooms. *Nat. Rev. Microbiol.* **2018**, *16*, 471–483. [CrossRef]
10. Cooke, G.D.; Welch, E.B.; Peterson, S.A.; Nichols, S. *Restoration and Management of Lakes and Reservoirs*, 3rd ed.; CRC Press-Taylor & Francis Group: Boca Raton, FL, USA, 2005; ISBN 1420032100.
11. Hilt, S.; Gross, E.M.; Hupfer, M.; Morscheid, H.; Mählmann, J.; Melzer, A.; Poltz, J.; Sandrock, S.; Scharf, E.-M.; Schneider, S.; et al. Restoration of submerged vegetation in shallow eutrophic lakes—A guideline and state of the art in Germany. *Limnol. Ecol. Manag. Inl. Waters* **2006**, *36*, 155–171. [CrossRef]
12. Paerl, H.W.; Barnard, M.A. Mitigating the global expansion of harmful cyanobacterial blooms: Moving targets in a human- and climatically-altered world. *Harmful Algae* **2020**, *96*, 101845. [CrossRef]
13. Huser, B.J.; Futter, M.; Lee, J.T.; Perniel, M. In-lake measures for phosphorus control: The most feasible and cost-effective solution for long-term management of water quality in urban lakes. *Water Res.* **2016**, *97*, 142–152. [CrossRef] [PubMed]

14. Lürling, M.; Van Oosterhout, F. Controlling eutrophication by combined bloom precipitation and sediment phosphorus inactivation. *Water Res.* **2013**, *47*, 6527–6537. [CrossRef] [PubMed]
15. Waajen, G.; van Oosterhout, F.; Douglas, G.; Lürling, M. Management of eutrophication in Lake De Kuil (The Netherlands) using combined flocculant—Lanthanum modified bentonite treatment. *Water Res.* **2016**, *97*, 83–95. [CrossRef] [PubMed]
16. Carpenter, S.R. Eutrophication of aquatic ecosystems: Bistability and soil phosphorus. *PNAS* **2005**, *102*, 10002–10005. [CrossRef] [PubMed]
17. Fastner, J.; Abella, S.; Litt, A.; Morabito, G.; Vörös, L.; Pálffy, K.; Straile, D.; Kümmerlin, R.; Matthews, D.; Phillips, M.G.; et al. Combating cyanobacterial proliferation by avoiding or treating inflows with high P load—experiences from eight case studies. *Aquat. Ecol.* **2016**, *50*, 367–383. [CrossRef]
18. Lürling, M.; Mucci, M. Mitigating eutrophication nuisance: In-lake measures are becoming inevitable in eutrophic waters in The Netherlands. *Hydrobiologia* **2020**. [CrossRef]
19. Lürling, M.; Kang, L.; Mucci, M.; van Oosterhout, F.; Noyma, N.P.; Miranda, M.; Huszar, V.L.M.; Waajen, G.; Manzi, M. Coagulation and precipitation of cyanobacterial blooms. *Ecol. Eng.* **2020**, *158*. [CrossRef]
20. Mucci, M.; Waajen, G.; van Oosterhout, F.; Yasseri, S.; Lürling, M. Whole lake application PAC-Phoslock treatment to manage eutrophication and cyanobacterial bloom. *Inl. Waters* **2020**, under review.
21. Li, L.; Pan, G. A Universal Method for Flocculating Harmful Algal Blooms in Marine and Fresh Waters Using Modified Sand. *Environ. Sci. Technol.* **2013**, *47*, 4555–4562. [CrossRef]
22. Yang, R.; Li, H.; Huang, M.; Yang, H.; Li, A. A review on chitosan-based flocculants and their applications in water treatment. *Water Res.* **2016**, *95*, 59–89. [CrossRef]
23. Chen, G.; Zhao, L.; Qi, Y.; Cui, Y. Chitosan and Its Derivatives Applied in Harvesting Microalgae for Biodiesel Production: An Outlook. *J. Nanomater.* **2014**, *2014*. [CrossRef]
24. Renault, F.; Sancey, B.; Badot, P.-M.; Crini, G. Chitosan for coagulation/flocculation processes—An eco-friendly approach. *Eur. Polym. J.* **2009**, *45*, 1337–1348. [CrossRef]
25. Pan, G.; Chen, J.; Anderson, D.M. Modified local sands for the mitigation of harmful algal blooms. *Harmful Algae* **2011**, *10*, 381–387. [CrossRef] [PubMed]
26. Pan, G.; Zou, H.; Chen, H.; Yuan, X. Removal of harmful cyanobacterial blooms in Taihu Lake using local soils III. Factors affecting the removal efficiency and an in situ field experiment using chitosan-modified local soils. *Environ. Pollut.* **2006**, *141*, 206–212. [CrossRef]
27. Allan, C.R.; Hadwigei, L.A. The Fungicidal Effect of Chitosan on Fungi of Varying Cell Wall Composition. *Exp. Mycol.* **1979**, *3*, 285–287. [CrossRef]
28. Kendra, D.F.; Hadwiger~, L.A.; Kendra, D.E. Characterization of the smallest chitosan oligomer that is maximally antifungal to fusarium solani and elicits pisatin formation in *Pisum sativum*. *Exp. Mycol.* **1984**, *8*, 276–281. [CrossRef]
29. Kong, M.; Chen, X.G.; Xing, K.; Park, H.J. Antimicrobial properties of chitosan and mode of action: A state of the art review. *Int. J. Food Microbiol.* **2010**, *144*, 51–63. [CrossRef] [PubMed]
30. No, H.K.; Young Park, N.; Ho Lee, S.; Meyers, S.P. Antibacterial activity of chitosans and chitosan oligomers with different molecular weights. *Int. J. Food Microbiol.* **2002**, *74*, 65–72. [CrossRef]
31. Sudarshan, N.R.; Hoover, D.G.; Knorr, D. Antibacterial action of chitosan. *Food Biotechnol.* **1992**, *6*, 257–272. [CrossRef]
32. Jeon, Y.-J.; Kamil, J.Y.V.A.; Shahidi, F. Chitosan as an Edible Invisible Film for Quality Preservation of Herring and Atlantic Cod. *J. Agric. food Chem.* **2002**, *50*, 5167–5178. [CrossRef]
33. Cao, Z.; Sun, Y. Chitosan-based rechargeable long-term antimicrobial and biofilm-controlling systems. *J. Biomed. Mater. Res. Part A* **2009**, *89A*, 960–967. [CrossRef]
34. Campaniello, D.; Bevilacqua, A.; Sinigaglia, M.; Corbo, M.R. Chitosan: Antimicrobial activity and potential applications for preserving minimally processed strawberries. *Food Microbiol.* **2008**, *25*, 992–1000. [CrossRef]
35. Li, H.; Pan, G. Simultaneous removal of harmful algal blooms and microcystins using microorganism- and chitosan-modified local soil. *Environ. Sci. Technol.* **2015**, *49*, 6249–6256. [CrossRef]
36. Pan, G.; Dai, L.; Li, L.; He, L.; Li, H.; Bi, L.; Gulati, R.D. Reducing the Recruitment of Sedimented Algae and Nutrient Release into the Overlying Water Using Modified Soil/Sand Flocculation- Capping in Eutrophic Lakes. *Environ. Sci. Technol.* **2012**, *46*, 5077–5084. [CrossRef]

37. Zou, H.; Pan, G.; Chen, H.; Yuan, X. Removal of cyanobacterial blooms in Taihu Lake using local soils. II. Effective removal of *Microcystis aeruginosa* using local soils and sediments modified by chitosan. *Environ. Pollut.* **2006**, *141*, 201–205. [CrossRef]
38. Wang, L.; Pan, G.; Shi, W.; Wang, Z.; Zhang, H. Manipulating nutrient limitation using modified local soils: A case study at Lake Taihu (China). *Water Res.* **2016**, *101*, 25–35. [CrossRef]
39. Peng, L.; Lei, L.; Xiao, L.; Han, B. Cyanobacterial removal by a red soil-based flocculant and its effect on zooplankton: An experiment with deep enclosures in a tropical reservoir in China. *Environ. Sci. Pollut. Res.* **2019**, *26*, 30663–30674. [CrossRef]
40. de Magalhães, L.; Noyma, N.P.; Furtado, L.L.; Mucci, M.; van Oosterhout, F.; Huszar, V.L.M.; Marinho, M.M.; Lürling, M. Efficacy of Coagulants and Ballast Compounds in Removal of Cyanobacteria (Microcystis) from Water of the Tropical Lagoon Jacarepaguá (Rio de Janeiro, Brazil). *Estuaries Coasts* **2016**, 1–13. [CrossRef]
41. Miranda, M.; Noyma, N.; Pacheco, F.S.; de Magalhaes, L.; Pinto, E.; Santos, S.; Soares, M.F.A.; Huszar, V.L.; Lürling, M.; Marinho, M.M. The efficiency of combined coagulant and ballast to remove harmful cyanobacterial blooms in a tropical shallow system. *Harmful Algae* **2017**, *65*, 27–39. [CrossRef]
42. Noyma, N.P.; De Magalhães, L.; Miranda, M.; Mucci, M.; Van Oosterhout, F.; Huszar, V.L.M.; Marinho, M.M.; Lima, E.R.A.; Lürling, M. Coagulant plus ballast technique provides a rapid mitigation of cyanobacterial nuisance. *PLoS ONE* **2017**, *12*. [CrossRef]
43. de Lucena-Silva, D.; Molozzi, J.; dos Santos Severiano, J.; Becker, V.; de Lucena Barbosa, J.E. Removal efficiency of phosphorus, cyanobacteria and cyanotoxins by the "flock & sink" mitigation technique in semi-arid eutrophic waters. *Water Res.* **2019**, *159*, 262–273. [CrossRef]
44. Park, Y.H.; Kim, S.; Kim, H.S.; Park, C.; Choi, Y.-E. Adsorption Strategy for Removal of Harmful Cyanobacterial Species *Microcystis aeruginosa* Using Chitosan Fiber. *Sustainability* **2020**, *12*, 4587. [CrossRef]
45. Noyma, N.P.; de Magalhães, L.; Furtado, L.L.; Mucci, M.; van Oosterhout, F.; Huszar, V.L.M.; Marinho, M.M.; Lürling, M. Controlling cyanobacterial blooms through effective flocculation and sedimentation with combined use of flocculants and phosphorus adsorbing natural soil and modified clay. *Water Res.* **2016**, 1–13. [CrossRef]
46. Mucci, M.; Noyma, N.P.; de Magalhães, L.; Miranda, M.; van Oosterhout, F.; Guedes, I.A.; Huszar, V.L.M.; Marinho, M.M.; Lürling, M. Chitosan as coagulant on cyanobacteria in lake restoration management may cause rapid cell lysis. *Water Res.* **2017**, *118*, 121–130. [CrossRef]
47. Pei, H.-Y.; Ma, C.-X.; Hu, W.-R.; Sun, F. The behaviors of *Microcystis aeruginosa* cells and extracellular microcystins during chitosan flocculation and flocs storage processes. *Bioresour. Technol.* **2014**, *151*, 314–322. [CrossRef]
48. Wang, Z.; Wang, C.; Wang, P.; Qian, J.; Hou, J.; Ao, Y.; Wu, B. The performance of chitosan/montmorillonite nanocomposite during the flocculation and floc storage processes of *Microcystis aeruginosa* cells. *Environ. Sci Pollut Res* **2015**, *22*, 11148–11161. [CrossRef]
49. Pan, G.; Yang, B.; Wang, D.; Chen, H.; Tian, B.; Zhang, M.; Yuan, X.; Chen, J. In-lake algal bloom removal and submerged vegetation restoration using modified local soils. *Ecol. Eng.* **2011**, *37*, 302–308. [CrossRef]
50. Lürling, M.; Mackay, E.; Reitzel, K.; Spears, B. Editorial—A critical perspective on geo-engineering for eutrophication management in lakes. *Water Res.* **2016**, *97*, 1–10. [CrossRef]
51. Merel, S.; Walker, D.; Chicana, R.; Snyder, S.; Baurès, E.; Thomas, O. State of knowledge and concerns on cyanobacterial blooms and cyanotoxins. *Environ. Int.* **2013**, *59*, 303–327. [CrossRef]
52. Harke, M.J.; Steffen, M.M.; Otten, T.G.; Wilhelm, S.W.; Wood, S.A.; Paerl, H.W. A review of the global ecology, genomics, and biogeography of the toxic cyanobacterium, *Microcystis* spp. *Harmful Algae* **2016**, *54*, 4–20. [CrossRef]
53. O'Neil, J.M.; Davis, T.W.; Burford, M.A.; Gobler, C.J. The rise of harmful cyanobacteria blooms: The potential roles of eutrophication and climate change. *Harmful Algae* **2012**, *14*, 313–334. [CrossRef]
54. Srivastava, A.; Singh, S.; Ahn, C.-Y.; Oh, H.-M.; Asthana, R.K. Monitoring Approaches for a Toxic Cyanobacterial Bloom. *Environ. Sci. Technol.* **2013**, *47*, 8999–9013. [CrossRef]
55. Lürling, M.; Noyma, N.P.; de Magalhães, L.; Miranda, M.; Mucci, M.; van Oosterhout, F.; Huszar, V.L.M.; Marinho, M.M. Critical assessment of chitosan as coagulant to remove cyanobacteria. *Harmful Algae* **2017**, *66*, 1–12. [CrossRef]
56. Liu, H.; Du, Y.; Wang, X.; Sun, L. Chitosan kills bacteria through cell membrane damage. *Int. J. Food Microbiol.* **2004**, *95*, 147–155. [CrossRef]

57. Witt, H.T. Primary Reactions of Oxygenic Photosynthesis. *Ber. Bunsenges. Phys. Chem* **1996**, *100*, 1923–1942. [CrossRef]
58. Forni, C.; Telo', F.R.; Caiola, M.G. Comparative analysis of the polysaccharides produced by different species of *Microcystis* (Chroococcales, Cyanophyta). *Phycologia* **1997**, *36*, 181–185. [CrossRef]
59. Xu, H.; Cai, H.; Yu, G.; Jiang, H. Insights into extracellular polymeric substances of cyanobacterium *Microcystis aeruginosa* using fractionation procedure and parallel factor analysis. *Water Res.* **2013**, *47*, 2005–2014. [CrossRef]
60. Gao, L.; Pan, X.; Zhang, D.; Mu, S.; Lee, D.-J.; Halik, U.; Zhang, D. Extracellular polymeric substances buffer against the biocidal effect of H_2O_2 on the bloom-forming cyanobacterium *Microcystis aeruginosa*. *Water Res.* **2015**, *69*, 51–58. [CrossRef]
61. Zhang, D.; Wang, J.; Pan, X. Cadmium sorption by EPSs produced by anaerobic sludge under sulfate-reducing conditions. *J. Hazard. Mater.* **2006**, *138*, 589–593. [CrossRef]
62. Ozturk, S.; Aslim, B. Relationship between chromium(VI) resistance and extracellular polymeric substances (EPS) concentration by some cyanobacterial isolates. *Environ. Sci. Pollut. Res.* **2008**, *15*, 478–480. [CrossRef]
63. Bai, L.; Xu, H.; Wang, C.; Deng, J.; Jiang, H. Extracellular polymeric substances facilitate the biosorption of phenanthrene on cyanobacteria *Microcystis aeruginosa*. *Chemosphere* **2016**, *162*, 172–180. [CrossRef]
64. De Philippis, R.; Colica, G.; Micheletti, E. Exopolysaccharide-producing cyanobacteria in heavy metal removal from water: Molecular basis and practical applicability of the biosorption process. *Appl. Microbiol. Biotechnol.* **2011**, *92*, 697–708. [CrossRef] [PubMed]
65. Chung, Y.; Su, Y.; Chen, C.; Jia, G.; Wang, H.; Wu, J.C.G.; Lin, J. Relationship between antibacterial activity of chitosan and surface characteristics of cell wall. *Acta Pharmacol. Sin.* **2004**, *25*, 932–936. [PubMed]
66. Ma, C.; Hu, W.; Pei, H.; Xu, H.; Pei, R. Enhancing integrated removal of *Microcystis aeruginosa* and adsorption of microcystins using chitosan-aluminum chloride combined coagulants: Effect of chemical dosing orders and coagulation mechanisms. *Colloids Surfaces A Physicochem. Eng. Asp.* **2016**, *490*, 258–267. [CrossRef]
67. Lee, J.; Walker, H.W. Adsorption of microcystin-Lr onto iron oxide nanoparticles. *Colloids Surfaces A Physicochem. Eng. Asp.* **2011**, *373*, 94–100. [CrossRef]
68. Shimizu, Y.; Hsu, C.-P.; Genenah, A. Structure of Saxitoxin in Solutions and Stereochemistry of Dihydrosaxitoxins1. *J. Am. Chem. Soc* **1981**, *103*, 605–609. [CrossRef]
69. Chorus, I.; Bartram, J. *Toxic Cyanobacteria in Water: A Guide to Their Public Health Consequences, Monitoring and Management*; Chorus, I., Bartram, J., Eds.; E & FN Spon: London, UK, 1999; ISBN 0419239308.
70. Lürling, M.; Mucci, M.; Waajen, G. Removal of Positively Buoyant Planktothrix rubescens in Lake Restoration. *Toxins* **2020**, *12*, 700. [CrossRef]
71. Tashyreva, D.; Elster, J.; Billi, D. A Novel Staining Protocol for Multiparameter Assessment of Cell Heterogeneity in Phormidium Populations (Cyanobacteria) Employing Fluorescent Dyes. *PLoS ONE* **2013**, *8*. [CrossRef]
72. Geng, L.; Qin, B.; Yang, Z. Unicellular, *Microcystis aeruginosa* cannot revert back to colonial form after short-term exposure to natural conditions. *Biochem. Syst. Ecol.* **2013**, *51*, 104–108. [CrossRef]
73. Lürling, M.; Beekman, W. Palmelloids formation in Chlamydomonas reinhardtii: Defence against rotifer predators? *Ann. Limnol. Int. J. Limnol.* **2006**, *42*, 65–72. [CrossRef]
74. Marinho, M.M.; Souza, M.B.G.; Lürling, M. Light and phosphate competition between *Cylindrospermopsis raciborskii* and *Microcystis aeruginosa* is strain dependent. *Microb. Ecol.* **2013**, *66*, 479–488. [CrossRef] [PubMed]
75. Dittmann, E.; Neilan, B.A.; Erhard, M.; von Döhren, H.; Börner, T. Insertional mutagenesis of a peptide synthetase gene that is responsible for hepatotoxin production in the cyanobacterium *Microcystis aeruginosa* PCC 7806. *Mol. Microbiol.* **1997**, *26*, 779–787. [CrossRef] [PubMed]
76. Ger, K.A.; Faassen, E.J.; Pennino, M.G.; Lürling, M. Effect of the toxin (microcystin) content of *Microcystis* on copepod grazing. *Harmful Algae* **2016**, *52*, 34–45. [CrossRef] [PubMed]
77. Lürling, M.; Meng, D.; Faassen, E.J. Effects of hydrogen peroxide and ultrasound on biomass reduction and toxin release in the cyanobacterium, *Microcystis aeruginosa*. *Toxins* **2014**, *6*, 3260–3280. [CrossRef]
78. Shao, J.; Wang, Z.; Liu, Y.; Liu, H.; Peng, L.; Wei, X.; Lei, M.; Li, R. Physiological responses of *Microcystis aeruginosa* NIES-843 (cyanobacterium) under the stress of chitosan modified kaolinite (CMK) loading. *Ecotoxicology* **2012**, *21*, 698–704. [CrossRef]

79. Guo, P.; Liu, Y.; Liu, C. Effects of chitosan, gallic acid, and algicide on the physiological and biochemical properties of *Microcystis flos-aquae*. *Environ. Sci. Pollut. Res.* **2015**, *22*, 13514–13524. [CrossRef]
80. Rojsitthisak, P.; Burut-Archanai, S.; Pothipongsa, A.; Powtongsook, S. Repeated phosphate removal from recirculating aquaculture system using cyanobacterium remediation and chitosan flocculation. *Water Environ. J.* **2017**, *31*, 598–602. [CrossRef]
81. Capelete, B.C.; Brandão, C.C.S. Evaluation of trihalomethane formation in treatment of. *Water Sci. Technol. Water Supply* **2013**, *13.4*, 1167–1173. [CrossRef]
82. Huang, Y.; Xu, L.; Han, R.; Wang, G.; Wang, J.; Jia, J.; Zhang, P.; Pang, Y. Using chitosan-modified clays to control black-bloom-induced black suspended matter in Taihu Lake: Deposition and resuspension of black matter/clay flocs. *Harmful Algae* **2015**, *45*, 33–39. [CrossRef]
83. Lama, S.; Muylaert, K.; Karki, T.B.; Foubert, I.; Henderson, R.K.; Vandamme, D. Flocculation properties of several microalgae and a cyanobacterium species during ferric chloride, chitosan and alkaline flocculation. *Bioresour. Technol.* **2016**, *220*, 464–470. [CrossRef]
84. Li, L.; Pan, G. Cyanobacterial bloom mitigation using proteins with high isoelectric point and chitosan-modified soil. *J. Appl. Phycol.* **2016**, *28*, 357–363. [CrossRef]
85. Ma, C.; Pei, H.; Hu, W.; Cheng, J.; Xu, H.; Jin, Y. Significantly enhanced dewatering performance of drinking water sludge from a coagulation process using a novel chitosan–aluminum chloride composite coagulant in the treatment of cyanobacteria-laden source water. *RSC Adv.* **2016**, *6*, 61047–61056. [CrossRef]
86. Rakesh, S.; Saxena, S.; Dhar, D.W.; Prasanna, R.; Saxena, A.K. Comparative evaluation of inorganic and organic amendments for their flocculation efficiency of selected microalgae. *J. Appl. Phycol.* **2014**, *26*, 399–406. [CrossRef]
87. Yan, Q.; Yu, Y.; Feng, W.; Pan, G.; Chen, H.; Chen, J.; Yang, B.; Li, X.; Zhang, X. Plankton Community Succession in Artificial Systems Subjected to Cyanobacterial Blooms Removal using Chitosan-Modified Soils. *Microb. Ecol.* **2009**, *58*, 47–55. [CrossRef]
88. Yuan, Y.; Zhang, H.; Pan, G. Flocculation of cyanobacterial cells using coal fly ash modified chitosan. *Water Res.* **2016**, *97*, 11–18. [CrossRef]
89. Pandhal, J.; Choon, W.; Kapoore, R.; Russo, D.; Hanotu, J.; Wilson, I.; Desai, P.; Bailey, M.; Zimmerman, W.; Ferguson, A. Harvesting Environmental Microalgal Blooms for Remediation and Resource Recovery: A Laboratory Scale Investigation with Economic and Microbial Community Impact Assessment. *Biology* **2017**, *7*, 4. [CrossRef]

toxins

MDPI

Article

Removal of Positively Buoyant *Planktothrix rubescens* in Lake Restoration

Miquel Lürling [1,*], Maíra Mucci [1] and Guido Waajen [2]

[1] Aquatic Ecology and Water Quality Management Group, Department of Environmental Sciences, Wageningen University, Droevendaalsesteeg 3a, 6708 PB Wageningen, The Netherlands; maira.mucci@wur.nl
[2] Water Authority Brabantse Delta, Team Knowledge, P.O. Box 5520, 4801 DZ Breda, The Netherlands; g.waajen@brabantsedelta.nl
* Correspondence: miquel.lurling@wur.nl; Tel.: +31-317-489-838

Received: 12 October 2020; Accepted: 3 November 2020; Published: 5 November 2020

check for updates

Abstract: The combination of a low-dose coagulant (polyaluminium chloride—'Floc') and a ballast able to bind phosphate (lanthanum modified bentonite, LMB—'Sink/Lock') have been used successfully to manage cyanobacterial blooms and eutrophication. In a recent 'Floc and Lock' intervention in Lake de Kuil (the Netherlands), cyanobacterial chlorophyll-*a* was reduced by 90% but, surprisingly, after one week elevated cyanobacterial concentrations were observed again that faded away during following weeks. Hence, to better understand why and how to avoid an increase in cyanobacterial concentration, experiments with collected cyanobacteria from Lakes De Kuil and Rauwbraken were performed. We showed that the *Planktothrix rubescens* from Lake de Kuil could initially be precipitated using a coagulant and ballast but, after one day, most of the filaments resurfaced again, even using a higher ballast dose. By contrast, the *P. rubescens* from Lake Rauwbraken remained precipitated after the Floc and Sink/Lock treatment. We highlight the need to test selected measures for each lake as the same technique with similar species (*P. rubescens*) yielded different results. Moreover, we show that damaging the cells first with hydrogen peroxide before adding the coagulant and ballast (a 'Kill, Floc and Lock/Sink' approach) could be promising to keep *P. rubescens* precipitated.

Keywords: in-lake measures; lake restoration; Floc and Lock; Kill; Floc and sink; Hydrogen peroxide; Phoslock; PAC

Key Contribution: Use a low dose of coagulant together with ballast (Floc & Sink/Lock) can be enough to keep *P. rubescens* precipitated. If not; damaging the cell first with H_2O_2 might be needed (Kill; Floc & Sink/Lock) to avoid filaments resurfacing.

1. Introduction

Eutrophication—the over-enrichment of surface waters with nutrients—is the largest water quality issue worldwide [1]. It may result in a massive proliferation of cyanobacteria in lakes, ponds, and reservoirs [2,3]. Inasmuch as several strains of the most abundant, cosmopolite cyanobacteria may produce potent toxins, cyanobacterial blooms may impair ecosystems services, such as drinking water production, irrigation, recreation, aquaculture, and fisheries [4]. Clearly, managing eutrophication and reducing cyanobacterial blooms is a significant priority, but ongoing anthropogenic activities and climate change are predicted to aggravate further eutrophication and cyanobacterial blooms [4–9]. The Organisation for Economic Cooperation and Development (OECD) is already referring to eutrophication and harmful blooms as "becoming a global epidemic" with annual costs associated with nutrient pollution in Australia, Europe, and the USA at over 100 billion USD [10]. Hence, more effort of authorities is needed to control eutrophication and cyanobacterial blooms.

The most logical management strategy to mitigate eutrophication is to reduce the external nutrient inputs to surface water (e.g., [11–13]). However, this is easier said than done. About 70% of the point source nutrient pollution from municipal and industrial wastewater is treated in well-developed countries, while this is only 10% in low-income countries [14]. In OECD countries, nowadays, eutrophication is largely caused by legacies from the past and diffuse nutrient pollution from mostly agricultural activities [10,15]. Nutrient legacies in lake beds have built up over many years and will keep on fueling cyanobacterial blooms for years or decades after a successful reduction of the external nutrient load [16–18]. Recovery can be speeded up by targeting the legacy phosphorus pool [19]. Non-point source diffuse-nutrient pollution is more difficult to tackle and requires catchment-wide measures that may come with time lags of decades to centuries before water quality improves [20,21]. Consequently, in-lake measures are needed to bring real-time relief from either targeting cyanobacteria directly or indirectly via a strong reduction in nutrient availability [22]. A whole range of in-lake measures is proposed, several of which are not effective at all, but effective ones include algaecides, coagulants, and phosphate binders [22].

The combination of a low-dose coagulant (polyaluminium chloride, PAC—'Floc') and a phosphate binder (lanthanum modified bentonite, LMB—'Lock') was applied successfully in the Dutch stratifying Lake Rauwbraken [23]. This 'Floc and Lock' intervention effectively aggregated a developing bloom of *Aphanizomenon flos-aquae*, sedimented the aggregates out of the water column, reduced water column phosphate, strongly lowered sediment phosphate release, and improved water quality for more than 10 years after the intervention [23–25]. Likewise, in Lake De Kuil (the Netherlands) a system analysis revealed that around 95% of the phosphorus loading was released from the sediment, while the lake was also suffering from an *A. flos-aquae* bloom, which made the water authority opt for a "Floc and Lock" intervention too [26]. A low dose of iron chloride (as coagulant—'Floc') was added together with LMB to the lake in 2009 and successfully reduced phosphorus and chlorophyll-*a* concentrations, hampered the P release from the lake bed, and improved water quality [26]. However, continuing diffuse P-inputs undermined the strongly improved water quality and led to a *Planktothrix rubescens* bloom in early 2017, 8 years after the intervention in 2009.

To counteract the developing cyanobacterial bloom and to prevent nuisance during the swimming season, the water authority decided to perform a second 'Floc and Lock' intervention. To this end, Lake De Kuil received a combined PAC and LMB treatment on 8–10 May 2017. At the moment of the application, the lake still experienced water column-dispersed *P. rubescens*. The combination of LMB as ballast and PAC as coagulant was added to clear the water column of these cyanobacteria, whilst injection of LMB in the hypolimnion would control phosphorus release from the sediment. On the first day (8 May), eight tons of LMB were added; on the second day, six tons of PAC (Calflock P-14) was added, both at the surface of the lake; and on the third day, the majority of the LMB (23 tons) was injected 4 m deeper in the water column. The cyanobacterial chlorophyll-*a*, which comprised the vast majority of the total chlorophyll-*a*, was reduced by almost 90%, but after one week, some elevated concentrations were observed that faded away in subsequent weeks (Appendix A; Figure A1) [27]. Because increasing cyanobacteria concentrations one week after the intervention (Figure A1) had not been observed in the previous studies [23,26], laboratory experiments were conducted with *P. rubescens* collected and concentrated from both Lake De Kuil and additional experiments included *P. rubescens* concentrated from Lake Rauwbraken. The experiments tested the hypothesis that some of the entrapped *P. rubescens* could escape sedimented flocs. Additional experiments focused on the possibility to include hydrogen peroxide to kill *P. rubescens* prior to 'Floc and Lock' and tested the hypothesis that a 'Kill, Floc and Lock' technique would not only effectively keep *P. rubescens* down, but also strongly reduce extracellular microcystin concentrations compared to solo hydrogen peroxide treatments.

2. Results

2.1. Floc and Sink Experiment—Ballast Dose

One hour after *P. rubescens* suspensions were treated with PAC (2 mg Al L^{-1}) and different doses of ballast (50, 100 and 200 mg LMB L^{-1}), in each treatment, the cyanobacteria were effectively translocated to the bottom of the tubes, while they accumulated in the top in the controls (Figure 1a). The one-way analysis of variance (ANOVA) indicated significant differences in chlorophyll-*a* concentrations at the water surface ($F_{3,11}$ = 495.8; $p < 0.001$) and at the bottom ($F_{3,11}$ = 835.1; $p < 0.001$) of the test tubes. In the control, the chlorophyll-*a* concentration at the water surface was significantly higher than in the treatments, while at the bottom, it was the opposite (Figure 1a). Hence, controls were significantly different from treatments. Photosystem II efficiencies (Φ_{PSII}) in the top of the tube were similar ($F_{3,11}$ = 2.40; p = 0.143) in controls and treatments; in the bottom they differed ($F_{3,11}$ = 46.6; $p < 0.001$) and were significantly reduced in the 100 and 200 mg LMB L^{-1} treatments compared to control and 50 mg LMB L^{-1} treatment (Figure 1a). Notably, two hours later—first at the lowest LMB dose, and then followed by the 100 mg LMB L^{-1} treatments—settled flocks started to rise due to entrapped oxygen bubbles in the flocks (Appendix B).

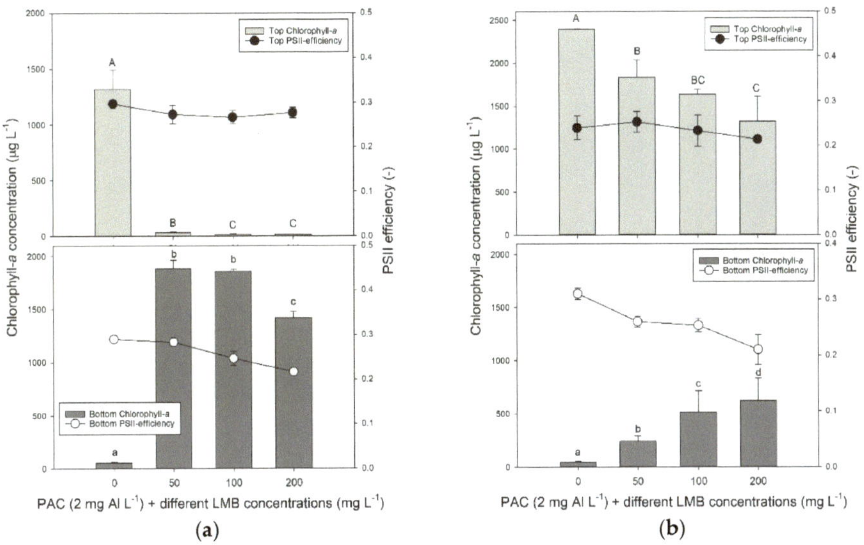

Figure 1. (**a**) chlorophyll-*a* concentrations (µg L^{-1}) in the top 2 mL (top light grey bars) and bottom 2 mL (lower dark grey bars) of 100 mL *P. rubescens* suspensions from De Kuil incubated for 1 h in the absence or presence of different concentrations ballast (50, 100, and 200 mg lanthanum modified bentonite (LMB) L^{-1}) combined with the flocculent polyaluminium chloride (PAC) (2 mg Al L^{-1}). Also included are the Photosystem II efficiencies (PSII) of the cyanobacteria collected at the surface of the tubes (filled circles) and at the bottom (open circles). Error bars indicate 1 standard deviation (SD, n = 3). Similar letters indicate homogeneous groups that are not different at the $p < 0.05$ level. (**b**) Similar to the panel (a), but now after 24 h incubation.

After 24 h, most *P. rubescens* had surfaced again (Figure 1b; Appendix B). A one-way ANOVA indicated significant differences in chlorophyll-*a* concentrations at the water surface ($F_{3,11}$ = 19.3; $p < 0.001$) and at the bottom (log-transformed data; $F_{3,11}$ = 45.8; $p < 0.001$) of the test tubes. The LMB dose had a significant effect on the amount of chlorophyll surfacing and remaining at the bottom, with more chlorophyll accumulating at the lowest LMB dose and the least at the highest dose (Figure 1b). The Φ_{PSII} in the top of the tube were similar ($F_{3,11}$ = 1.33; p = 0.330) in controls and treatments; in the bottom they differed ($F_{3,11} = 19.5$; $p < 0.001$) and were significantly reduced in the 100 and 200 mg LMB L^{-1} treatments compared to control and 50 mg LMB L^{-1} treatment (Figure 1b). The pH in the controls (pH = 8.38 ± 0.16) was significantly higher ($F_{3,11} = 12.8$; $p = 0.002$) than the pH in the three treatments, which were a pH of 8.10 (± 0.09) in the 50 mg LMB L^{-1} treatment, a pH of 7.98 (± 0.06) in the 100 mg LMB L^{-1} treatment, and of 7.93 (± 0.03) in the 200 mg LMB L^{-1} treatment.

It is obvious that, despite more cyanobacteria remaining precipitated with higher ballast dose, the majority of the settled flocks had risen again and accumulated at the water surface (Figures 1b and A2).

2.2. Floc and Sink Experiment—Cyanobacteria Concentration

A laboratory experiment using different concentrations of cyanobacteria and ballast was performed to examine how effective cyanobacterial biomass could be precipitated with PAC and LMB. The results after 1 h showed that 100 mg LMB L^{-1} sufficed to settle even the highest chlorophyll-*a* concentration used (200 μg chlorophyll L^{-1}), while lower biomass could still be precipitated effectively using around 50 mg LMB L^{-1} (Figure 2a). However, the results after 24 h showed much lower removal and strongly hampered efficiency; even at low biomass, high amounts of ballast could only maximally remove 76% of the chlorophyll-*a* concentration (Figure 2b).

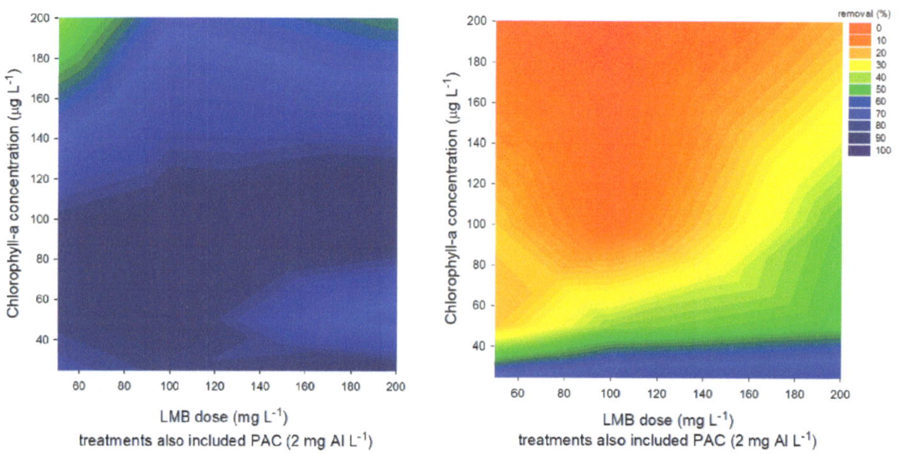

(a) **(b)**

Figure 2. (**a**) Percentage of chlorophyll-*a* removal in *P. rubescens* suspensions with different chlorophyll-*a* concentrations (25–200 μg L^{-1}) after 1 h exposure to different LMB concentrations mixed with the coagulant PAC (2 mg Al L^{-1}). (**b**) Percentage of chlorophyll-*a* removal in *P. rubescens* suspensions with different chlorophyll-*a* concentrations (25–200 μg L^{-1}) after 24 h exposure to different LMB concentrations mixed with the coagulant PAC (2 mg Al L^{-1}).

2.3. Effect of Hydrogen Peroxide on P. rubescens

The sensitivity of *P. rubescens* from Lake De Kuil to hydrogen peroxide was studied to test a possible strategy of killing/damaging *P. rubescens* before sweeping the water column clear of cyanobacteria with a coagulant and ballast. For comparison also, *P. rubescens* from Lake Rauwbraken was included.

P. rubescens from Lake De Kuil was less sensitive to hydrogen peroxide (H_2O_2) than *P. rubescens* from Lake Rauwbraken even though the latter was incubated at almost twice as high chlorophyll-*a* concentrations (Figure 3a). The Photosystem II efficiency (Φ_{PSII}) of *P. rubescens* from Lake Rauwbraken dropped to zero at 4 mg H_2O_2 L^{-1}, while this was reached at 10 mg H_2O_2 L^{-1} for *P. rubescens* from Lake De Kuil (Figure 3b). Nonetheless, Φ_{PSII} had dropped strongly from 2 mg H_2O_2 L^{-1} and higher for both, reflecting damages to the cells. Strongly increased chlorophyll-*a* concentrations further exemplify this damage. As evidenced by filtrate measurements, this was caused partly by dissolved fluorescent pigments (Figure 3a), which did not yield any Φ_{PSII} (Figure 3b).

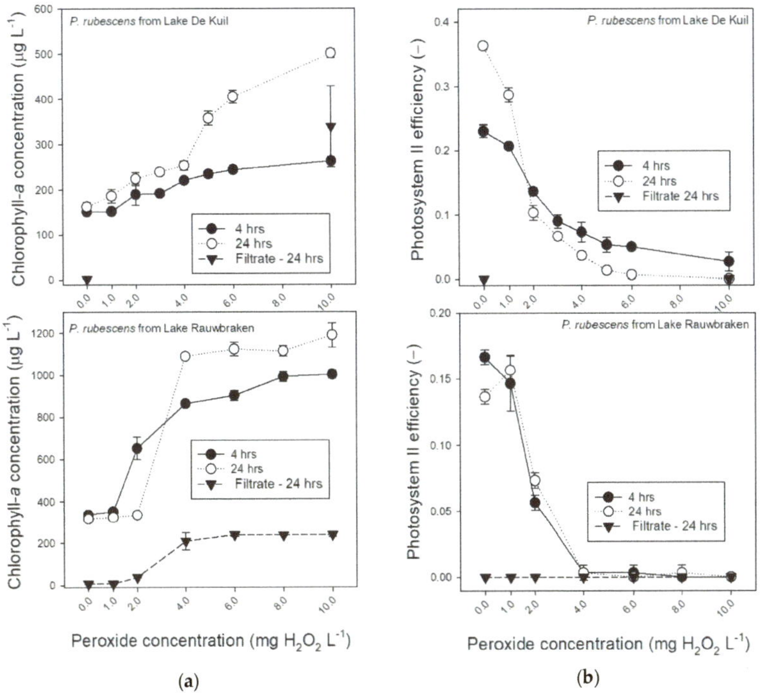

Figure 3. (**a**) chlorophyll-*a* concentrations (µg L^{-1}) in 25 mL *P. rubescens* suspensions from Lake De Kuil (top panel) and Lake Rauwbraken (bottom panel) after 4 h exposure (filled circles) and 24 h exposure (open circles) to different concentrations hydrogen peroxide (0–10 mg L^{-1}). Also included are the chlorophyll concentrations determined in 0.45 µm filtered samples after 24 h (triangles). Note that for Lake De Kuil these were only tested for the controls (0 mg L^{-1}) and the 10 mg H_2O_2 L^{-1} treatment. Error bars indicate 1 SD ($n = 3$). (**b**) Photosystem II efficiencies of *P. rubescens* suspensions from Lake De Kuil (top panel) and Lake Rauwbraken (bottom panel) after 4 h exposure (filled circles) and 24 h exposure (open circles) to different concentrations hydrogen peroxide (0–10 mg L^{-1}). Also included are the Photosystem II efficiencies determined in 0.45 µm filtered samples after 24 h (triangles). Error bars indicate 1 SD ($n = 3$).

2.4. *Efficacy of a Combined Hydrogen Peroxide and Floc and Sink Treatment on P. rubescens*

In the series with *P. rubescens* from Lake De Kuil, the chlorophyll-*a* concentrations in the top of the test tubes were significantly different ($F_{3,11} = 149.5$; $p < 0.001$) between controls and the various treatments after 24 h incubation (Figure 4a). In the control and sole peroxide treatment (5 mg H_2O_2 L^{-1}), chlorophyll-*a* concentrations were highest and not different from each other; in the Floc and Lock treatment (PAC, 2 mg Al L^{-1} and LMB, 200 mg L^{-1}), chlorophyll-*a* concentrations were significantly lower, but still 100 x higher than in the combined treatment (H_2O_2, 5 mg L^{-1}, PAC, 2 mg Al L^{-1} and LMB, 200 mg L^{-1}) in which virtually all chlorophyll-*a* had remained at the bottom of the test tubes (Figure 4a). In the bottom of the tubes, chlorophyll-*a* concentrations were lowest in the control and sole peroxide treatment, significantly higher (log-transformed data; $F_{3,11} = 401.8$; $p < 0.001$) in the Floc and Lock treatment, and highest in the combined treatment (Figure 4a). The Φ_{PSII} in the top of the test tubes was significantly different among treatments ($F_{3,11} = 206.9$; $p < 0.001$), as was Φ_{PSII} in the bottom samples ($F_{3,11} = 70.2$; $p < 0.001$). The Holm–Sidak post hoc pairwise comparison revealed for both top and bottom water samples two homogenous groups: (1) the controls and the Floc and Lock treatments, and (2) both peroxide treatments (solo and combined).

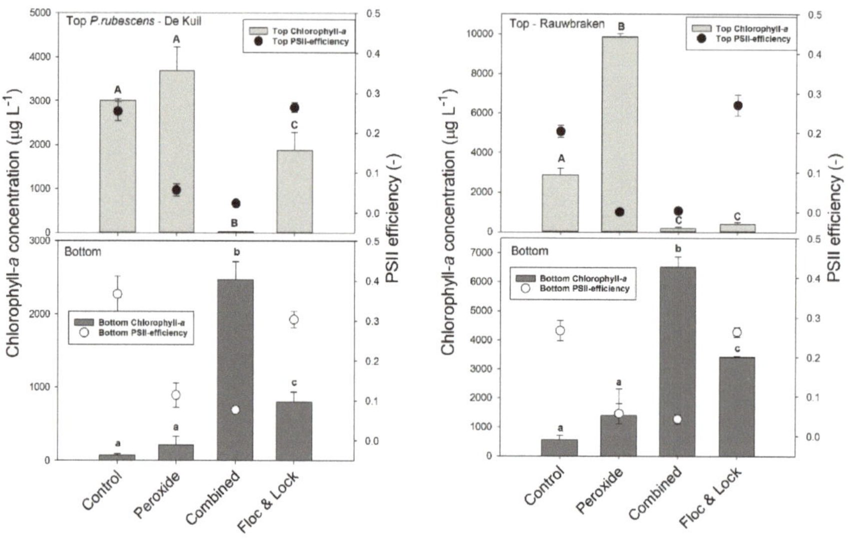

(a) (b)

Figure 4. chlorophyll-*a* concentrations (µg L^{-1}) in the top 2 mL (top light grey bars) and bottom 2 mL (lower dark grey bars) of 100 mL cyanobacteria suspension from (**a**) Lake De Kuil and (**b**) Lake Rauwbraken after 24 h exposure to hydrogen peroxide (5 mg L^{-1}), peroxide + coagulant (2 mg Al L^{-1}) and ballast (200 mg LMB L^{-1}) (combined) or only coagulant (2 mg Al L^{-1}) and LMB (Floc & Lock). Also included are the Photosystem II efficiencies (PSII) of the cyanobacteria collected at the water surface (filled circles) and at the bottom (open circles). Error bars indicate 1 SD ($n = 3$). Similar letters indicate homogeneous groups that are not different at the $p < 0.05$ level.

In the series with *P. rubescens* from Lake Rauwbraken in both the controls and the sole peroxide treatments, the vast majority of the filaments aggregated at the water surface (Figure 4b). In contrast, in both the Floc and Lock treatment and the combined treatment, most cyanobacteria were at the bottom of the tube (Figure 4b). One-way ANOVA indicated significant differences in chlorophyll-*a* concentrations at the water surface ($F_{3,11} = 1428$; $p < 0.001$) and at the bottom ($F_{3,11} = 86.4$; $p < 0.001$) of the test tubes. The Φ_{PSII} in the top of the test tubes was significantly different among treatments ($F_{3,11} = 237.7$; $p < 0.001$), where Holm–Sidak post hoc pairwise comparison revealed that Φ_{PSII} in both peroxide treatments (sole and combined) were significantly lower than in control and the Floc and Lock treatment (Figure 4b). The Φ_{PSII} in the bottom of the test tubes was also significantly different among treatments ($F_{3,11} = 120.8$; $p < 0.001$), and Holm–Sidak post hoc pairwise comparison revealed two homogenous groups: 1) the controls and Floc and Lock treatments, and 2) both peroxide treatments.

The exposure of *P. rubescens* suspensions collected from Lake De Kuil to hydrogen peroxide caused a sharp increase in the concentration of extracellular microcystins (MCs) of which the variant dmMC-RR was most abundant (Figure 5a). The one-way ANOVA indicated significant differences ($F_{3,11} = 226.1$; $p < 0.001$); extracellular MC was lowest in control and Floc and Lock treatments and the highest in the solely peroxide treatment (Figure 5a). In the combined treatment, extracellular MC concentration was about 60% lower than in the sole peroxide treatment (Figure 5a). In suspensions with *P. rubescens* from Lake Rauwbraken only exposure to solely hydrogen peroxide caused significantly ($F_{3,11} = 78.0$; $p < 0.001$) elevated extracellular MC concentrations (Figure 5b).

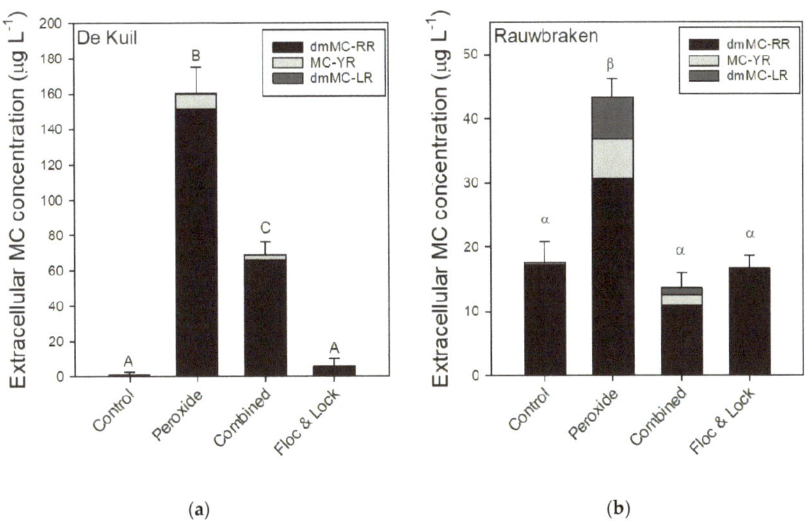

Figure 5. Extracellular microcystin (MC) concentrations (µg L^{-1}) of three MC variants quantified in samples from *P. rubescens* suspensions from (**a**) Lake De Kuil and (**b**) Lake Rauwbraken after 24 h exposure to hydrogen peroxide (5 mg L^{-1}), peroxide + coagulant (2 mg Al L^{-1}) and ballast (200 mg LMB L^{-1}) (combined) or only coagulant (2 mg Al L^{-1}) and LMB (Floc & Lock). Error bars indicate 1 SD (n = 3). Similar letters indicate homogeneous groups that are not different at the $p < 0.05$ level.

3. Discussion

The experiments provided clear evidence that *P. rubescens* from Lake De Kuil could initially be precipitated using a coagulant and a ballast, but that after 24 h, most filaments had resurfaced again. Those results should be a warning when it comes to the use of short-term (1–2 h) tests to determine the efficacy of a coagulant and ballast in so-called 'Floc and Sink' assays [28–31]. The reason for this study was based on field observations that showed reoccurring *P. rubescens* after a 'Floc and Lock' treatment of Lake De Kuil in May 2017 [27]. However, in other whole lake 'Floc and Lock' interventions [23], including one in Lake De Kuil in 2009 [26], no such reappearance had been observed. In those lakes, another cyanobacterium (*Aphanizomenon flas-aquae*) was dominating at the time of intervention. Likewise, in an experiment in which sediment cores and over-standing water infested with *Microcystis aeruginosa* were treated with PAC + LMB, chlorophyll-*a* concentrations were within 1.5 h more than 90% lower than in the control, which remained low during the entire 13 days of the experiment [32]. Clearly, the outcome of those experiments in which the entrapped cyanobacteria stay alive is influenced by species/strain-specific characteristics.

One important feature of *Planktothrix* as a member of the Oscillatoriales is its motility, which is a gliding movement or positive phototactic orientation; an oriented movement towards light [33]. This movement could allow the filaments to crawl out of flocs, as flocs are composed of aggregates with differently sized pores [34]. Another characteristic of *P. rubescens* is that it is highly adapted to low light conditions and can even grow using low amounts of green light prevailing at depth [35,36]. Considering the relative shallowness of Lake De Kuil (maximum water depth ~9 m, average depth ~4 m), ongoing photosynthesis on the sediment with a cleared water column would have been very likely. Consequently, flotation of flocs by oxygen bubbles generated by photosynthesis may occur [37], which could lead to resurfacing of some of the flocs.

Despite the fact that higher ballast doses kept more cyanobacteria at the bottom of the tubes (see Figure 1b), even a ballast dose of 200 mg LMB L^{-1} was insufficient to keep most of the *Planktothrix* at the bottom of the test tubes. In line with previous findings [29], more ballast was needed to remove higher cyanobacterial biomass, but only low cyanobacterial biomass could be kept at the bottom of the tubes for at least 24 h (see Figure 2b). During the application in Lake de Kuil, the biomass in the first 4 m was around 24 μg chlorophyll L^{-1}, and around 30 mg LMB L^{-1} (based on the whole lake volume was applied at the surface as ballast before the PAC application [27]. This means the ballast dose has been higher than 30 mg LMB L^{-1} in the upper one–two meters of the water column. The results of our experiments suggested that at 25 μg chlorophyll L^{-1}, when 50 mg LMB L^{-1} was added, the removal efficiency was maximally 68%. Extrapolating this to the field implies that even when the cyanobacterial biomass had been reduced by two-thirds over time, the reoccurring biomass is large enough to accumulate in relatively high densities at the shore. Hence, biomass plays a role in determining the amount of ballast needed to remove the cells efficiently. However, in this case even using the highest amount of ballast at the lowest chlorophyll-*a* concentrations could not prevent a return to the water column after 24 h. Adding even more ballast would probably not have kept all the biomass at the bottom for reasons of motility and ongoing photosynthesis.

Consequently, additional measures to kill or damage the cyanobacteria and then remove them from the water column seem a strategy to control the nuisance. Such "Kill, Floc and Sink" combination [25] has already been tested with hydrogen peroxide as cyanobacteriocide combined with the coagulant polymeric ferric sulphate (PFS) and lake sediment [38]. In a 91 m^2 enclosure, a *Microcystis* bloom was treated with 60 mg H_2O_2 L^{-1}, followed 2 h later by combined 20 mg PFS L^{-1} and 2 g sediment L^{-1} as ballast [38]. Because effective hydrogen peroxide doses against *Planktothrix* sp. (e.g., [39–42]), are much lower than the high concentration used by Wang et al. [38], hydrogen peroxide was tested in a lower dosage range, which also implies limited side effects on non-target organisms [39,43]. Inasmuch as sensitivity to hydrogen peroxide might differ between cyanobacteria [42] and between strains [44], the sensitivity to hydrogen peroxide of *P. rubescens* from Lake De kuil was compared to that of *P. rubescens* concentrated from Lake Rauwbraken. Photosystem II efficiency (Φ_{PSII}) was chosen as an endpoint because it reflects the fitness of photosynthetic organisms and can be used to demonstrate the damage of H_2O_2 to the photosystem of cyanobacteria [45–47]. The *P. rubescens* from Lake Rauwbraken was more sensitive than *P. rubescens* from Lake De Kuil as its Φ_{PSII} was already zero at 4 mg H_2O_2 L^{-1}, while Φ_{PSII} of *P. rubescens* from Lake De Kuil dropped to zero at 10 mg H_2O_2 L^{-1}, but a strong decline was already observed at much lower concentrations of 2 mg L^{-1}, which is comparable to findings with other cyanobacteria [39,46].

At these H_2O_2 concentrations, the chlorophyll-*a* concentrations (in μg L^{-1}) determined by the PHYTO-PAM were also elevated. This is caused by the detachment of pigments from the thylakoid membranes [46] and leakage of them into the water. Those water soluble extracellular pigments from cyanobacteria can contribute considerably to the detected fluorescence signal, which does not reflect an increase of biomass [48]. The increase in the filterable chlorophyll-*a* without any Φ_{PSII} is a clear indicator of this cell leakage as in general the release of intracellular components is an indication of membrane damage [49]. Given that these extracellular pigments were still elevated 24 h after application, the oxidizing power of the introduced H_2O_2 was not enough to destroy released cell constituents. Likewise, in the combined hydrogen peroxide and Floc and Sink experiment, the H_2O_2 treatments had higher chlorophyll-*a* concentrations than their corresponding controls (the water surface for sole peroxide and control; see Figure 4).

In the combined hydrogen peroxide and Floc and Sink experiment, we chose a dose of 5 mg H_2O_2 L^{-1}, which was sufficient to damage *P. rubescens* from Lake De Kuil for a period of three hours after which the coagulant and ballast were added. This was sufficient to reduce the viability of the filaments to such an extent that they remained precipitated after 24 h incubation. The chlorophyll-*a* concentration in the top of Lake De Kuil tubes was 0.8 (± 0.3) % of that in the bottom, while in the Lake Raubraken combined treated tubes, it was 2.8 (± 1.4) %. When only coagulant and ballast were used, a considerable part of the *P. rubescens* from Lake De Kuil had resurfaced after 24 h, just as observed in the previous experiments. In contrast, *P. rubescens* from Lake Rauwbraken remained precipitated after a sole Floc and Sink treatment. In the Lake Rauwbraken series the chlorophyll-*a* concentration in the top of the sole Floc and Sink tubes was 11 (± 3) % of that in the bottom, while in the Lake De Kuil series, it was 294 (± 96) %. Evidently, the preceding H_2O_2 treatment was effective in keeping *P. rubescens* at the bottom of the tubes. A side effect of the H_2O_2 treatment was leakage of cell constituents, such as pigments and toxins (microcystins, MC). In both *P. rubescens*, exposure to H_2O_2 led to strongly elevated extracellular MC concentrations, which has also been observed for *Microcystis aeruginosa* exposed to H_2O_2 [50]. However, when followed by coagulant and ballast, the extracellular MC concentrations were strongly reduced and for Lake Rauwbraken, even similar to the controls. These results match with the recently reported capacity of LMB to lower dissolved MC concentrations; LMB dosed at 50, 100, and 150 ppm decreased MC concentrations by 61.2%, 86.0%, and 75.4% relative the controls, respectively [51]. However, in the sole Floc and Sink treatments, no further reduction of extracellular MC concentrations was observed. Inasmuch as a Floc and Sink treatment will not damage filaments or cells, and thus will not liberate MCs rapidly, its strongest effect on MC concentration will be via precipitation of cyanobacteria and thereby the removal of particulate MCs from the water column. Nonetheless, concomitant reduction of extracellular MCs is possible, as was shown by a combined chitosan-nano scale montmorillonite treatment that effectively precipitated *Microcystis aeruginosa* (94% removal) and removed 90% of the extracellular MCs within one hour [52]. Clearly, concomitant measurements of cyanobacterial biomass and cyanotoxin concentrations during an intervention are strongly recommended.

The effective precipitation of *P. rubescens* using a combined hydrogen peroxide and Floc and Lock treatment also indicates that despite the fact that H_2O_2 will decay and produce oxygen, this is not leading to the surfacing of flocs. Wang et al. [38] reported that "the floc of *Microcystis* bloom was oxygen-rich ... ", but also that the flocs were deposited on the sediment. However, when calcium peroxide was used as cyanobacteriocide combined with chitosan as a coagulant and red soil as ballast, part of the settled cyanobacteria/ballast flocs migrated upwards again [28]. Hence, the separation of the oxidizing agent (added 3 h before the coagulant and ballast) prevents entrapment of oxygen bubbles inside the aggregates formed; and seems more effective than including a granular formulation together with the coagulant and ballast.

Our experiments yielded insight that a combined hydrogen peroxide and 'Floc and Lock' treatment could be effective in keeping *P. rubescens* precipitated and showed that similar species (*P. rubescens*), but from two different lakes, yielded dissimilar results. This further underpins the necessity to test selected measures for each lake first. Not a single lake is unique, and this also holds for the target organisms, even when belonging to the same species.

4. Conclusions

Short-term (1–2 h) tests to determine the efficacy of precipitation of cyanobacteria by a coagulant and ballast in so-called 'Floc and Sink' assays should be extended to at least 24 h. Motile or low-light adapted cyanobacteria, such as *P. rubescens*, may cause resurfacing of initially settled flocs within 24 h. Using hydrogen peroxide preceding the 'Floc and Sink' treatment seems effective in keeping the cyanobacteria precipitated and thus out of the water column. Moreover, the coagulant and ballast reduce extracellular MCs liberated from damaging the cyanobacteria by H_2O_2. Up-scaled experiments are needed to test the proposed "Kill, Floc and Sink/Lock" approach under more realistic (field) conditions prior to field applications.

5. Materials and Methods

On 30 May 2017, samples were taken from Lake De Kuil (the Netherlands). The lake had orange-colored, odorous surface scums accumulated in some shore regions. Microscopy revealed it consisted of *Planktothrix rubescens*. A large volume (10 L) surface accumulated material was collected to have some higher biomass samples to be used in the experiments. Water samples over the vertical, as well as samples from different sites were collected. In the laboratory, accumulated material and collected water were mixed to create suspensions that were used in experiments to test combined treatments of a coagulant (Floc) and a ballast (Sink) as well as treatments that include hydrogen peroxide (Kill).

5.1. Floc and Sink Experiment—Ballast Dose

The total dose of PAC applied to Lake De Kuil was 6 tons of Calflock P-14 (Caldic Belgium N.V., Hemiksem, Belgium), which contains 7.2% Al, and has a specific gravity of 1.31 kg L^{-1}. An average volume of 268,000 m^3 yielded a PAC dose of 2.1 mg Al L^{-1}. The lanthanum modified bentonite (LMB) Phoslock® (Phoslock Europe Ltd., Manchester, U.K.) was dosed at 31 tons yielding around 116 mg L^{-1}. The LMB ballast, however, will surely have been less as only part of the LMB had been added the first day. Therefore, the effect of different ballast doses on the efficiency of *P. rubescens* removal was studied. The PAC dose was kept at 2 mg Al L^{-1} (Floc), while LMB as ballast (Sink) was dosed at 50, 100, and 200 mg L^{-1}. The experiment was conducted in 125 mL glass tubes filled with 100 mL of a *P. rubescens* suspension created by mixing surface accumulated *P. rubescens* and water collected from Lake De Kuil. This *P. rubescens* infested test water had a chlorophyll-*a* concentration of 114 (± 5) μg L^{-1}. Treatments were run in triplicate, while three non-treated tubes served as controls. A slurry of LMB was added to the treatments, immediately followed by the addition of PAC. All suspensions—controls and treatments—were stirred using a metal rod and left untouched for one hour. Then, 2 mL samples were taken from both the top as well as the bottom of each tube using a 10 mL Eppendorf Varipette pipette (Eppendorf Nederland B.V., Nijmegen, the Netherlands). These samples were analyzed on their chlorophyll-*a* concentrations and Photosystem II efficiencies (Φ_{PSII}) using a PHYTO-PAM phytoplankton analyzer (HeinzWalz GmbH, Effeltrich, Germany). After measuring, the samples were gently placed back in the region from where they were taken. After 24 h, again, 2 mL samples were taken from both the top as well as the bottom of each tube and measured as indicated before. The pH was measured using a WTW multi 340i meter (WTW GmbH and Co. KG, Weilheim, Germany).

The chlorophyll-*a* concentrations after 1 h and after 24 h in the top of the test tubes and in the bottom were evaluated running separate one-way ANOVAs using the toolpack SigmaPlot version 14.0 (Systat Software Inc., San Jose, CA, USA, 2017). The same was done for the Φ_{PSII} and for the pH. Significant differences were distinguished using Holm–Sidak post-hoc pairwise comparisons. Normality (normality test: Shapiro-Wilk) and homogeneity of variance (equal variance test: Brown–Forsythe) were checked prior to running the one-way ANOVAs.

5.2. Floc and Sink Experiment—Cyanobacteria Concentration

A second 'Floc and Sink' experiment was performed to test the hypothesis that the cyanobacterial biomass was too high to be removed. To this end, different combinations of cyanobacteria and ballast concentrations were tested: LMB (50, 100 and 200 mg L^{-1}) in combination with PAC (2 mg Al L^{-1}) were added to various cyanobacteria concentrations (25, 50, 100 and 200 μg chlorophyll-*a* L^{-1}) that were determined by PHYTO-PAM. One tube per cyanobacteria concentration was left untreated. After 1 h, samples from top and bottom were taken. In addition, an extra series was left standing for 24 h and measured then as described above. Hence, this experiment comprised 4 chlorophyll-*a* concentrations $\times$ 4 LMB doses $\times$ 2 series = 32 experimental tubes.

5.3. Effect of Hydrogen Peroxide on P. rubescens

The Floc and Sink experiments yielded insight that additional measures to kill/damage *P. rubescens* prior to adding a coagulant and ballast seem a possible strategy to control the nuisance. Hereto, the sensitivity of *P. rubescens* to hydrogen peroxide was tested. For comparison also *P. rubescens* from Lake Rauwbraken was included, which was sampled using a plankton net from deeper water layers where a low biomass resided [53]. *P. rubescens* from Lake Rauwbraken was further concentrated by pipetting off the surface-accumulated material and transferring it into filtered lake water. The material was visually inspected under the light microscope, no other cyanobacteria were observed.

Aliquots of 25 mL *P. rubescens* suspensions from Lake Rauwbraken (318 ± 34 μg chlorophyll-*a* L^{-1}; Φ_{PSII} = 0.31 ± 0.01) and from Lake De Kuil (318 ± 34 μg chlorophyll-*a* L^{-1}; Φ_{PSII} = 0.30 ± 0.01) were transferred into transparent polystyrene vials (VWR® vials with cap, VWR International B.V., Amsterdam, the Netherlands). Hydrogen peroxide (H_2O_2 30%, 1.07209.0500, Merck KGaA, Darmstadt, Germany) was tested in triplicate concentrations of 0, 1, 2, 4, 6, 8 and 10 mg L^{-1} for Lake Rauwbraken and at 0, 1, 2, 3, 4, 5, 6, and 10 mg L^{-1} for Lake De Kuil. It was pipetted from a 100× diluted stock, where after the vials were closed with a lid and gently shaken. After 4 h, the vials were shaken, a 2 mL subsample analyzed on their chlorophyll-*a* concentrations and Photosystem II efficiencies (Φ_{PSII}) using the PHYTO-PAM phytoplankton analyzer and pipetted back into the vial. After 24 h, the measurement was repeated.

5.4. *Efficacy of a Combined Hydrogen Peroxide and Floc and Sink Treatment on P. rubescens*

The efficiency of a combined H_2O_2, PAC and LMB treatment ('Kill, Floc and Sink') on removing *P. rubescens* from the water column was examined. This experiment tested the hypothesis that H_2O_2 would damage *P. rubescens* cells enough to strongly hamper photosynthesis and buoyancy, which would keep filaments aggregated in flocs at the bottom of the test units. Based on the peroxide exposure experiment (described in Section 5.3) a working dose of 5 mg H_2O_2 L^{-1} was chosen. Aliquots of 100 mL concentrated *P. rubescens* from Lake Rauwbraken (202 ± 5 μg chlorophyll-*a* L^{-1}; Φ_{PSII} = 0.35 ± 0.03) was transferred to 12 glass tubes of 125 mL. Similarly, samples of 100 mL from Lake De Kuil concentrate (156 ± 4 μg chlorophyll-*a* L^{-1}; Φ_{PSII} = 0.32 ± 0.03) was brought into 12 other tubes. Three tubes of each series remained untreated (controls), six tubes of each series were treated with peroxide (5 mg H_2O_2 L^{-1}). After three hours, three of the peroxide treated tubes were treated with a coagulant (PAC, 2 mg Al L^{-1}) and ballast (LMB, 200 mg L^{-1}), while the remaining three tubes per series were treated with an only coagulant (PAC, 2 mg Al L^{-1}) and ballast (LMB, 200 mg L^{-1}), which is referred to as a Floc and Lock treatment. The tubes were incubated for 24 h in the laboratory, where after 2 mL samples from the top of the test tubes and from the bottom of each tube were collected. These samples were analyzed on their chlorophyll-*a* concentrations and the Φ_{PSII}.

From the middle of each tube, a 5 mL sample was taken with a syringe and filtered through a 0.45 μm unit filter (Aqua 30/0.45CA, Whatman, Germany). The filtrates were collected in 8 mL glass tubes and evaporated to dryness in a Speedvac (Thermo Scientific Savant SPD121P, Waltham, MA, USA). The dried filtrates were reconstituted with 800 μL methanol and transferred to 2-mL Eppendorf vials with a cellulose-acetate filter (0.2 μm, Grace Davison Discovery Sciences, Deerfield, IL, USA) and centrifuged for 5 min at 16,000× *g* (VWR Galaxy 16DH, VWR International, Buffalo Grove, IL, USA). Filtrates were transferred to amber glass vials and analyzed for eight microcystins (MC) variants (dm-7-MC-RR, MC-RR, MC-YR, dm-7-MC-LR, MC-LR, MC-LY, MC-LW, and MC-LF) and nodularin (NOD) by LC-MS/MS. The variants dm-7-MC-RR, MC-YR, dm-7-MC-LR, MC-LR, MC-LY and NOD were obtained from DHI Lab products (Hørsholm, Denmark), the variants MC-RR, MC-LF and MC-LW were obtained from Novakits (Nantes, France). The LC-MS/MS was performed as described in Lürling and Faassen [54]. The concentration of each MC variant in the samples was calculated against a calibration curve of each standard and subsequently corrected for recovery, which had been determined for each variant by spiking a cyanobacterial matrix [54].

The chlorophyll-*a* concentrations in the top of the test tubes and at the bottom were evaluated by running separate one-way ANOVAs using the toolpack SigmaPlot version 14.0. Likewise, the Φ_{PSII} in the top of the test tubes and the bottom were evaluated by running separate one-way ANOVAs. Also the total MC concentrations were evaluated with a one-way ANOVA. Holm–Sidak post-hoc pairwise comparisons tests were run to distinguish significant differences. Normality (normality test: Shapiro–Wilk) and homogeneity of variance (equal variance test: Brown–Forsythe) was checked before running the one-way ANOVAs.

Author Contributions: Conceptualization, M.L., M.M. and G.W.; methodology, M.L. and M.M.; validation, M.L., and M.M.; formal analysis, M.L., M.M. and G.W. writing—original draft preparation, M.L.; writing—review and editing, M.L.; M.M. and G.W.; visualization, M.L.; funding acquisition, G.W. All authors have read and agreed to the published version of the manuscript.

Funding: The application of Floc and Lock to Lake De Kuil in May 2017 was funded by Water Authority Brabantse Delta and the Municipality of Breda. The monitoring of the water quality in Lake De Kuil before, during and after the application of May 2017 was funded by Water Authority Brabantse Delta and M.M. was funded by SWB/ CNPq (201328/2014-3).

Acknowledgments: Leonardo de Magalhães is cordially thanked for help during the samplings.

Conflicts of Interest: The authors declare no conflict of interest. The funders had no role in the design of the study; in the collection, analyses, or interpretation of data; in the writing of the manuscript; or in the decision to publish the results.

Appendix A

The Floc and Lock intervention in Lake De Kuil (8–10 May 2017) effectively precipitated a bloom of the cyanobacterium *P. rubescens*. Total chlorophyll-*a* concentration was reduced by 80% and the blue signal—indicative for phycocyanin-containing cyanobacteria—was reduced by more than 90% (Figure A1). The PHYTO-PAM uses four different excitation wavelengths, provided by light-emitting diodes (LEDs) peaking at 470, 535, 620, and 659 nm that allow a separation between cyanobacteria, green algae and diatoms/dinoflagellates [55]. However, the cyanobacteria that can be distinguished are those containing mostly phycocyanin, while phycoerythrin-containing species, such as *P. rubescens*, will not be accurately placed in the "blue channel". A LED with a wavelength around 570 nm would be needed for the excitation of phycoerythrin, which would allow identification of phycoerythrin containing cyanobacteria and cryptophyta [56]. Hence, it is likely part of the signal picked up in the "brown channel" originated from *P. rubescens*. Nonetheless, the intervention reduced chlorophyll-*a* concentrations strongly, but after one week (18 May) a slight increase in the blue channel was detected (Figure A1).

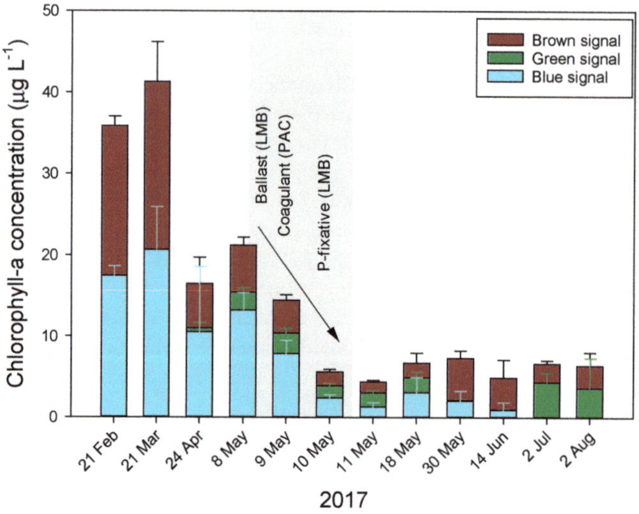

Figure A1. The course of water column (0–9 m) averaged chlorophyll-*a* concentrations (μg L^{-1}) in Lake De Kuil in 2017 before, during (gray plane), and after a Floc and Lock intervention. chlorophyll-*a* was determined with a PHYTO-PAM and separated in a blue-, green- and brown signal, based on different excitation wavelengths.

Appendix B

P. rubescens could initially be settled to the bottom of the test tubes rise due to entrapped oxygen bubbles in the flocks (see pictures below).

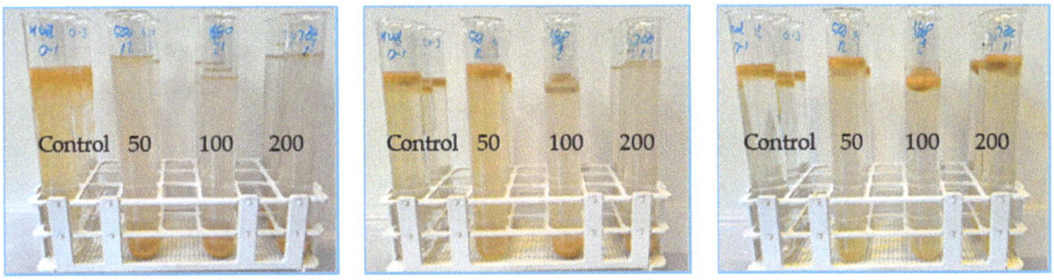

(**a**) (**b**) (**c**)

Figure A2. Pictures of test tubes with *P. rubescens* from lake De Kuil after one hour (**a**), three hours (**b**) and 24 h (**c**) exposure to 2 mg Al L^{-1} PAC + different doses of ballast; 50, 100 or 200 mg LMB L^{-1}. Note that after three hours in the lower ballast doses, cyanobacteria started to float again. After 24 h, most have surfaced again.

References

1. Downing, J.A. Limnology and oceanography: Two estranged twins reuniting by global change. *Inland Waters* **2014**, *4*, 215–232. [CrossRef]
2. Smith, V.; Tilman, G.; Nekola, J. Eutrophication: Impacts of excess nutrient inputs on freshwater, marine, and terrestrial ecosystems. *Environ. Pollut.* **1999**, *100*, 179–196. [CrossRef]
3. Paerl, H.W.; Xu, H.; McCarthy, M.J.; Zhu, G.; Qin, B.; Li, Y.; Gardner, W.S. Controlling harmful cyanobacterial blooms in a hyper-eutrophic lake (Lake Taihu, China): The need for a dual nutrient (N & P) management strategy. *Water Res.* **2011**, *45*, 1973–1983. [CrossRef] [PubMed]
4. Paerl, H.W.; Paul, V.J. Climate change: Links to global expansion of harmful cyanobacteria. *Water Res.* **2012**, *46*, 1349–1363. [CrossRef] [PubMed]
5. Cordell, D.; Drangert, J.-O.; White, S. The story of phosphorus: Global food security and food for thought. *Glob. Environ. Chang.* **2009**, *19*, 292–305. [CrossRef]
6. Sinha, E.; Michalak, A.M.; Balaji, V. Eutrophication will increase during the 21st century as a result of precipitation changes. *Science* **2017**, *357*, 405–408. [CrossRef] [PubMed]
7. O'Neil, J.; Davis, T.W.; Burford, M.; Gobler, C.J. The rise of harmful cyanobacteria blooms: The potential roles of eutrophication and climate change. *Harmful Algae* **2012**, *14*, 313–334. [CrossRef]
8. Jeppesen, E.; Kronvang, B.; Meerhoff, M.; Søndergaard, M.; Hansen, K.M.; Andersen, H.E.; Lauridsen, T.L.; Liboriussen, L.; Beklioglu, M.; Özen, A.; et al. Climate Change Effects on Runoff, Catchment Phosphorus Loading and Lake Ecological State, and Potential Adaptations. *J. Environ. Qual.* **2009**, *38*, 1930–1941. [CrossRef]
9. Beaulieu, J.J.; DelSontro, T.; Downing, J.A. Eutrophication will increase methane emissions from lakes and impoundments during the 21st century. *Nat. Commun.* **2019**, *10*, 1–5. [CrossRef] [PubMed]
10. OECD. *Diffuse Pollution, Degraded Waters*; OECD: Paris, France, 2017; ISBN 9789264269057.
11. Hamilton, D.P.; Salmaso, N.; Paerl, H.W. Mitigating harmful cyanobacterial blooms: Strategies for control of nitrogen and phosphorus loads. *Aquat. Ecol.* **2016**, *50*, 351–366. [CrossRef]
12. Paerl, H.W.; Gardner, W.S.; Havens, K.E.; Joyner, A.R.; McCarthy, M.J.; Newell, S.E.; Qin, B.; Scott, J.T. Mitigating cyanobacterial harmful algal blooms in aquatic ecosystems impacted by climate change and anthropogenic nutrients. *Harmful Algae* **2016**, *54*, 213–222. [CrossRef]
13. Huisman, J.; Codd, G.A.; Paerl, H.W.; Ibelings, B.W.; Verspagen, J.M.H.; Visser, P.M. Cyanobacterial blooms. *Nat. Rev. Genet.* **2018**, *16*, 471–483. [CrossRef] [PubMed]
14. WWAP. *United Nations World Water Assessment Programme the United Nations World Water Development Report 2017*; Wastewater: Paris, France, 2017.

15. OECD. *Water Governance in the Netherlands: Fit for the Future?* OECD: Paris, France, 2014; ISBN 9789264208940.
16. Fastner, J.; Abella, S.; Litt, A.; Morabito, G.; Vörös, L.; Pálffy, K.; Straile, D.; Kümmerlin, R.; Matthews, D.; Phillips, M.G.; et al. Combating cyanobacterial proliferation by avoiding or treating inflows with high P load—experiences from eight case studies. *Aquat. Ecol.* **2015**, *50*, 367–383. [CrossRef]
17. Cullen, P.; Forsberg, C. Experiences with reducing point sources of phosphorus to lakes. *Hydrobiologia* **1988**, *170*, 321–336. [CrossRef]
18. O'Connell, D.W.; Ansems, N.; Kukkadapu, R.K.; Jaisi, D.; Orihel, D.M.; Cade-Menun, B.J.; Hu, Y.; Wiklund, J.; Hall, R.I.; Chessell, H.; et al. Changes in Sedimentary Phosphorus Burial Following Artificial Eutrophication of Lake 227, Experimental Lakes Area, Ontario, Canada. *J. Geophys. Res. Biogeosci.* **2020**, *125*, 125. [CrossRef]
19. Lürling, M.; Mackay, E.; Reitzel, K.; Spears, B.M. Editorial—A critical perspective on geo-engineering for eutrophication management in lakes. *Water Res.* **2016**, *97*, 1–10. [CrossRef]
20. Jarvie, H.P.; Sharpley, A.N.; Spears, B.; Buda, A.R.; May, L.; Kleinman, P.J.A. Water Quality Remediation Faces Unprecedented Challenges from "Legacy Phosphorus". *Environ. Sci. Technol.* **2013**, *47*, 8997–8998. [CrossRef]
21. Goyette, J.-O.; Bennett, E.M.; Maranger, R. Low buffering capacity and slow recovery of anthropogenic phosphorus pollution in watersheds. *Nat. Geosci.* **2018**, *11*, 921–925. [CrossRef]
22. Lürling, M.; Mucci, M. Mitigating eutrophication nuisance: In-lake measures are becoming inevitable in eutrophic waters in the Netherlands. *Hydrobiololgia* **2020**, 1–21. [CrossRef]
23. Lürling, M.; Van Oosterhout, F. Controlling eutrophication by combined bloom precipitation and sediment phosphorus inactivation. *Water Res.* **2013**, *47*, 6527–6537. [CrossRef] [PubMed]
24. Van Oosterhout, F.; Waajen, G.; Yasseri, S.; Marinho, M.M.; Noyma, N.P.; Mucci, M.; Douglas, G.; Waajen, M.L.G. Lanthanum in Water, Sediment, Macrophytes and chironomid larvae following application of Lanthanum modified bentonite to lake Rauwbraken (The Netherlands). *Sci. Total. Environ.* **2020**, *706*, 135–188. [CrossRef]
25. Lürling, M.; Kang, L.; Mucci, M.; Van Oosterhout, F.; Noyma, N.P.; Miranda, M.; Huszar, V.L.; Waajen, G.; Marinho, M.M. Coagulation and precipitation of cyanobacterial blooms. *Ecol. Eng.* **2020**, *158*, 106032. [CrossRef]
26. Waajen, G.; Van Oosterhout, F.; Douglas, G.; Lürling, M. Management of eutrophication in Lake De Kuil (The Netherlands) using combined flocculant—Lanthanum modified bentonite treatment. *Water Res.* **2016**, *97*, 83–95. [CrossRef]
27. Mucci, M.; Waajen, G.; van Oosterhout, F.; Yasseri, S.; Lürling, M. Whole lake application PAC-Phoslock treatment to manage eutrophication and cyanobacterial bloom. *Inland Waters*. (under review).
28. Noyma, N.P.; De Magalhães, L.; Furtado, L.L.; Mucci, M.; Van Oosterhout, F.; Huszar, V.L.; Marinho, M.M.; Lürling, M. Controlling cyanobacterial blooms through effective flocculation and sedimentation with combined use of flocculants and phosphorus adsorbing natural soil and modified clay. *Water Res.* **2016**, *97*, 26–38. [CrossRef]
29. Noyma, N.P.; De Magalhães, L.; Miranda, M.; Mucci, M.; Van Oosterhout, F.; Huszar, V.L.M.; Marinho, M.M.; Lima, E.R.A.; Lürling, M. Coagulant plus ballast technique provides a rapid mitigation of cyanobacterial nuisance. *PLoS ONE* **2017**, *12*, e0178976. [CrossRef]
30. Miranda, M.; Noyma, N.; Pacheco, F.S.; De Magalhães, L.; Pinto, E.; Santos, S.; Soares, M.F.A.; Huszar, V.L.; Lürling, M.; Marinho, M.M. The efficiency of combined coagulant and ballast to remove harmful cyanobacterial blooms in a tropical shallow system. *Harmful Algae* **2017**, *65*, 27–39. [CrossRef] [PubMed]
31. De Lucena-Silva, D.; Molozzi, J.; Severiano, J.D.S.; Becker, V.; Barbosa, J.E.D.L. Removal efficiency of phosphorus, cyanobacteria and cyanotoxins by the "flock & sink" mitigation technique in semi-arid eutrophic waters. *Water Res.* **2019**, *159*, 262–273. [CrossRef]
32. De Magalhães, L.; Noyma, N.P.; Furtado, L.L.; Drummond, E.; Leite, V.B.G.; Mucci, M.; Van Oosterhout, F.; Huszar, V.L.D.M.; Lürling, M.; Marinho, M.M. Managing Eutrophication in a Tropical Brackish Water Lagoon: Testing Lanthanum-Modified Clay and Coagulant for Internal Load Reduction and Cyanobacteria Bloom Removal. *Estuar. Coast.* **2019**, *42*, 390–402. [CrossRef]
33. Häder, D.P. *Photomovement*; Springer Science and Business Media LLC: Berlin, Germany, 1984; pp. 435–443.
34. Gorczyca, B.; Ganczarczyk, J. Fractal Analysis of Pore Distributions in Alum Coagulation and Activated Sludge Flocs. *Water Qual. Res. J.* **2001**, *36*, 687–700. [CrossRef]

35. Kromkamp, J.C.; Domin, A.; Dubinsky, Z.; Lehmann, C.; Schanz, F. Changes in photosynthetic properties measured by oxygen evolution and variable chlorophyll fluorescence in a simulated entrainment experiment with the cyanobacterium *Planktothrix rubescens*. *Aquat. Sci.* **2001**, *63*, 363–382. [CrossRef]
36. Oberhaus, L.; Briand, J.; Leboulanger, C.; Jacquet, S.; Humbert, J.F. Comparative effects of the quality and quantity of light and temperature on the growth of *Planktothrix agardhii* and *P. rubescens*. *J. Phycol.* **2007**, *43*, 1191–1199. [CrossRef]
37. Eldridge, R.J.; A Hill, D.R.; Gladman, B. A comparative study of the coagulation behaviour of marine microalgae. *Environ. Boil. Fishes* **2012**, *24*, 1667–1679. [CrossRef]
38. Wang, Z.; Li, D.; Qin, H.; Li, Y. An integrated method for removal of harmful cyanobacterial blooms in eutrophic lakes. *Environ. Pollut.* **2012**, *160*, 34–41. [CrossRef]
39. Matthijs, H.C.; Visser, P.M.; Reeze, B.; Meeuse, J.; Slot, P.C.; Wijn, G.; Talens, R.; Huisman, J. Selective suppression of harmful cyanobacteria in an entire lake with hydrogen peroxide. *Water Res.* **2012**, *46*, 1460–1472. [CrossRef] [PubMed]
40. Bauzá, L.; Aguilera, A.; Echenique, R.; Andrinolo, D.; Giannuzzi, L. Application of Hydrogen Peroxide to the Control of Eutrophic Lake Systems in Laboratory Assays. *Toxins* **2014**, *6*, 2657–2675. [CrossRef]
41. Dziga, D.; Tokodi, N.; Drobac, D.; Kokociński, M.; Antosiak, A.; Puchalski, J.; Strzałka, W.; Madej, M.; Svirčev, Z.; Meriluoto, J.; et al. The Effect of a Combined Hydrogen Peroxide-MlrA Treatment on the Phytoplankton Community and Microcystin Concentrations in a Mesocosm Experiment in Lake Ludoš. *Toxins* **2019**, *11*, 725. [CrossRef]
42. Lusty, M.W.; Gobler, C.J. The Efficacy of Hydrogen Peroxide in Mitigating Cyanobacterial Blooms and Altering Microbial Communities across Four Lakes in NY, USA. *Toxins* **2020**, *12*, 428. [CrossRef]
43. Matthijs, H.C.P.; Jančula, D.; Visser, P.M.; Maršálek, B. Existing and emerging cyanocidal compounds: New perspectives for cyanobacterial bloom mitigation. *Aquat. Ecol.* **2016**, *50*, 443–460. [CrossRef]
44. Dziallas, C.; Grossart, H.-P. Increasing Oxygen Radicals and Water Temperature Select for Toxic Microcystis sp. *PLoS ONE* **2011**, *6*, e25569. [CrossRef]
45. Drábková, M.; Admiraal, W.; Maršálek, B. Combined Exposure to Hydrogen Peroxide and Light Selective Effects on Cyanobacteria, Green Algae, and Diatoms. *Environ. Sci. Technol.* **2007**, *41*, 309–314. [CrossRef]
46. Drábková, M.; Matthijs, H.C.P.; Admiraal, W.; Marsalek, B. Selective effects of H_2O_2 on cyanobacterial photosynthesis. *Photosynthetica* **2007**, *45*, 363–369. [CrossRef]
47. Piel, T.; Sandrini, G.; White, E.; Xu, T.; Schuurmans, J.M.; Huisman, J.; Visser, P.M. Suppressing Cyanobacteria with Hydrogen Peroxide Is More Effective at High Light Intensities. *Toxins* **2019**, *12*, 18. [CrossRef]
48. Bastien, C.; Cardin, R.; Veilleux, É.; Deblois, C.; Warren, A.; Laurion, I. Performance evaluation of phycocyanin probes for the monitoring of cyanobacteria. *J. Environ. Monit.* **2011**, *13*, 110–118. [CrossRef]
49. Liu, H.; Du, Y.; Wang, X.; Sun, L. Chitosan kills bacteria through cell membrane damage. *Int. J. Food Microbiol.* **2004**, *95*, 147–155. [CrossRef]
50. Lürling, M.; Meng, D.; Faassen, E. Effects of Hydrogen Peroxide and Ultrasound on Biomass Reduction and Toxin Release in the Cyanobacterium, *Microcystis aeruginosa*. *Toxins* **2014**, *6*, 3260–3280. [CrossRef]
51. Laughinghouse, H.; Lefler, F.W.; Berthold, D.E.; Bishop, W.M. Sorption of dissolved microcystin using lanthanum-modified bentonite clay. *J. Aquat. Plant Manag.* **2020**, *58*, 72–75.
52. Wang, Z.; Wang, C.; Wang, P.; Qian, J.; Hou, J.; Ao, Y.; Wu, B. The performance of chitosan/montmorillonite nanocomposite during the flocculation and floc storage processes of *Microcystis aeruginosa* cells. *Environ. Sci. Pollut. Res.* **2015**, *22*, 11148–11161. [CrossRef] [PubMed]
53. van Oosterhout, F.; Yasseri, S.; Noyma, N.; Huszar, V.; Marinho, M.M.; Mucci, M.; Waajen, G.; Lurling, M. Evaluation of a whole lake eutrophication management technique using combined flocculation and in-situ phosphorus immobilization. *Inland Waters*. (under review).
54. Lürling, M.; Faassen, E.J. Dog Poisonings Associated with a *Microcystis aeruginosa* Bloom in the Netherlands. *Toxins* **2013**, *5*, 556–567. [CrossRef] [PubMed]

55. Xie, J.; Zhang, W.; Mei, J. A Data Grid System Oriented Biologic Data. In Proceedings of the 2007 IEEE/WIC/ACM International Conferences on Web Intelligence and Intelligent Agent Technology—Workshops, Silicon Valley, CA, USA, 5–12 November 2007.
56. Beutler, M.; Wiltshire, K.H.; Meyer, B.; Moldaenke, C.; Lüring, C.; Meyerhöfer, M.; Hansen, U.-P.; Dau, H. A fluorometric method for the differentiation of algal populations in vivo and in situ. *Photosynth. Res.* **2002**, *72*, 39–53. [CrossRef]

Publisher's Note: MDPI stays neutral with regard to jurisdictional claims in published maps and institutional affiliations.

toxins

MDPI

Article

Potential Impacts on Treated Water Quality of Recycling Dewatered Sludge Supernatant during Harmful Cyanobacterial Blooms

Kanarat Pinkanjananavee [1], Swee J. Teh [2], Tomofumi Kurobe [2], Chelsea H. Lam [2], Franklin Tran [2] and Thomas M. Young [1,*]

[1] Department of Civil & Environmental Engineering, University of California Davis, Davis, CA 95616, USA; kanpin@ucdavis.edu

[2] Department of Anatomy, Physiology, and Cell Biology, School of Veterinary Medicine, University of California Davis, Davis, CA 95616, USA; sjteh@ucdavis.edu (S.J.T.); tkurobe@ucdavis.edu (T.K.); chylam@ucdavis.edu (C.H.L.); fdtran@ucdavis.edu (F.T.)

* Correspondence: tyoung@ucdavis.edu

Abstract: Cyanobacterial blooms and the associated release of cyanotoxins pose problems for many conventional water treatment plants due to their limited removal by typical unit operations. In this study, a conventional water treatment process consisting of coagulation, flocculation, sedimentation, filtration, and sludge dewatering was assessed in lab-scale experiments to measure the removal of microcystin-LR and *Microcystis aeruginosa* cells using liquid chromatography with mass spectrometer (LC-MS) and a hemacytometer, respectively. The overall goal was to determine the effect of recycling cyanotoxin-laden dewatered sludge supernatant on treated water quality. The lab-scale experimental system was able to maintain the effluent water quality below relevant the United States Environmental Protection Agency (US EPA) and World Health Organisation (WHO) standards for every parameter analyzed at influent concentrations of *M. aeruginosa* above 10^6 cells/mL. However, substantial increases of 0.171 NTU (Nephelometric Turbidity Unit), 7×10^4 cells/L, and 0.26 µg/L in turbidity, cyanobacteria cell counts, and microcystin-LR concentration were observed at the time of dewatered supernatant injection. Microcystin-LR concentrations of 1.55 µg/L and 0.25 µg/L were still observed in the dewatering process over 24 and 48 h, respectively, after the initial addition of *M. aeruginosa* cells, suggesting the possibility that a single cyanobacterial bloom may affect the filtered water quality long after the bloom has dissipated when sludge supernatant recycling is practiced.

Keywords: harmful cyanobacteria; cyanotoxins; conventional water treatment

Key Contribution: The effects of recycling dewatered sludge supernatant on the turbidity, *Microcystis* cell count, and microcystin-LR concentration were determined. Recycling supernatant extended the effects of the cyanobacterial bloom on treated water quality well beyond the event.

check for updates

Citation: Pinkanjananavee, K.; Teh, S.J.; Kurobe, T.; Lam, C.H.; Tran, F.; Young, T.M. Potential Impacts on Treated Water Quality of Recycling Dewatered Sludge Supernatant during Harmful Cyanobacterial Blooms. *Toxins* **2021**, *13*, 99. https://doi.org/10.3390/toxins13020099

Received: 29 November 2020
Accepted: 27 January 2021
Published: 29 January 2021

1. Introduction

An increase in the frequency and severity of harmful cyanobacterial blooms has been observed worldwide [1–6]. Cyanobacteria are prokaryotes, but have been widely referred to as "blue-green algae" because of their ability to perform photosynthesis and similarity in size and color to many algal species. Multiple cyanobacteria species grow in a typical bloom, but the most abundant and common cyanobacteria found in harmful algal blooms across the globe is *M. aeruginosa* [7–9]. There are many important factors controlling the growth *of M. aeruginosa* in an aquatic environment including physical disturbance, light, temperature, nutrients, and grazing. Eutrophication linked to the increase of nutrients including nitrogen (N) and phosphorus (P) in water systems is generally considered as one of the major factors that favor the development of planktonic cyanobacterial blooms [10,11]

A critical problem caused by *M. aeruginosa* is associated with their potential to release cyanotoxins [11–14]. These toxins have the ability to cause human health problems such as liver cancer or neurotoxic effects [15]. The most common and intensively identified toxin class is microcystins (MCs), especially the microcystin-LR (MC-LR) form, in which L and R stand for the distinguishing amino acids leucine and arginine, respectively [12–14]. Due to the toxicity of MCs, the WHO has established a drinking water guideline of 1.0 μg/L of MC-LR equivalents [16]. The United States has not established a national MCs drinking water standard; consequently, MCs standards vary by state.

To protect people from MCs exposure, physical and chemical treatment methods are usually applied to remove algal cells and associated toxins during water treatment. Coagulation, flocculation, and sedimentation is a commonly used treatment sequence during drinking water treatment, supplying many benefits such as ease of operation, low cost, rapid reaction rates, and high efficiency [4]. These preliminary treatments have been shown to achieve 78% to 92% removal of cyanobacteria cells [17,18]. Metal salt coagulants such as alum ($Al(SO_4)_3 \cdot 18H_2O$) are effective because they can neutralize the negative surface charges present on cyanobacterial cells at natural water pH levels [19]. Charge neutralization promotes floc formation, allowing cyanobacteria to settle out during subsequent sedimentation.

Although the conventional treatment method of coagulation-flocculation and sedimentation is considered effective in removing cyanobacteria cells [4,12,20], the process only neutralizes the charges on the surface of the cells but does not deactivate cellular functions [21–23]. Thus, it is possible that the cells that were not removed by the sedimentation process could repopulate in downstream filtration media and negatively impact treated water quality [24]. A rapid increase of cyanobacteria cell counts caused by incomplete deactivation of cyanobacteria in a water treatment plant has been reported by Gad and El-Tawel, (2016) [23].

Cyanobacterial regrowth in the filter media may not be the only mechanism for the persistence of, and a possible increase in, the cyanobacterial populations in water treatment plants. Many water treatment plants recycle settling basin supernatant to the plant headworks, potentially allowing cyanobacterial regrowth and cyanotoxin redistribution [25–27]. The accumulation and subsequent regrowth of cyanobacteria and the excretion of cyanotoxins in sludge from the dewatering process may lead to prolonged effects of cyanobacterial blooms in drinking water treatment operations [26], particularly since cyanobacteria cell lysis may occur and result in the release of additional cyanotoxins.

Therefore, the objectives of this study were to: (1) investigate the accumulation of cyanobacteria cells and microcystins in conventional filter media during operation of a lab-scale drinking water treatment system experiencing a simulated bloom of toxigenic *M. aeruginosa* cells, (2) explore the possibility of cyanobacteria repopulation and microcystin excretion during the dewatering process, and (3) examine the effect of dewatered supernatant recycling on the quality of filtered water. The lab-scale treatment system, depicted in Figure 1, includes a conventional process train of coagulation, flocculation, sedimentation, and filtration.

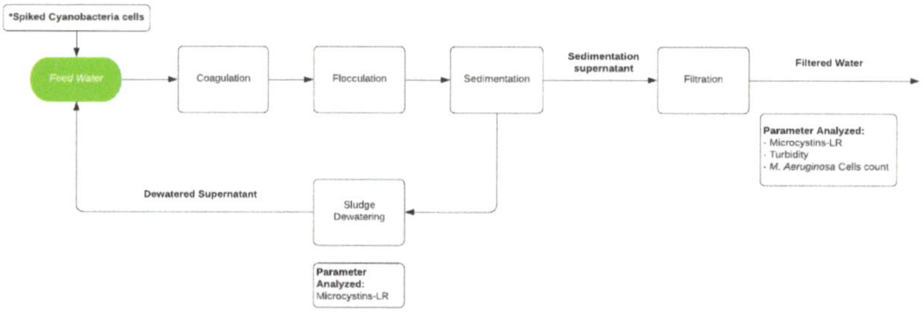

Figure 1. Schematic of the lab-scale water treatment system. * Only for experiments AE (A-Spiked) and BE (B-Spiked).

2. Results

2.1. Feed Sample Characteristics

Source water obtained from the intake to the Woodland and Davis Clean Water Association (WDCWA) treatment plant on two collection dates featured relatively low turbidity (<32 NTU), with a pH of 7.6 and 8.2 for experiment A and experiment B, respectively, and no detectable concentrations of MC-LR (Table 1). Experiments A and B are two pilot-scale experiments conducted under the same operating parameters at different times to ensure the reproducibility of the results.

Table 1. Raw water characteristics at the WDCWA treatment plant intake.

Experiment	Collection Date	Turbidity	Cell Counts	MC-LR	pH
		NTU	Cells/mL	μg/L	
A	13 February 2020	31.8	<1 × 10^4	<LOD *	7.6
B	27 February 2020	13.2	<1 × 10^4	<LOD *	8.2

* Limit of detection (LOD): 0.005 μg/L.

The lab-scale treatment system used in these experiments (Figure 1) is a batch system designed to mimic the behavior of a full-scale treatment process that includes coagulation, filtration, sludge dewatering, and supernatant recycle to the filters. Feed water supplied to the system consisted of WDCWA raw water with (spiked experiment, E) or without (control experiment, C) the addition of laboratory cultured *M. aeruginosa* cells. The water quality of the spiking solutions and the spiked feed water for each experiment are reported in Table 2. The turbidity of the feed water is considered to be the same as for the raw water since the change in turbidity was insignificant (<0.1 NTU). The *M. aeruginosa* cell counts in the experimental water samples are controlled to be 1 × 10^6–1 × 10^7 cells/mL, which is within the range of *M. aeruginosa* cell counts commonly observed during bloom events [4,18]. Also shown in Table 2 are the flow rates, dewatered sludge supernatant injection times, and optimal coagulant dosages applied (as determined by separate jar tests) for each experiment.

Table 2. Feed water and spiking solution characteristics and Lab-scale experimental conditions.

Experiment	Optimal Coagulant Dosage	Cyanobacteria Solution		Cell Count	MC-LR Extracellular	MC-LR Total	Feed Water Average Flow Rate	Time of Injection of Dewatered Sludge Supernatant
	mg/L	Cells/mL	(MC-LR) μg/L	Cells/mL	μg/L		mL/s	Hours
A-Control (AC)	30	-	-	-	<LOD *	-	13	24
A-Spiked (AE)	30	2.00 × 10^8	11.41	1.0×10^7	0.29	-	13	24
B-Control (BC)	30	-	-	-	<LOD *	-	14	25
B-Spiked (BE)	30	6.50 × 10^7	10.17	1.6×10^6	0.53	50.57	14	25

* Limit of detection (LOD): 0.005 μg/L.

After the process of coagulation-flocculation and sedimentation was completed, the supernatant of the settled sample was transferred into the filtration process and the settled sludge was concentrated in the dewatering system. After 24 h of dewatering process, the dewatered process supernatant (dewatered supernatant) was removed and mixed with WDCWA raw water to serve as experimental feed water during the second day of the experiment; this simulates the process of recycling liquids from dewatering operations to the feed water as practiced at many treatment facilities. MC-LR concentrations and *M.*

aeruginosa cell counts for dewatered supernatant for experiments AE (A-Spiked) and BE (B-Spiked) are shown in Figure 2.

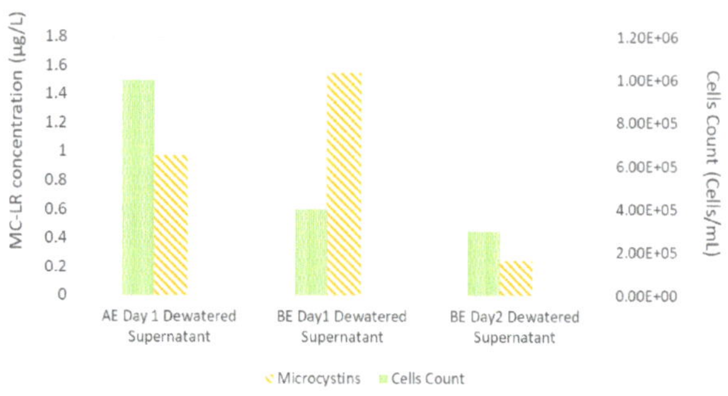

Figure 2. Experiment A (AE) and B (BE) dewatered supernatant composition.

Data shown in Figure 2 follow the same trend as the data shown in Table 2 with higher MC-LR and lower cell counts for experiment B than experiment A. The MC-LR concentrations of dewatered supernatant for experiments AE and BE (Figure 2) are higher than the feed water MC-LR concentrations (Table 2), this result highlights the potential for *M. aeruginosa* cells to be lysed during the coagulation and flocculation process and also during the dewatering process.

2.2. Filtration Performance

Overall, during these experiments, including following the injection of dewatered supernatant, the filtration system maintained an effluent turbidity level below the 1 NTU limit required under U.S. EPA regulations (Figure 3). The only exception was at the initial time point for experiment B, which slightly exceeded the 1 NTU limit. During experiment A, the system removed 99.0% of the initial 31.8 NTU (Table 2) turbidity for the control feed solution and 98.0% for the spiked feed solution (Figure 3a). During experiment B, the filtration system removed 96.7% of the initial 13.2 NTU turbidity from the control feed solution and 94.7% from the spiked feed solution (Figure 3b). During both experiments, the recycling of dewatered supernatant produced a sudden increase in the effluent turbidity, but the increase was not sufficient to cause the system to exceed the 1 NTU effluent guideline.

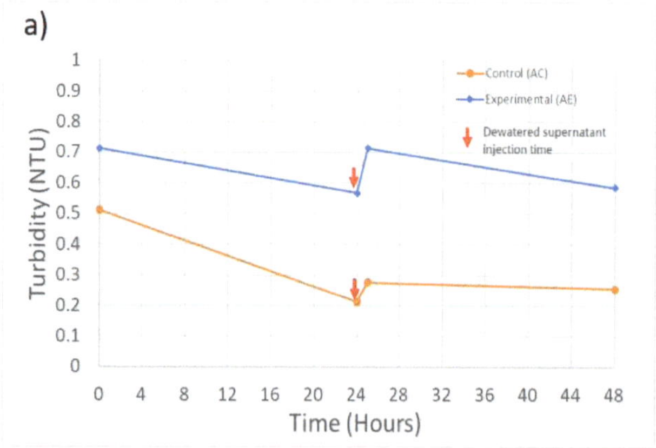

Figure 3. *Cont.*

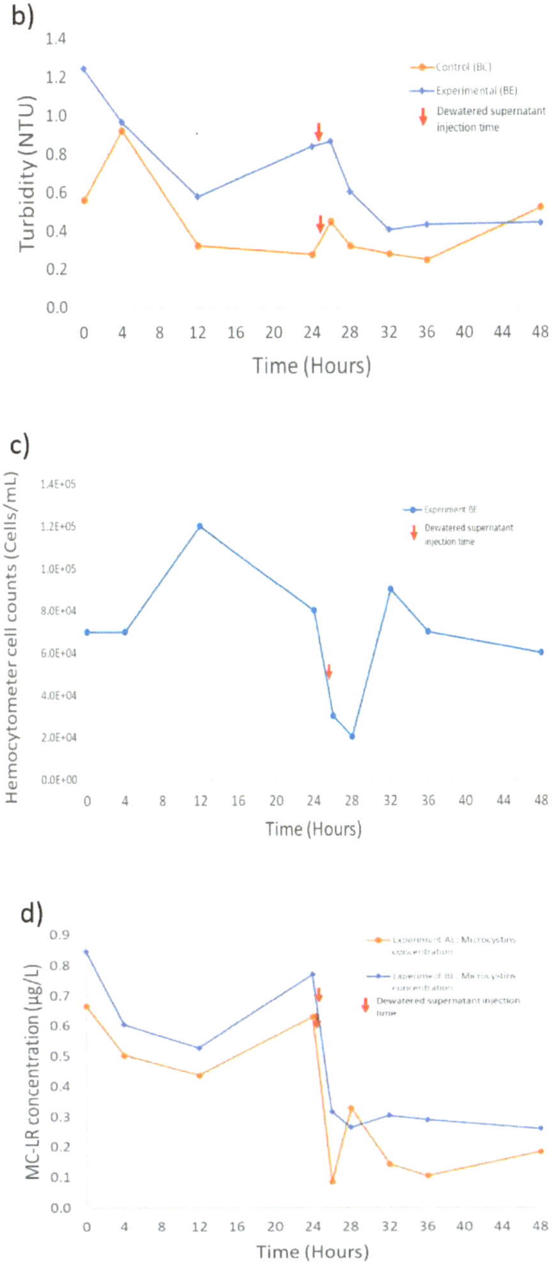

Figure 3. Experiment A and B filtration process effluent quality. (**a**) Experiment A turbidity (n = 3), initial concentration: 31.7 NTU; (**b**) Experiment B turbidity (n = 3), initial concentration: 13.2 NTU; (**c**) Experiment B *M. aeruginosa* cell counts (n = 1), initial concentration: 1.6×10^6 cells/Land; (**d**) Experiment A and B MC-LR concentrations (n = 1), initial concentration: 0.29 and 0.53 µg/L, respectively. Error bars represent standard deviation from measurement replicates.

M. aeruginosa cell counts for the filtration process effluent in experiment B are shown in Figure 3c (effluent cell counts are not available for experiment A). The treatment process in experiment B was able to remove 95.8% of the initial 1.6×10^6 cells/mL (Table 2). An increase in the cell counts following injection of dewatered supernatant can also be observed, but the timing of the increase is delayed in comparison with the time of the turbidity increase. The cause of this difference in timing is not clear, but it could mean that the prior turbidity increase may not be solely due to the influence of the injected *M. aeruginosa* cells.

MC-LR concentrations in filter effluents are shown in Figure 3d. The treatment process removed 55.6% and 38.5% of the MC-LR present in the feed solutions during experiments A and B, respectively. However, it can also be seen in Figure 3d that, at some time steps, the MC-LR concentration exceeded the concentration of MC-LR in the feed solution (Table 2). The additional MC-LR may have been released by cell lysis during the coagulation and flocculation steps, releasing the intracellular MC-LR into the dissolved phase, thereby increasing the extracellular MC-LR that was being measured. A further increase in the MC-LR filter effluent concentrations during both experiments can also be seen after the injection of the dewatered supernatant at the 28- and 32-h time points for experiments AE and BE, respectively. Although the MC-LR concentrations did not exceed the WHO guideline of 1 µg/L at any point in time during these experiments, the result reinforces previous findings that conventional water treatment systems are not highly efficient in removing extracellular MCs [28].

2.3. MC Retention in Filter Media

It is challenging to determine the number of *M. aeruginosa* cells retained within the filter media because of the large amount of other suspended solids also retained. Consequently, only the total MC-LR concentration within the sand media was investigated. After a full cycle of filter operation, the sand media had retained a total MC-LR concentration (extracellular and intracellular) of 14.89 µg/L (1.04 µg/L extracellular MC-LR and 13.85 µg/L calculated intracellular MC-LR). To put this value in context, this represents approximately 29.5% of the total amount of MC-LR (Table 2) delivered to the filter during experiment BE. This data supports a conceptual model in which intracellular MC-LR from the *M. aeruginosa* cells retained is the primary component of the total MC-LR concentration within the sand filter; this is consistent with the relatively low removal efficiency for extracellular MC-LR and the simultaneous high level of removal of cyanobacteria cells by these sand filters.

2.4. Effects of Dewatered Supernatant Recycling

A key goal of these experiments was to determine whether the recycling of dewatered supernatant could serve to prolong the detrimental effects of cyanobacterial bloom events on treated water quality. The concentration of extracellular MC-LR across the treatment system during experiment B is shown in Figure 4. After the feed water enters the pilot system and undergoes coagulation, flocculation, and sedimentation, the MC-LR concentration increases, presumably due to the rupture of cells caused by the mixing processes; this effect is also reflected in the dewatered supernatant MC-LR concentration shown in Figure 4.

The MC-LR originally retained primarily in an intracellular form are now able to move through the process and enter the filtration process. However, as mentioned previously, the filtration process exhibits relatively low removal of extracellular MC-LR, resulting in its release in the filtrate solution. Further, examining the MC-LR concentration in day 2 dewatered supernatant at the end of the operation, it can be seen that MC-LR remains in the supernatant waiting to be recycled back into the system during the next operation period along with the MC-LR accumulated in the sand media for the previous 48 h of operation time. The retained MC-LR could be released from the filter media during the backwashing operations. Thus, even for a cyanobacterial bloom that affected feed water for only 24 h, the effects of the toxins and cyanobacteria from the bloom may persist much longer than one treatment system residence time if the plant recycles dewatered supernatant and/or the filter media backwash water.

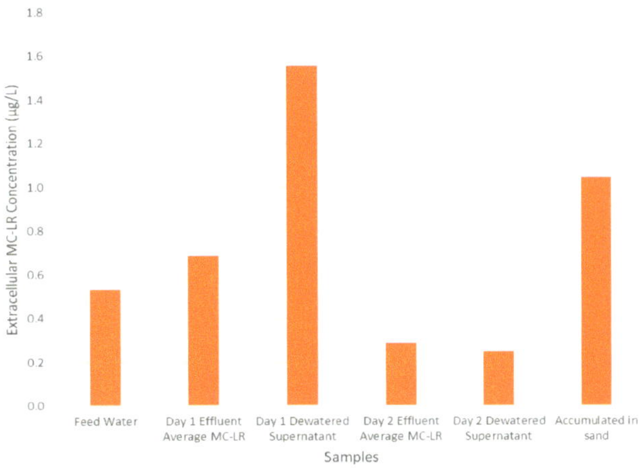

Figure 4. Experiment BE, extracellular MC-LR concentration throughout the water treatment process.

Even though the MC-LR concentration in the effluent water is below the WHO guideline, the measured concentration only considers the extracellular fraction of the water's total MC-LR content. To consider the potential amount of MC-LR that could be contributed to the solution if the intracellular portion was released to the solution, intracellular and extracellular MC-LR concentrations were measured for the dewatered supernatant and spiked feed water from experiments AE and BE, respectively (Figure 5). Most of the total MC-LR concentration (90%–98%) in these two samples is intracellular. Although MC-LR within intact cells can be efficiently removed by conventional treatment operations, the presence of high fractions of MC-LR in retained solids makes the permanence of this removal dependent on subsequent solids handling decisions.

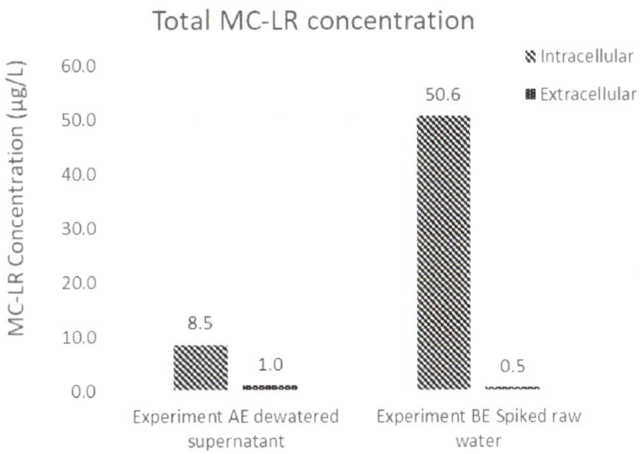

Figure 5. Extracellular and intracellular MC-LR concentration of experiment AE dewatered supernatant and experiment BE spiked raw water.

3. Discussion

3.1. *The Effectiveness of the Conventional Treatment Process*

The results from the water quality analysis demonstrate the capability of the pilot experimental system to remove turbidity, *M. aeruginosa* cells, and MC-LR to levels below the EPA and WHO regulations. Our finding that *M. aeruginosa* cells were removed with greater than 95% efficiency from the feed water by the processes of coagulation, flocculation, and sedimentation are similar to results reported by Chow et al. 1999 [29] but lower than found in a pilot-plant scale study by Zamyadi et al. 2013 [18]. Though the removal efficiency is high, the total number of *M. aeruginosa* cells remaining in the effluent stream might still cause problems in downstream unit operations, such as disinfection. It is worth noting that the microscopic observation method employed here using a hemacytometer is tedious and challenging. Implementation of a more precise cell counts method, such as via polymerase chain reaction (PCR) or flow cytometry is recommended.

Removal of the extracellular MC-LR cyanotoxin, however, was far less efficient, with overall removals below 50%. Although low, this is still higher than the results shown previously by Ho et al. 2006 [24]. The higher removal efficiency of intracellular, particle-associated cyanotoxins in comparison with MC-LR in its dissolved form is to be expected during conventional water treatment operations such as coagulation-flocculation and filtration, which are targeted at the removal of particulate matter. Other filter configurations may be more successful in removing MC-LR and related compounds; previous research indicates that the presence of biofilms on filter media, for example, can assist in degrading cyanotoxins; however, the bacterial biofilm required at least 4 days to form and be functional [13,24].

Regarding the extracellular MC-LR concentration, it can be seen that at the initial state of the treatment process (0 h) the MC-LR concentration in the filtered effluent is higher than the initial concentration in both experiments AE and BE. The increase in the MC-LR concentration observed here differs from findings reported previously by Chow et al. 1999 [29], who found no significant increase in MC-LR concentration following coagulation-flocculation. It is possible that the rapid mixing step used here caused additional cell lysis and contributed to the increased extracellular MC-LR in the effluent water.

3.2. *The Effects of Dewatering Supernatant Recycling*

Due to the high removal of *M. aeruginosa* during the coagulation-flocculation process, cyanobacteria cells accumulate in the sludge, which is subsequently sent to the dewatering system. The dewatering experiment revealed that intracellular MC-LR are released to the solution during this phase, exhibiting a 192% increase in extracellular MC-LR concentration compared with the concentration in the feed solution (Figure 4). If this extracellular MC-LR is recycled into the treatment system, it has the potential to extend the effects of a cyanobacterial bloom event on the water treatment system. The increased duration may be longer than the hydraulic residence time of the treatment system as both the dewatered supernatant from the 2nd-day operation and the MC-LR that may be released from the filter media through the backwashing process could be recycled back into the treatment system.

As mentioned previously, the treatment simulations performed in this study only addressed the effects of a cyanobacterial bloom on the coagulation-flocculation and filtration processes and do not address possible impacts on the disinfection process. If the disinfection process were included, a lower concentration of *M. aeruginosa* cells and microcystins might be achieved, as shown by previous studies [13,14,30–32] Even though the disinfection process may oxidize the cyanobacterial membrane and initiate the release of intracellular microcystins, this may produce other hazards in the form of increased concentrations of disinfection by-products including trihalomethanes (THM), haloacetic acids (HAA), and N-nitrosodimethylamine (NDMA) [14,29,33–36].

3.3. Future Extensions of the Research

Future extensions of this research would include upgrading the batch simulation approach used in these experiments to a continuous treatment process, which may result in a more accurate estimation of the MC-LR concentrations in effluents from conventional drinking water processes. The addition of a chlorination process to the simulated treatment train would allow the effects of recycling cyanobacteria-laden dewatered supernatant on the amount and product distribution of disinfection-by-products to be explored. Another worthwhile avenue for future research aimed at reducing cyanotoxin impacts is to explore possible origins of cyanobacterial blooms in various water sources; this could inform watershed management approaches that minimize cyanobacterial blooms to reduce the probability that cyanotoxins and/or elevated levels of disinfection-by-products will impact final drinking water quality.

Though a pilot-scale experiment might achieve the desired removal efficiency and better imitate the actual water treatment process, conducting a plant-scale experiment is ultimately recommended. Investigating these processes in a real water treatment plant will alleviate problems associated with limited water sample size and allow more robust analysis of the performance of each individual process. However, if a plant-scale experiment were conducted, the operators must ensure that the effluent from the experiment does not enter the drinking water distribution system, as it might be harmful for the community surrounding the plant.

4. Conclusions

Overall, the results of the current study have shown the possibility of an increase in turbidity, cell density, and MC-LR concentration caused by the recycling of dewatered supernatant and the possible extension of the water quality effects from seasonal cyanobacteria blooms. Thus, we strongly advise against the recycling of dewatered supernatant during cyanobacterial bloom periods, redirecting these materials instead to the sludge treatment process. These steps will help minimize the duration of cyanobacterial bloom impacts on finished water quality. However, a plant scale study to confirm these results in field-scale water treatment facilities is recommended.

5. Materials and Methods

5.1. Materials and Reagents

The microcystin-producing strain *M. aeruginosa* was obtained from the University of Texas Culture Collection of Algae (UTEX, https://utex.org/, UTEX Culture ID: LB2385, Austin, USA) and cultured in a synthetic cyanobacteria growth media, CB media (adapted from Shirai et al. 1989 [37]) under fluorescent lamps in an incubator with a controlled temperature of 25 °C. The cyanobacteria cells were harvested at 2 weeks of incubation during the logarithmic growth phase of the cyanobacteria and the initial cell counts were analyzed using Hausser Scientific Hemacytometer (Fisher Scientific, Pittsburgh, PA, USA). The MC-LR standard was purchased from Enzo Life Sciences (Farming Dale, NY, USA). The coagulant aluminum sulfate hydrate ($Al_2(SO_4)_3 \cdot 18H_2O$) was purchased from Sigma-Aldrich (St. Louis, MI, USA) and the solutions were prepared with deionized (Milli-Q) water prior to the experiments; new solutions were prepared for each experiment. The water source chosen for the study was the Sacramento River; samples were collected twice from the WDCWA treatment plant raw water intake pipe on 13 February and 27 February 2020. The raw water samples were then stored in a refrigerator at 4 °C. Prior to starting experiments each day, the water samples were taken out and allowed to attain room temperature.

5.2. Experimental Setup

5.2.1. Optimum Dosage Jar Test Experiment

In order to determine the optimum coagulant concentration for the lab-pilot scale experiment, a series of jar test experiments were conducted using a Programmable Jar tester

(Phipps &Bird Model PB-900, Fisher Scientific, Pittsburgh, PA, USA) with a 2000 mL beaker at room temperature (23 ± 1 °C). The coagulant was added to 1000 mL of *M. aeruginosa* spiked feed water with a rapid mixing speed of 250 rpm for 60 s [17] After the coagulation process, the sample underwent slow mixing at 40 rpm for 30 min. Then, the samples were allowed to settle for 30 min. The supernatant from the jar test experiments were collected to determine the turbidity and *M. aeruginosa* cell counts.

5.2.2. Filtration Column Preparation

The glass filtration column (length 20 cm, internal diameter 3 cm) was packed with an autoclaved sand filter media with a bed height of 15 cm. Milli-Q water was pumped through the bed in a downflow configuration until steady state was achieved at the designated flow rate. At that time, the Milli-Q water was replaced with the supernatant from the preliminary treatment process.5.2.3. Lab-Scale Experiments

The controlled sample without any spiking of *M. aeruginosa* cells (control, C) and the feed water with spiked *M. aeruginosa* to imitate cyanobacterial blooms entering the water treatment system influent (experiment, E) were prepared for both experiments A and B.

Stock *M. aeruginosa* cells suspensions with a density of 20 × 10^9 cells/mL were prepared and resuspended in raw water to achieve a cell suspension of 1 × 10^6 cells/mL for experiments AE and BE. No *M. aeruginosa* cells were added to the feed water for experiments AC and BC. After the resuspension of the *M. aeruginosa* stock solution, both experiments underwent the same procedures. The spiking of cyanobacteria cells into the raw water was only done on the first day of the experiment.

Jar test experiments were performed to represent the coagulation-flocculation and sedimentation processes during water treatment using the same configuration as the jar tests conducted to determine optimum coagulant dosages.

The supernatant from the jar test experiment was then fed to the filtration process. The filtration system was expected to be able to operate for 48 h without backwashing. A 1 L sample was collected from both experimental conditions (C and E) at 0, 4, 12, 24, 26, 28, 32, 36, and 48 h after filtration and stored in the refrigerator at 4 °C for a maximum of 48 h, before undergoing the microcystin analysis. These water samples were analyzed for turbidity and cyanobacteria cell counts immediately after being collected.

All the sludge from the jar test was collected, combined, and transferred to a separate dewatering container for each of the experiments and allowed to settle for 24 h. After 24 h of dewatering, the supernatant from the process was separated into two parts, 500 mL of the sample were used for MC-LR concentration analysis and the remainder was transferred to the second-day feed water to imitate the recycling of dewatered supernatant in a conventional water treatment system. The combined water was processed through the jar test experiment again to simulate the preliminary treatment.

5.3. Water Quality Analysis

5.3.1. Turbidity

The water effluent samples from each time step were analyzed for turbidity using a HACH 2100AN turbidity meter (HACH, Loveland, CO, USA). The instrument was calibrated according to the manufacture's specifications for every 24-h time step.

5.3.2. Analysis of Microcystins

In the absence of suitable standards for most microcystin variants, the microcystin concentration in this experiment for all samples was expressed as equivalent of microcystin-LR (MC-LR) [38–40].

Prior to the liquid chromatography MC-LR analysis, the samples were concentrated using solid-phase extraction cartridges (Waters, Oasis SPE, Milford, CT, USA). The MC-LR concentration was analyzed by using a Liquid Chromatography-Quadrupole Time-of-Flight-Mass Spectrometer (Agilent 1260 Infinity HPLC coupled with an Agilent 6530 QTOF-MS, LC-QTOF-MS, Santa Clara, USA). Sample volumes of 10 µL were injected into the column (Agilent Zorbax Eclipse C18, Santa Clara, CA, USA) at a flow rate of 0.35 mL/min. Negative Electrospray Ionization (-ESI) mode was selected for the analysis. The concentration of the MC-LR was determined by the calibration of the peak areas.

5.3.3. Analyzing the Sand Filter Media after the Experiment

After the breakthrough point for either the turbidity or water height was reached in both control and experimental filter beds, the filter media was removed from the glass column and transferred into separate Erlenmeyer flasks. Deionized water (Milli-Q, 500 mL) was then added to the flasks, and flasks were mechanically shaken for 24 h to remove deposited cyanobacteria cells and MC-LR from the media surface. Then, the supernatant from the process was filtered through a glass filter module. The filter paper was collected and cut into 1 mm × 1 mm pieces and sonicated in 20 mL of Milli-Q water for 2 h, after the completion of the sonication the sample was centrifuged at 3500 rpm for 5 min. Both the filtered and the centrifuged samples were combined and analyzed for total MC-LR concentration.

Author Contributions: Conceptualization: K.P. and T.M.Y.; investigation, K.P.; methodology, K.P., S.J.T., T.K., C.H.L., F.T., and T.M.Y.; writing—original draft, K.P.; writing—review and editing, T.K. and T.M.Y. All authors have read and agreed to the published version of the manuscript.

Funding: Research reported in this publication was supported in part by the National Institute of Environmental Health Sciences of the National Institutes of Health under award number P42 ES004699. The content is solely the responsibility of the authors and does not necessarily represent the official views of the National Institutes of Health.

Institutional Review Board Statement: Not applicable.

Informed Consent Statement: Not applicable.

Data Availability Statement: Data available upon request.

Acknowledgments: The authors appreciate assistance analyzing microcystin-LR provided by Luann Wong and the help with water sampling at WDCWA provided by Brian Frank.

Conflicts of Interest: The authors declare no conflict of interest.

References

1. Abrantes, N.; Antunes, S.; Pereira, M.; Gonçalves, F. Seasonal succession of cladocerans and phytoplankton and their interactions in a shallow eutrophic lake (Lake Vela, Portugal). *Acta Oecol.* **2006**, *29*, 54–64. [CrossRef]
2. Almuhtaram, H.; Cui, Y.; Zamyadi, A.; Hofmann, R. Cyanotoxins and Cyanobacteria Cell Accumulations in Drinking Water Treatment Plants with a Low Risk of Bloom Formation at the Source. *Toxins* **2018**, *10*, 430. [CrossRef] [PubMed]
3. Bláha, L.; Babica, P.; Maršálek, B. Toxins produced in cyanobacterial water blooms—Toxicity and risks. *Interdiscip. Toxicol.* **2009**, *2*, 36–41. [CrossRef] [PubMed]
4. Chintalapati, P.; Mohseni, M. Degradation of cyanotoxin microcystin-LR in synthetic and natural waters by chemical-free UV/VUV radiation. *J. Hazard. Mater.* **2020**, *381*, 120921. [CrossRef]
5. Chow, C.W.; Drikas, M.; House, J.; Burch, M.D.; Velzeboer, R.M. The impact of conventional water treatment processes on cells of the cyanobacterium Microcystis aeruginosa. *Water Res.* **1999**, *33*, 3253–3262. [CrossRef]
6. Coral, L.A.; Zamyadi, A.; Barbeau, B.; Bassetti, F.D.J.; Lapolli, F.R.; Prévost, M. Oxidation of Microcystis aeruginosa and Anabaena flos-aquae by ozone: Impacts on cell integrity and chlorination by-product formation. *Water Res.* **2013**, *47*, 2983–2994. [CrossRef]
7. Daly, R.I.; Ho, L.; Brookes, J.D. Effect of Chlorination onMicrocystis aeruginosaCell Integrity and Subsequent Microcystin Release and Degradation. *Environ. Sci. Technol.* **2007**, *41*, 4447–4453. [CrossRef]
8. Drikas, M.; Chow, C.W.K.; House, J.; Burch, M.D. Using Coagulation, and Settling to Remove Toxic Cyanobacte-ria. *J. Am. Water Work. Assoc.* **2001**, *93*, 100–111. [CrossRef]
9. Fan, J.; Hobson, P.; Ho, L.; Daly, R.; Brookes, J.D. The effects of various control and water treatment processes on the membrane integrity and toxin fate of cyanobacteria. *J. Hazard. Mater.* **2014**, *264*, 313–322. [CrossRef]
10. Fan, J.; Rao, L.; Chiu, Y.-T.; Lin, T.-F. Impact of chlorine on the cell integrity and toxin release and degradation of colonial Microcystis. *Water Res.* **2016**, *102*, 394–404. [CrossRef]
11. Gad, A.A.; El-Tawel, S. Effect of pre-oxidation by chlorine/permanganate on surface water characteristics and algal toxins. *Desalin. Water Treat.* **2015**, *57*, 17922–17934. [CrossRef]

12. Ghernaout, B.; Ghernaout, D.; Saiba, A. Algae and cyanotoxins removal by coagulation/flocculation: A review. *Desalin. Water Treat.* **2010**, *20*, 133–143. [CrossRef]
13. Han, J.; Jeon, B.-S.; Park, H.-D. Cyanobacteria cell damage and cyanotoxin release in response to alum treatment. *Water Supply* **2012**, *12*, 549–555. [CrossRef]
14. He, X.; Liu, Y.-L.; Conklin, A.; Westrick, J.; Weavers, L.K.; Dionysios, D.; Lenhart, J.J.; Mouser, P.J.; Szlag, D.; Walker, H.W. Toxic cyanobacteria and drinking water: Impacts, detection, and treatment. *Harmful Algae* **2016**, *54*, 174–193. [CrossRef]
15. Henderson, R.; Parsons, S.A.; Jefferson, B. The impact of algal properties and pre-oxidation on solid–liquid separation of algae. *Water Res.* **2008**, *42*, 1827–1845. [CrossRef]
16. Ho, L.; Dreyfus, J.; Boyer, J.; Lowe, T.; Bustamante, H.; Duker, P.; Meli, T.; Newcombe, G. Fate of cyanobacteria and their metabolites during water treatment sludge management processes. *Sci. Total. Environ.* **2012**, *424*, 232–238. [CrossRef]
17. Ho, L.; Meyn, T.; Keegan, A.; Hoefel, D.; Brookes, J.; Saint, C.P.; Newcombe, G. Bacterial degradation of microcystin toxins within a biologically active sand filter. *Water Res.* **2006**, *40*, 768–774. [CrossRef]
18. Hou, C.-R.; Hu, W.; Jia, R.-B.; Liu, P.-Q. The Mechanism of Cyanobacterium (M.aeruginosa) Microcystins Releasing by Chemical Oxidation in Drinking Water Treatment. In Proceedings of the 2008 2nd International Conference on Bioinformatics and Biomedical Engineering, Shanghai, China, 16–18 May 2008; IEEE: Piscataway, NJ, USA, 2008; pp. 3734–3737.
19. Jiang, J.Q.; Graham, N.J. Preliminary evaluation of the performance of new pre-polymerised inorganic coagulants for lowland surface water treatment. *Water Sci. Technol.* **1998**, *37*, 121–128. [CrossRef]
20. Jiang, J.-Q.; Graham, N.J.D.; Harward, C. Comparison of Polyferric Sulphate with Other Coagulants for the Removal of Algae and Algae-Derived Organic Matter. *Water Sci. Technol.* **1993**, *27*, 221–230. [CrossRef]
21. Kemp, A.; John, J. Microcystins associated withMicrocystis dominated blooms in the Southwest wetlands, Western Australia. *Environ. Toxicol.* **2006**, *21*, 125–130. [CrossRef]
22. Lin, D.; Ralph, S.; Evans, L.; Beuscher, D.B. *Algal Removal by Alum Coagulation*; Illinois State Water Survey: Champaign, IL, USA, 1971; Available online: https://www.ideals.illinois.edu/handle/2142/102010 (accessed on 28 January 2021).
23. Ma, M.; Liu, R.; Liu, H.; Qu, J. Chlorination of Microcystis aeruginosa suspension: Cell lysis, toxin release and degradation. *J. Hazard. Mater.* **2012**, *217*, 279–285. [CrossRef]
24. Pestana, C.J.; Reeve, P.J.; Sawade, E.; Voldoire, C.F.; Newton, K.; Praptiwi, R.; Collingnon, L.; Dreyfus, J.; Hobson, P.; Gaget, V.; et al. Fate of cyanobacteria in drinking water treatment plant lagoon supernatant and sludge. *Sci. Total Environ.* **2016**, *565*, 1192–1200. [CrossRef]
25. Pivokonsky, M.; Naceradska, J.; Kopecka, I.; Baresova, M.; Jefferson, B.; Li, X.; Henderson, R. The impact of algogenic organic matter on water treatment plant operation and water quality: A review. *Crit. Rev. Environ. Sci. Technol.* **2015**, *46*, 291–335. [CrossRef]
26. Qin, B.; Li, W.; Zhu, G.; Zhang, Y.; Wu, T.; Gao, G. Cyanobacterial bloom management through integrated monitoring and forecasting in large shallow eutrophic Lake Taihu (China). *J. Hazard. Mater.* **2015**, *287*, 356–363. [CrossRef]
27. Qin, B.; Zhu, G.; Gao, G.; Zhang, Y.; Li, W.; Paerl, H.W.; Carmichael, W.W. A Drinking Water Crisis in Lake Taihu, China: Linkage to Climatic Variability and Lake Management. *Environ. Manag.* **2010**, *45*, 105–112. [CrossRef]
28. Richardson, J.; Miller, C.; Maberly, S.C.; Taylor, P.; Globevnik, L.; Hunter, P.D.; Jeppesen, E.; Mischke, U.; Moe, S.J.; Pasztaleniec, A.; et al. Effects of multiple stressors on cyanobacteria abundance vary with lake type. *Glob. Chang. Biol.* **2018**, *24*, 5044–5055. [CrossRef]
29. Scott, J.T.; Marcarelli, A.M. Cyanobacteria in Freshwater Benthic Environments. In *Ecology of Cyanobacteria II*; Springer: Berlin/Heidelberg, Germany, 2012; pp. 271–289.
30. Shi, X.; Bi, R.; Yuan, B.; Liao, X.; Zhou, Z.; Li, F.; Sun, W. A comparison of trichloromethane formation from two algae species during two pre-oxidation-coagulation-chlorination processes. *Sci. Total Environ.* **2019**, *656*, 1063–1070. [CrossRef]
31. Shirai, M.; Matumaru, K.; Ohotake, A.; Takamura, Y.; Aida, T.; Nakano, M. Development of a Solid Medium for Growth and Isolation of Axenic Microcystis Strains (Cyanobacteria). *Appl. Environ. Microbiol.* **1989**, *55*, 2569–2571. [CrossRef]
32. Sun, J.; Bu, L.; Deng, L.; Shi, Z.; Zhou, S. Removal of Microcystis aeruginosa by UV/chlorine process: Inactivation mechanism and microcystins degradation. *Chem. Eng. J.* **2018**, *349*, 408–415. [CrossRef]
33. World Health Organization. Guidelines for Safe Recreational Water Environments. Volume 1. Coastal and Fresh Waters. 2013. Available online: https://apps.who.int/iris/handle/10665/42591 (accessed on 28 January 2021).
34. World Health Organization. *Guidelines for Drinking-Water Quality*; World Health Organization: Geneva, Switzerland, 2017.
35. Xu, H.; Qi, F.; Jin, Y.; Xiao, H.; Ma, C.; Sun, J.; Li, H. Characteristics of water obtained by dewatering cyanobacteria-containing sludge formed during drinking water treatment, including C-, N-disinfection byproduct formation. *Water Res.* **2017**, *111*, 382–392. [CrossRef]
36. Yuan, B.; Qu, J.-H.; Fu, M.-L. Removal of cyanobacterial microcystin-LR by ferrate oxidation–coagulation. *Toxicon* **2002**, *40*, 1129–1134. [CrossRef]
37. Zamyadi, A.; Dorner, S.; Sauvé, S.; Ellis, D.; Bolduc, A.; Bastien, C.; Prévost, M. Species-dependence of cyanobacteria removal efficiency by different drinking water treatment processes. *Water Res.* **2013**, *47*, 2689–2700. [CrossRef] [PubMed]
38. Zamyadi, A.; Ho, L.; Newcombe, G.; Bustamante, H.; Prévost, M. Fate of toxic cyanobacterial cells and disinfection by-products formation after chlorination. *Water Res.* **2012**, *46*, 1524–1535. [CrossRef] [PubMed]

39. Zamyadi, A.; MacLeod, S.L.; Fan, Y.; McQuaid, N.; Dorner, S.; Sauvé, S.; Prévost, M. Toxic cyanobacterial breakthrough and accumulation in a drinking water plant: A monitoring and treatment challenge. *Water Res.* **2012**, *46*, 1511–1523. [CrossRef] [PubMed]
40. Zhou, S.; Shao, Y.; Gao, N.; Li, L.; Deng, J.; Zhu, M.; Zhu, S. Effect of chlorine dioxide on cyanobacterial cell integrity, toxin degradation and disinfection by-product formation. *Sci. Total. Environ.* **2014**, 208–213. [CrossRef]

toxins

MDPI

Article

Can Cyanobacterial Diversity in the Source Predict the Diversity in Sludge and the Risk of Toxin Release in a Drinking Water Treatment Plant?

Farhad Jalili [1,*], Hana Trigui [1], Juan Francisco Guerra Maldonado [1], Sarah Dorner [1], Arash Zamyadi [2], B. Jesse Shapiro [3,4,5], Yves Terrat [3], Nathalie Fortin [6], Sébastien Sauvé [7] and Michèle Prévost [1]

1 Department of Civil, Geological and Mining Engineering, Polytechnique Montréal, Montréal, QC H3C 3A7, Canada; hana.trigui@polymtl.ca (H.T.); juan-francisco.guerra-maldonado@polymtl.ca (J.F.G.M.); sarah.dorner@polymtl.ca (S.D.); michele.prevost@polymtl.ca (M.P.)
2 Water Research Australia, Adelaide SA 5001, Australia; arash.zamyadi@waterra.com.au
3 Department of Biological Sciences, University of Montréal, Montréal, QC H2V 0B3, Canada; jesse.shapiro@umontreal.ca (B.J.S.); yves.terrat@umontreal.ca (Y.T.)
4 Department of Microbiology and Immunology, McGill University, Montréal, QC H3A 2B4, Canada
5 McGill Genome Center, McGill University, Montréal, QC H3A 0G1, Canada
6 National Research Council Canada, Energy, Mining and Environment, Montréal, QC H4P 2R2, Canada; nathalie.fortin@cnrc-nrc.gc.ca
7 Department of Chemistry, University of Montréal, Montréal, QC H3C 3J7, Canada; sebastien.sauve@umontreal.ca
* Correspondence: farhad.jalili@polymtl.ca

Citation: Jalili, F.; Trigui, H.; Guerra Maldonado, J.F.; Dorner, S.; Zamyadi, A.; Shapiro, B.J.; Terrat, Y.; Fortin, N.; Sauvé, S.; Prévost, M. Can Cyanobacterial Diversity in the Source Predict the Diversity in Sludge and the Risk of Toxin Release in a Drinking Water Treatment Plant?. *Toxins* **2021**, *13*, 25. https://doi.org/10.3390/toxins13010025

Received: 30 November 2020
Accepted: 29 December 2020
Published: 1 January 2021

Abstract: Conventional processes (coagulation, flocculation, sedimentation, and filtration) are widely used in drinking water treatment plants and are considered a good treatment strategy to eliminate cyanobacterial cells and cell-bound cyanotoxins. The diversity of cyanobacteria was investigated using taxonomic cell counts and shotgun metagenomics over two seasons in a drinking water treatment plant before, during, and after the bloom. Changes in the community structure over time at the phylum, genus, and species levels were monitored in samples retrieved from raw water (RW), sludge in the holding tank (ST), and sludge supernatant (SST). *Aphanothece clathrata brevis, Microcystis aeruginosa, Dolichospermum spiroides,* and *Chroococcus minimus* were predominant species detected in RW by taxonomic cell counts. Shotgun metagenomics revealed that Proteobacteria was the predominant phylum in RW before and after the cyanobacterial bloom. Taxonomic cell counts and shotgun metagenomic showed that the *Dolichospermum* bloom occurred inside the plant. Cyanobacteria and Bacteroidetes were the major bacterial phyla during the bloom. Shotgun metagenomics also showed that *Synechococcus, Microcystis,* and *Dolichospermum* were the predominant detected cyanobacterial genera in the samples. Conventional treatment removed more than 92% of cyanobacterial cells but led to cell accumulation in the sludge up to 31 times more than in the RW influx. Coagulation/sedimentation selectively removed more than 96% of *Microcystis* and *Dolichospermum*. Cyanobacterial community in the sludge varied from raw water to sludge during sludge storage (1–13 days). This variation was due to the selective removal of coagulation/sedimentation as well as the accumulation of captured cells over the period of storage time. However, the prediction of the cyanobacterial community composition in the SST remained a challenge. Among nutrient parameters, orthophosphate availability was related to community profile in RW samples, whereas communities in ST were influenced by total nitrogen, Kjeldahl nitrogen (N- Kjeldahl), total and particulate phosphorous, and total organic carbon (TOC). No trend was observed on the impact of nutrients on SST communities. This study profiled new health-related, environmental, and technical challenges for the production of drinking water due to the complex fate of cyanobacteria in cyanobacteria-laden sludge and supernatant.

Keywords: cyanobacteria; microcystins (MCs); water treatment; sludge; shotgun metagenomics; cyanobacterial community; high-throughput sequencing

Key Contribution: High-throughput sequencing was applied to investigate cyanobacterial community in raw water, stored sludge, and supernatant of sludge holding tanks in a drinking water treatment plant.

1. Introduction

Cyanobacterial cells and their associated cyanotoxins are considered to represent an important challenge due to (1) their health threat to humans and animals; (2) their negative aesthetic impacts with respect to taste, odor, and color; (3) the implications of extra water treatment requirements for ozonation and membrane filtration; and (4) the increased consumption of coagulants, flocculants, and activated carbon [1–3].

Conventional treatment using coagulation, flocculation, sedimentation, and filtration is a common approach for cyanobacterial removal from intake water [1,3,4]. However, conventional treatment is not efficient at removing dissolved cyanotoxins [5]. In addition, hydraulic and chemical stresses during treatment may cause damage to cells and trichomes, leading to the release of cyanotoxins [6,7]. Another challenge of conventional treatment is the increase in cyanobacteria and cyanotoxin concentrations in the clarifiers and filters of water treatment plants (WTPs) and their accumulation in the sludge of clarifiers [3,4,8,9]. Cyanobacterial cells can be present at concentrations 10–100 times higher in the sludge than in intake water, even in plants with low cyanobacterial flux (<1000 cells/mL) [10,11]. Moreover, some investigations have shown that coagulated cells can stay viable in the sludge for 2 to 10 days [4,9,12–14]. More recently, cell viability in the sludge was observed for more than 20 days [15]. During this period, microcystin-LR (MCLR) and cylindrospermopsin concentrations increased to 3–7 times their initial levels. The authors of [16] reported that metabolite concentrations in the sludge supernatant after storage were up to five times greater than those within the sludge before storage. This shows a new challenge in sludge management during storage and when the sludge supernatant is recycled to the head of the plant [16,17]. The fate of cyanobacteria and cyanotoxins during and after coagulation and in the sludge is not fully understood. The impacts of coagulation on cyanobacterial cells are still controversial. Although some studies have demonstrated that coagulation depends on cyanobacterial species [3,17,18], another study showed that cells are not selectively captured by coagulation [15]. It has been shown that cell damage and metabolites released in sludge are associated with various environmental conditions; however, due to the complex interactions of cyanobacteria with treatment processes, the primary factors behind this complex behavior are still not determined [16,19]. Additionally, although the positive impact of powdered activated carbon (PAC) on cyanotoxin degradation in raw water (RW) has been widely studied [5,20–23], there are no data about the role of injected PAC in RW in the degradation of accumulated cyanotoxins within sludge.

Recently, high-throughput sequencing and metagenomics techniques have been successfully applied to describe microbial communities in the water resources to predict the occurrence of cyanobacterial blooms [24,25]. During the last decade, several studies have investigated bacterial communities in WTPs and have demonstrated that while microbial communities in the water treatment chain are represented by water intake, treatment processes have an impact on the microbial community structure through WTPs [17,26–32]. Few studies have investigated bacterial communities within sludge in WTPs [33,34]. The authors of [33] reported similar bacterial communities in sludge samples collected from six different Chinese WTPs with the same treatment processes. They reported similar bacterial communities in sludge samples, suggesting that bacterial communities in sludge might be shaped by RW communities; however, they did not compare the bacterial composition in sludge with that in RW. The authors of [34] studied the impact of different coagulants on bacterial communities and metabolite release in sludge. They found that the relative abundance of the dominant genera *Microcystis, Rhodobacter, Phenylobacterium,* and *Hydrogenophaga* decreased, reflecting their damage and the subsequent release of extracellular microcystin and organic matter. They suggest that the sludge should be treated or disposed of within 4 days to avoid the proliferation of the pathogens.

There are basically no studies exploring the impact of RW cyanobacterial communities on cyanobacteria-laden sludge and its supernatant in WTPs. Moreover, previous studies considered sludge as a batch samples, while in WTPs cyanobacteria-laden sludge might be dynamically affected by different parameters such as RW characteristics, treatment process functionality, and sludge storage time. The impact of these parameters on the cyanobacterial community structure of sludge has not been investigated. No comparative analysis has been carried out on bacterial/cyanobacterial community composition within RW, sludge, and sludge supernatant. Due to the knowledge gaps related to the fate of cyanobacteria and cyanotoxins in sludge, high-throughput sequencing techniques could be useful to better understand the community dynamics of cyanobacteria-laden sludge through a WTP. This study is the first to use both shotgun metagenomic sequencing and taxonomic cell count approaches to provide an overview of cyanobacterial composition in RW, a sludge holding tank (ST), and the corresponding sludge supernatant (SST) in a WTP.

The general objective of this research was to study the fate of cyanobacteria and its associated cyanotoxins in a WTP. The specific objectives were to: (1) diagnose critical points of WTPs where cyanobacteria cells and their associated cyanotoxins accumulate; (2) determine the relationship between cyanobacterial communities in RW, sludge, and its supernatant; (3) determine the impact of nutrients on cyanobacterial community shifts in RW, the sludge holding tank, and its supernatant; and (4) compare taxonomic cell counts with shotgun metagenomic sequencing results.

2. Results and Discussion

2.1. Impact of Conventional Treatment on Cyanobacteria and Cyanotoxins

During the period prior to the bloom (1 July to 30 August 2017), taxonomic cell counts in RW were below 5.0×10^4 cells/mL and *Aphanothece clathrata brevis* was the most dominant species, representing 65–100% of total cell counts. *Dolichospermum spiroides* (0–26%), *Chroococcus minimus* (0–34%), *Microcystis aeruginosa* (0–2%), and *Dolichospermum circinale* (<1%) were detected frequently (Figure 1). During this period, low concentrations of MCs were detected, with dissolved microcystin (MC) levels below 90 ng/L and cell-bound MC levels below 10 ng/L (Figure 2, Table S2).

A cyanobacterial bloom appeared in Missisquoi Bay in late August, and total taxonomic cell counts increased in RW to 3.1×10^5 cells/mL on 1 September. The dominant species was *D. spiroides*, representing 52% of total cell counts with a concentration of 1.6×10^5 cells/mL. Other identified species were *A. clathrata brevis* (1.0×10^5 cells/mL, 33%), *M. aeruginosa* (4.0×10^4 cells/mL, 13%), and *Coelosphaerium kuetzingianum* (1.7×10^3 cells/mL, 1%) (Figure 1). The low total MC level increased to 260.1 ng/L, of which 191.9 ng/L were dissolved (Figure 2). On the sampling dates following the bloom (5 September and 27 October), cell counts decreased to around 3.9×10^4 cells/mL and remained constant. The species *A. clathrata brevis* was dominant during those two dates (68–85%). *C. minimus* (31%) was also found on 5 September and *M. aeruginosa* (31%) and *Aphanizomenon gracile* (1%) were found on 27 October. MCs remained below the detection limit (DL) on 5 September. On 27 October, the total MC concentration increased to 142.6 ng/L, of which 121.9 ng/L were dissolved (Figure 2).

Through the treatment process, 86–99% of total cyanobacterial cells were removed by the clarifier. In particular, 85–100% of *M. aeruginosa, A. clathrata brevis, C. minimus, A. gracile,* and *D. spiroides* were eliminated. These results are in agreement with previous studies documenting coagulation efficiency of between 62% and 99% [4,8,14,35,36]. Meanwhile, 14–71% of the escaped cells from the clarifier were removed by filtration (Figure 1). Overall, 92–99% of cells were eliminated by conventional treatment. A previous study on this plant found a similar reduction [3]. Furthermore, taxonomic cell counts increased from 25% to 120% in treated water (TW) on all sampling dates. *A. clathrata brevis* (72–99%) remained dominant in TW, except on the bloom date (1 September). On 1 September, the cyanobacterial composition in TW consisted of *D. spiroides* (62%), *A. clathrata brevis* (29%), and *Pseudanabaena mucicola* (9%) (Figure 1). Cell counts in TW (after chlorination) were 1.3–2 times greater than cell counts in filtered water (FW) (before chlorination). This accumulation can be problematic if accumulated cells produce cyanotoxins.

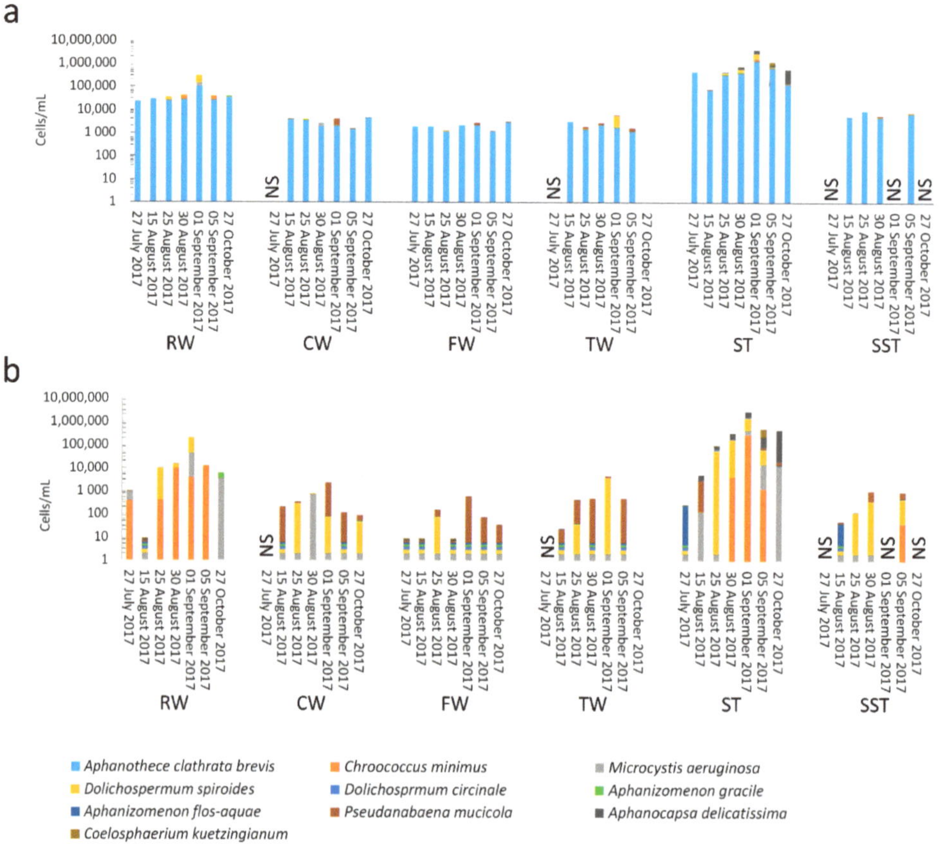

Figure 1. Distribution of cyanobacterial species in the water treatment plant (WTP) by taxonomic cell counts: (**a**) all species, (**b**) speciation of species other than *Aphanothece clathrata brevis*. RW: raw water, CW: clarified water, FW: filtered water, TW: treated water, ST: sludge holding tank, SST: sludge holding tank supernatant. Sludge storage times—27 July 2017: 8 days; 15 August 2017: 3 days; 25 August 2017: 8 days; 30 August 2017: 13 days; 1, 5 September 2017: 1 day; 27 October 2017: 2 days. NS: sample not taken.

Total taxonomic cell counts in ST remained around 3–31 times greater than in RW (Figure 1). The cell percentage of *A. clathrata brevis* in ST decreased from ~100% on 27 July to 32% on 1 September (bloom date). In contrast, during this period, the percentages of *Aphanocapsa delicatissima* and *D. spiroides* increased from 0% to 29% and 27%, respectively. On 1 September, *M. aeruginosa* counts in ST were four times greater than in RW. Furthermore, *A. delicatissima, P. mucicola, Aphanizomenon flos-aquae, D. circinale, and C. kuetzingianum* were detected in small amounts (0–1.7 $\times$ 10^3 cells/mL) in RW, whereas their cell counts increased to between 2.5 $\times$ 10^2 to 1.2 $\times$ 10^6 cells/mL in ST on the corresponding dates (Figure 1).

Cell counts in SST remained around 94–98% lower than those in ST and 69–97% lower than those in RW. *A. clathrata brevis* was dominant (83–99%) in SST, with small percentages of *D. spiroides* (1–6%) and *P. mucicola* (5–11%).

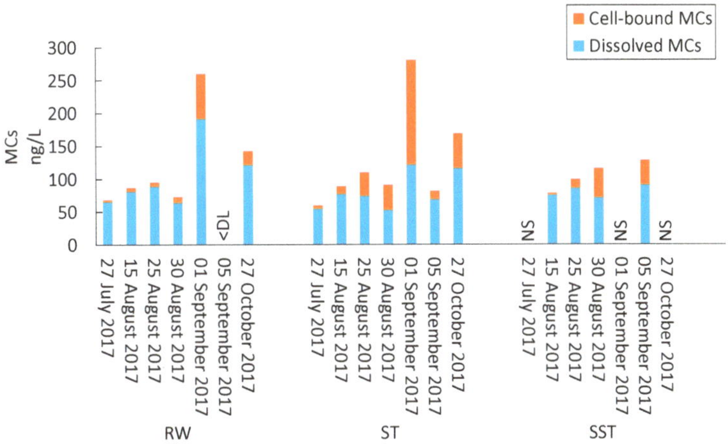

Figure 2. Concentration of dissolved and cell-bound microcystins (MCs) in raw water (RW), in the sludge holding tank (ST), and in sludge holding tank supernatant (SST). NS: sample not taken, DL: below detection limit.

Total MCs in ST and SST remained below 281 and 128 ng/L, respectively, during the sampling campaign (Figure 2). These MC trends are inconsistent with previous investigations which reported MCLR concentrations in the clarifier sludge to be around 10 times greater than in RW [3,8]. This low concentration of MCs measured in the ST also contradicts the results of [16], where it was shown that cyanobacterial metabolites in lagoon supernatant were 2 to 5 times greater than the initial concentrations. One reason behind the low MC concentrations in the ST as well as in the SST in our study may be the impact of injected PAC in RW on accumulated MCs in stored sludge. Indeed, levels of dissolved MCs were reduced from 121.7 ng/L (1 September) to below DL on 5 September when the PAC dose increased from 9.2 to 27.3 mg/L. In contrast, the concentration of dissolved MCs increased to 116.7 ng/L on 27 October when the PAC dose decreased to 7.0 mg/L (Figures 2 and 3). A second reason may be the biodegradation of MCs during sludge storage. However, the authors of [37] documented biodegradation of MCs as being very low compared to individual microcystin analogues.

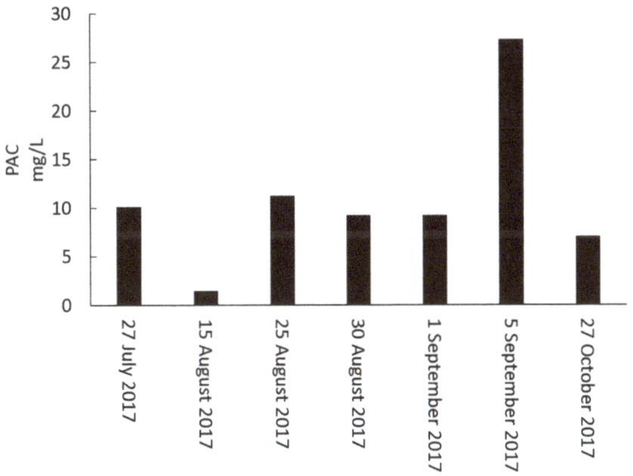

Figure 3. Powder activated carbon (wood-based PAC) doses injected into raw water (RW) during the 2017 sampling campaigns.

2.2. Cyanobacterial Diversity in Sludge and Supernatant Assessed by Shotgun Metagenomic Sequencing

From 27 July to 25 August 2017, Proteobacteria remained dominant in RW and their relative abundance increased from 26% to 56%. Actinobacteria (12–26%) and Bacteroidetes (14%) were the following dominant phyla in RW. During this period, Cyanobacteria, Verrucomicrobia, and Firmicutes had small relative abundances below 6%, 5%, and 3%, respectively. On 30 August, the abundance of Proteobacteria and Actinobacteria decreased to 35% and 14.6%, respectively, while that of Cyanobacteria and Bacteroidetes increased to 19%. On 1 September, the community profile was associated with high cyanobacterial levels (38%) and was distinct from those of other sampling dates where there were lower cyanobacterial levels. This is coherent with trends observed in taxonomic cell count results on 1 September. Similarly, the relative abundance of Bacteroidetes reached its highest level (32%) on that date (Figure 4), as supported by previous reports linking cyanobacterial blooms with Bacteroidetes [38,39]. Indeed, Bacteroidetes is associated with nutrient loadings which promote the growth of Cyanobacteria [40]. In this work, we observed that total (TN) and dissolved nitrogen (DN) were significantly associated with the Bacteroidetes community (Table S1, Figure S1). On the next sampling dates (5 September and 27 October), the abundance of Cyanobacteria and Bacteroidetes decreased to 12.4% and 4.5% on 5 September, and 19% and 14% on 27 October, respectively. The abundance of Proteobacteria increased from 37% to 48% (Figure 4). On 30 August (before the bloom) and 5 September (after the bloom) Proteobacteria and Actinobacteria were the two dominant phyla in the ST (56–57% and 14–17%, respectively) and SST (56–68% and 18–20%, respectively) (Figure 4). During this period, the relative cyanobacterial abundance in the ST and SST was about 7% and 4%, respectively. Interestingly, Bacteroidetes was also found at low levels, fluctuating from 5% to 12% and from 7% to 18% in the ST and SST, respectively.

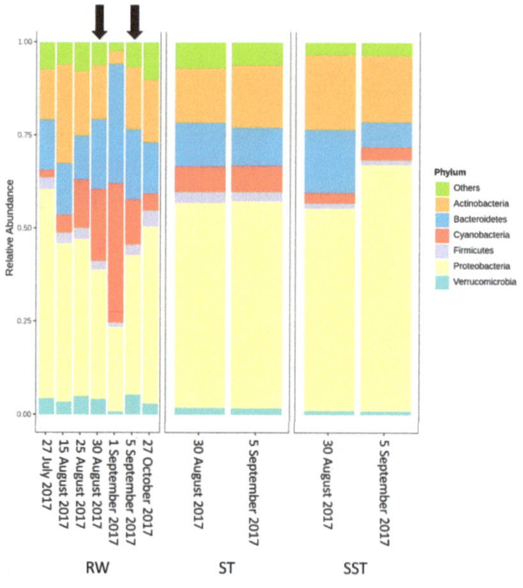

Figure 4. Bacterial community at the phylum level in raw water (RW) (27 July, 15, 25, and 30 August, 1 and 5 September, and 27 October 2017), in the sludge holding tank (ST), and in sludge holding tank supernatant (SST) (30 August, 5 September 2017). The black arrows show the corresponding dates with the ST and SST samples.

At the genus level within Cyanobacteria in RW, *Synechococcus* and *Microcystis* were predominant on 27 July and 15 August (Figure 5). In late August, the relative abundance of *Synechococcus* and *Microcystis*, declined, while that of *Dolichospermum* and *Nostoc*, increased. The relative abundance of *Dolichospermum* reached its maximum level on 30 August and 1 September (bloom date). After the bloom date (5 September and 27 October), the relative abundance of *Dolichospermum* decreased, while that of *Microcystis* and *Synechococcus* increased, and the diversity of cyanobacterial communities almost returned to pre-bloom conditions (Figure 5). A previous investigation in Missisquoi Bay documented that the relative abundance of *Dolichospermum* and *Microcystis* repeatedly alternated in bloom and non-bloom events [24], while our study showed that *Synechococcus* also shifted from being a highly abundant taxa before and after the bloom to being present with very low abundance during the bloom.

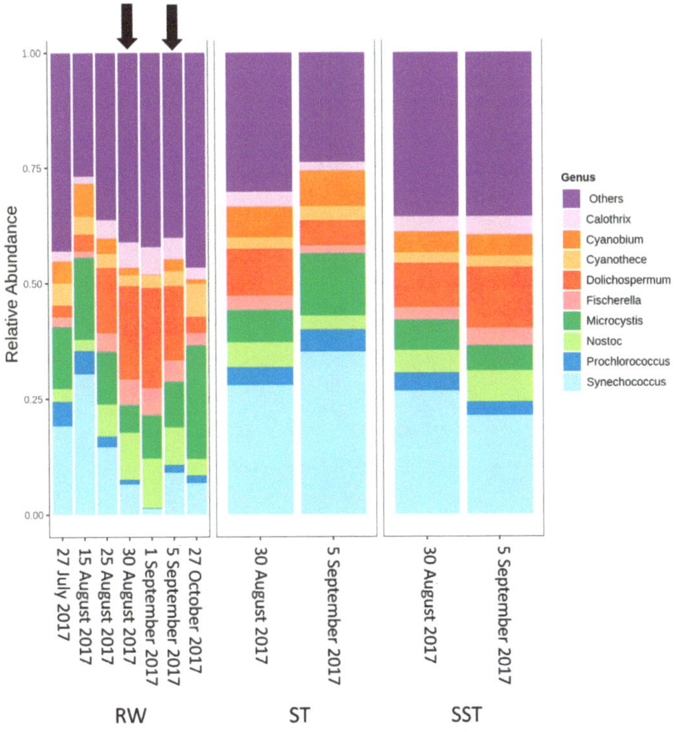

Figure 5. Cyanobacterial community at the genus level in raw water (RW) (27 July, 15, 25, and 30 August, 1 and 5 September, and 27 October 2017), in the sludge holding tank (ST), and in sludge holding tank supernatant (SST) (30 August, 5 September 2017). The black arrows show the corresponding dates for the ST and SST samples.

The relative abundance of the genera and species changed between RW and ST/SST stages (Figure 5). It is important to note that the structural composition of the sludge communities was not expected to match because sludge is the result of several days of cyanobacterial cell accumulation in the holding tank. For example, when considering samples from 25 August 2017, the sludge holding time was estimated as 13 days, while that of 5 September was 1 day. A comparison between RW (25 and 30 August) and ST (30 August) showed that there was a higher abundance of *Synechococcus* and a lower abundance of *Dolichospermum* in the ST as compared to RW. This trend was also observed on 5 September. On 25 August, the relative abundance of *Microcystis* was lower in the ST in comparison to RW, while the opposite trend was observed on 30 August. On 5 September, the abundance of *Microcystis* was higher in the ST than in the RW. The abundance of *Synechococcus* in SST (30 August) was higher than in RW (25 and 30 August). This trend was also observed on 5 September. The opposite trend of *Synechococcus* was observed within that of *Dolichospermum*. The abundance of *Microcystis* in SST (30 August) was lower than in RW on 25 August but higher than in RW on 30 August. On 5 September, the abundance of *Microcystis* in SST was lower than that in RW. The relative abundance of *Synechococcus* in the ST was higher than that in SST on both sampling dates (30 August and 5 September). On 30 August, the relative abundances of *Microcystis* and *Dolichospermum* were similar in the ST and SST. Interestingly, on 5 September, the abundance of *Microcystis* decreased in SST compared to the ST, while *Dolichospermum* showed the opposite trend (Figure 5). At the species level, similar trends were observed in *M. aeruginosa*, *Dolichospermum sp. 90*, and *Synechococcus sp. CB0101* (Figure 6). Additionally, other genera with lower relative abundance (<6%) were detected in the samples. For example, *Prochlorococcus*, *Cyanobium*, *Fischerella*, *Calothrix*, and *Cyanothece* did not show significant changes between the RW, ST, and SST (Figure 6, Figure S2 and S3). The richness of cyanobacterial species (Chao1 index) remained approximately constant in the RW (578 and 598 on 30 August and 5 September, respectively) and the ST (599 and 620 on 30 August and 5 September, respectively), while it decreased in the SST (275 and 475 on 30 August and 5 September, respectively) (Figure 7a). The difference in richness between RW and ST with respect to SST suggests that the cyanobacterial communities in SST and ST were different but that there were similarities between the ST and RW.

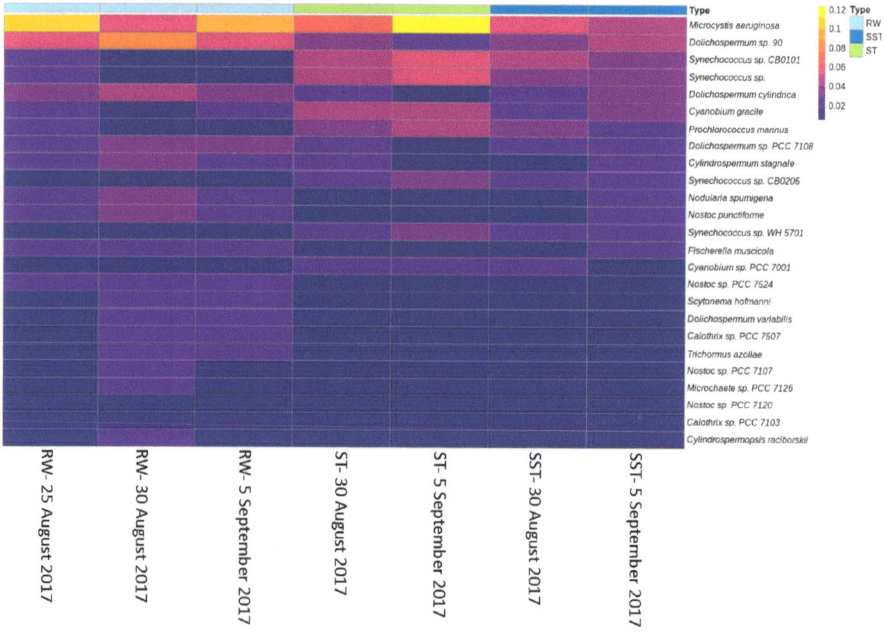

Figure 6. Relative abundance of the top 25 major abundant species in raw water (RW) (25 and 30 August, 5 September 2017), in the sludge holding tank (ST), and in sludge holding tank supernatant (SST) (30 August, 5 September 2017).

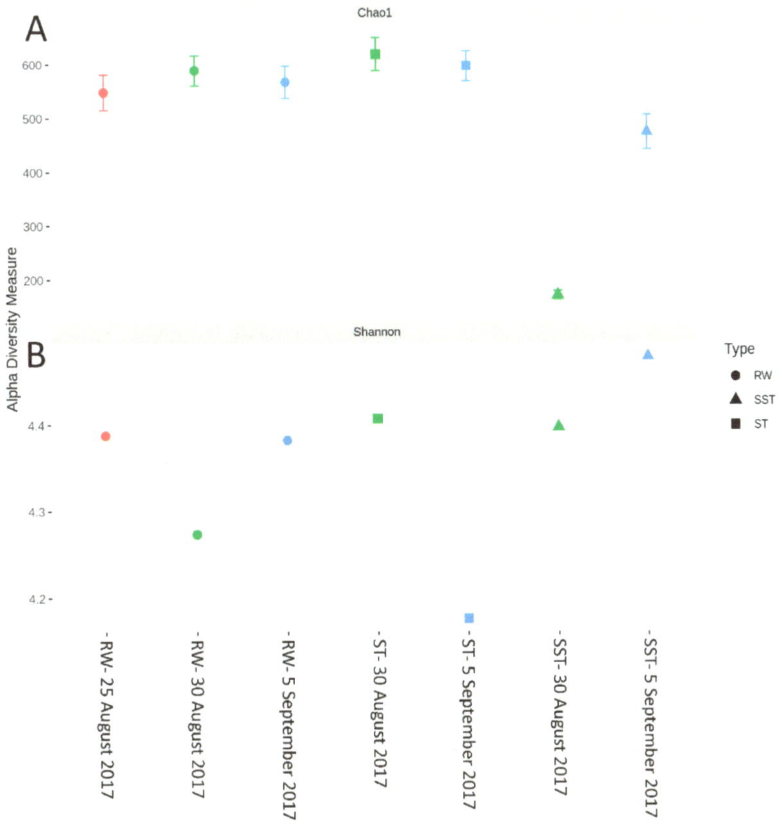

Figure 7. Evaluation of cyanobacterial richness and diversity in raw water (RW) (25 and 30 August and 5 September 2017), in the sludge holding tank (ST), and in sludge holding tank supernatant (SST) on 30 August and 5 September 2017 using (**A**) Chao1 and (**B**) the Shannon index.

Between 27 July and 30 August, the Shannon index increased from 4.18 to 4.51 in RW, while it decreased to 4.06 on bloom on 1 September (Figure S4). Our results showed that the diversity decreased during the bloom, in agreement with [24]. The Shannon index in the ST on 30 August (4.4) indicated similar diversity profiles to RW on 25 August (4.39) and 30 August (4.28). Due to 13 days of sludge storage time on 30 August, the diversity in the ST was affected by the bacterial populations in RW in samples on both 25 August and 30 August. The Shannon index in the sludge decreased to 4.16 on 5 September, which is a lower value than the Shannon index in the RW (4.38) on the same date (Figure 7b).

Overall, when relating structural composition of the communities found in raw water (RW), in the sludge holding tank (ST), and in sludge holding tank supernatant (SST), several factors should be considered:

- As expected, changes in composition of RW communities were observed and affected the microbial populations in the ST. Since sludge accumulates over the period of time (1–13 days in this case), the ST profile is expected to reflect accumulative diversity considering both the relative abundance and biomass. Furthermore, the efficacy of coagulation and settling is species-dependent, as shown by [3,17]. Previous investigations also demonstrated that 96–100% of *Dolichospermum* and *Microcystis* cells were more likely to be captured by the clarifier [3,8], and that the coagulation efficiency for these genera was twice the value observed for *Synechococcus* [41].

- The communities found in the ST and SST showed different trends at the phylum and genus level as shown by the Shannon index (Figure S5, Figures 4 and 5). In fact, at the phylum level, Cyanobacteria was selectively removed and retained within sludge (Figure 4). The cyanobacterial community distribution in the supernatant reflects the incoming sludge and the subsequent buoyancy of the community in the sludge. Storage in the holding tank of the sludge may cause cell breakage, leading to vesicle damage [42] and interruption of buoyancy regulation [43]. This would affect the profile of the supernatant in our work. The increase in cyanobacterial richness in SST on 5 September might be due to the longer sludge storage time in the 30 August sample (13 days) compared to that of 5 September (1 day), providing more time for cell damage.
- Cell survival, re-growth, and damage might have occurred in ST during sludge storage. The longer storage of the sludge might have led to cell lysis in the sludge. These phenomena were documented in several studies for various dominant genera including the most dominant genera in this study (*Microcystis* and *Dolichospermum*) after 2 days of sludge storage [4,15,16,44]. Furthermore, trichome damage of *Dolichospermum* due to the treatment stress has been already reported [7]. However, there are no data on the fate of *Synechococcus* in stored sludge.

Understanding the community structure and dynamics in the ST and SST is important for quantitative cyanobacterial risk assessment. Water operators need to be able to predict the exchanges between the sludge and the supernatant. Supernatant (SST) can be discharged into water resources or recycled to the head of the WTP and could constitute a risk for the water intake or an additional burden on the plant treatment processes. Sludge (ST) can be disposed of in wastewater collectors, processed as sludge in lagoons or sludge facilities, or land-applied.

Other studies have shown that environmental conditions can impact sludge communities [16,19]. Redundancy analysis (RDA) analyses were performed to evaluate the relationship between nutrients (Table S1) and cyanobacterial communities. Orthophosphate (OP), total nitrogen (TN, sum of Kjeldahl nitrogen (N- Kjeldahl), organic nitrogen, nitrite, and nitrate), N-Kjeldahl, total phosphorous (TP), particulate phosphorous (PP), and total organic carbon (TOC) exerted significant effects ($p < 0.05$) on community profiles in different ways (Figure 8). A clear correlation was observed between OP in RW, with *Nostocales* (reported to have a 4.5 times higher relative abundance of genes related to phosphorous metabolism than *Chroococcales*) found at low concentrations of phosphorous [45]. Other studies have demonstrated that higher concentrations of nitrogen, phosphorous, and carbon resulted in better conditions for bacterial communities and led to an increase in microbial growth [33,34,46]. In our study, RDA analyses showed that OP was more available in RW than in the ST and SST and that TN, N- Kjeldahl, TP, PP, and TOC had a strong impact on the cyanobacterial population in the ST, which mostly contained *Chroococcales*. None of these nutrient parameters seemed to affect the SST. This is in accordance with our previous observations on the different patterns of cell accumulation in SST. However, it must be noted that the mass of nutrients measured in sludge (2.0–32.8 mg/L of TN and 0.48–5.9 mg/L of TP), was not associated with cyanobacterial cell-bound nutrients nor with dissolved nutrients. Using reference values for cell nutrient content, cell-bound nutrients consist of less than 0.8% nitrogen and 0.4% phosphorous [47]. The persistence of *Chroococcales* in the sludge environment could be the result of the high environmental resistance and the ability to thrive in the presence of elevated levels of nutrients (Figure 8).

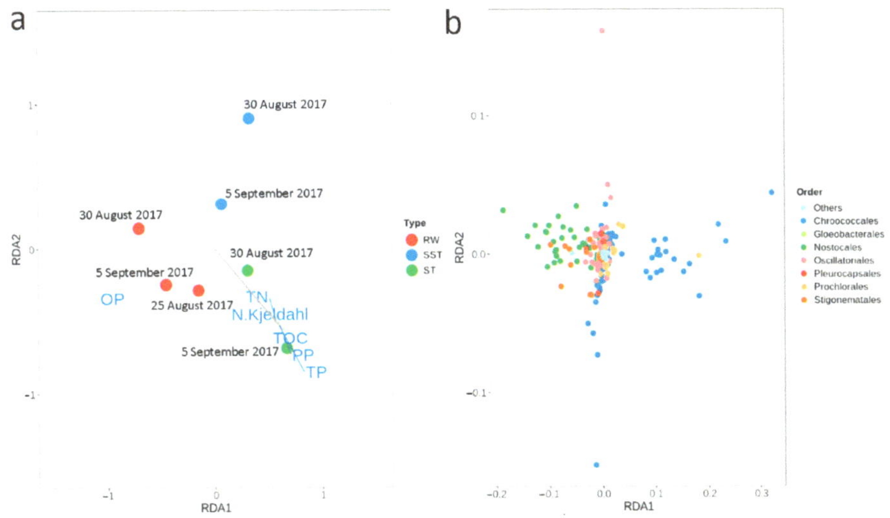

Figure 8. Redundancy analysis (RDA) of cyanobacterial communities with respect to nutrient parameters in (**a**) raw water (RW), in the sludge holding tank (ST), and in sludge holding tank supernatant (SST); (**b**) cyanobacterial distribution at the order level. RDA1: 65.6%, RDA2: 8.7%. Only significant parameters ($p < 0.05$) are shown.

2.3. Comparison between Shotgun Metagenomic Sequencing and Taxonomic Cell Counts

The observed genera from microscopic cell counts do not completely match the high-throughput sequencing results, as *Aphanothece*, *Chroococcus*, *Aphanocapsa* and *Coelosphaerium* were not detected by shotgun metagenomics. In RW, *M. aeruginosa*, *D. circinale*, *D. spiroides*, *A. gracile*, and *P. mucicola* were detected through both taxonomic cell counts and metagenomics (data not shown). In contrast, *A. clathrata brevis*, *C. minimus*, and *C. kuetzingianum* were detected only by taxonomic cell counts and not by metagenomics (data not shown). In ST, only *M. aeruginosa*, *D. spiroides*, and *D. circinale* were detected by both approaches, while *A. clathrata brevis*, *C. minimus*, *P. mucicola*, *A. delicatissima A. flos-aquae*, and *C. kuetzingianum* were only detected in taxonomic cell counts (data not shown). Overall, 76% and 88% of the detected species by metagenomics were not detected by taxonomic cell counts in RW and ST, respectively (Figures 1 and 6). As recently discussed [48], taxonomic cell counts and high-throughput sequencing can yield different community profiles because of the limitations inherent to each of these methods. Physical and chemical stress in WTPs may cause damaged cells and affect taxonomic cell counts [7], while DNA can be extracted from lysed and dead cells and provide metagenomics shotgun reads [49]. Despite the advantages of taxonomic cell counts, measurement bias such as misidentification of morphologically similar species, the impact of the conservation agent on biovolume, and the complexity of counting species in aggregates should be considered [50–52]. In sludge samples, the presence of debris, sediments, and a high number of cells might increase the probability of cross interferences. New metagenomic approaches based on direct cloning and shotgun sequencing of environmental DNA represent a powerful tool for species classification and the evaluation of community dynamics through water treatment processes. However, use of metagenomics also represents some challenges such as: (1) an inadequate recovery rate of DNA [53]; (2) contamination of DNA during extraction [54,55]; and (3) a lack of a standard identification pipeline that includes all species [56].

3. Conclusions

- Bacterial communities shifted before and after the cyanobacterial bloom. Proteobacteria was the predominant phylum in RW before the bloom. Levels of Cyanobacteria

and Bacteroidetes progressively increased to reach their greatest abundance on the bloom date. This high abundance of Bacteroidetes was associated with nutrient-rich conditions which occurred during the cyanobacterial bloom. After the bloom, bacterial communities returned to almost the same composition as prior to the bloom.
- Conventional treatment eliminated 92–97% of the cyanobacterial cells, as revealed by cell counts. Overall, 96% of *Microcystis* and *Dolichospermum* were eliminated by this process. At first glance, this is an effective approach to controlling the cyanobacterial flux. However, coagulation leads to accumulation of cyanobacterial cells in the sludge. Even a low cell number in intake water (3.9×10^4 cells/mL) led to 31 times as much cell accumulation in the sludge.
- Selective removal of cyanobacteria at the genus and species levels by coagulation/sedimentation has been highlighted by both metagenomic shotgun sequencing and taxonomic cell counts. Sludge (ST) cyanobacterial composition differs from RW if only samples from the same day are considered. Sludge diversity reflects both selective removal by coagulation/sedimentation and the accumulation of captured cells over a period of time as determined by sludge age.
- Monitoring strategies focusing on sporadic measurement of the diversity in raw water cannot capture the risk associated with the storage and disposal of the sludge. Sludge community profiles also appear to be a better indicator for evaluating the influx of cyanobacterial communities in WTPs. Indeed, the sludge profile reflects a cumulative community in terms of relative abundance and biomass.
- Bacterial and cyanobacterial communities of sludge in the holding tank (ST) markedly differed from those measured in sludge supernatant (SST). The communities found in the ST and SST showed different trends at the phylum and genus level as shown by the Shannon index. The prediction of cyanobacterial communities in the supernatant remains a challenge as it is often recycled, possibly adding cyanobacteria and cyanotoxins to the intake water.
- Considering environmental parameters monitored, nutrients were the most discriminating factors affecting cyanobacterial communities. Cyanobacterial communities in RW were influenced by OP, while the sludge communities were correlated with TN, N- Kjeldahl, TP, PP, and TOC.
- Storage, management, and disposal of the cyanobacteria-laden sludge are technical and health-related challenges. By adjusting the storage time and adding PAC, risk assessment of supernatant recycling can be applied to minimize the impact of cyanobacteria and cyanotoxin accumulation.

4. Materials and Methods

4.1. Description of the Studied Water Body and Plant, Including Treatment Schematics

A plant located on the Canadian side of Missisquoi Bay, Lake Champlain, located South East of Montreal was monitored from 27 July to 27 October 2017. The plant intake water is situated 180 m within the bay. The treatment chain is presented in Figure 9. Briefly, powdered activated carbon (PAC) injection followed by conventional treatment (coagulation, flocculation, sedimentation) and a post-chlorination step are applied as treatment. The clarifier sludge is stored in a sludge holding tank. The supernatant of this sludge is discharged into the lake and the sludge is transferred to the local wastewater treatment plant.

Overall, seven sampling campaigns were performed on the following dates: 27 July, 15, 25, and 30 August, 1 and 5 September, and 27 October 2017. Specifications of the plant and treatment are summarized in Table S2.

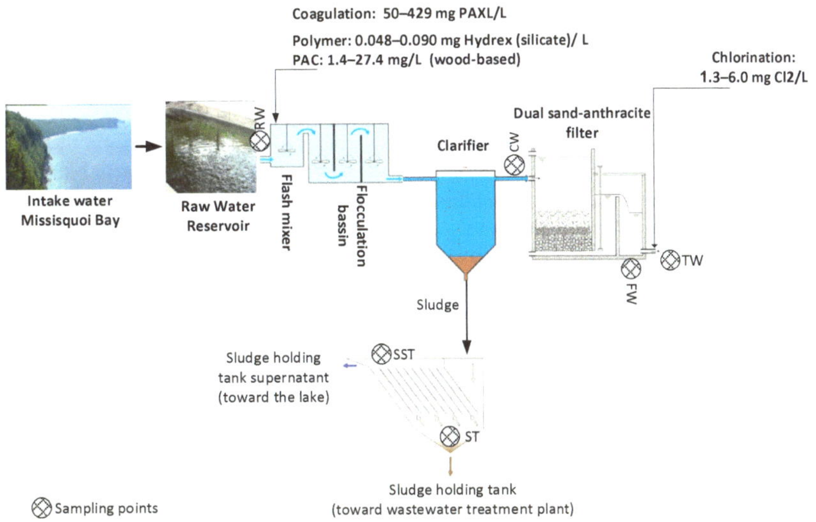

Figure 9. The treatment chain of the WTP and sampling points. The water intake is from Missisquoi Bay. Sampling points are indicated by . RW: raw water; CW: clarified water; FW: filtered water; TW: treated water; ST: sludge holding tank; SST: supernatant of the sludge holding tank.

4.2. Description of the Studied Water Body and Plant Including Treatment Schematics

Samples were taken following each treatment step from raw water (RW), clarified water (CW), filtered water (FW), treated water (TW), the sludge holding tank (ST), and sludge holding tank supernatant (SST).

Autoclaved 1-L polypropylene bottles, 5-L polypropylene containers, and 40-mL glass vials were used for each sampling point. Before sampling, all containers and vials were rinsed with the water from the sampling point. The 40-mL vials were used for taxonomic cell counts. The taxonomic samples were preserved with Lugol's iodine. Subsamples were taken for metagenomics from the 1-L bottles; the 5-L containers were used for cell-bound and dissolved microcystins (MCs) and nutrient samples.

Genomic subsamples were prepared by sample filtration via 0.2-μm polyethersulfone hydrophilic membranes (Millipore Sigma, Oakville, ON). Membranes were stored in the sterile Falcon tube at −80 °C. Cell-bound and dissolved microcystin subsamples were prepared by sample filtration using pre-weighted 0.45-μm GHP membranes (Pall, Mississauga, ON). The filters were kept in the petri dish and the filtrate was kept in 125-mL graduated polyethylene terephthalate glycol (PETG) amber bottles (Thermo Fisher, Mississauga, ON). Subsamples for total nitrogen (TN), total phosphorous (TP), and total organic carbon (TOC) were aliquoted directly. Dissolved nitrogen (DN), Kjeldahl nitrogen (N- Kjeldahl), ammonium nitrogen (NH_4), nitrite/nitrate (NO_2/NO_3) and dissolved phosphorous (DP) subsamples were filtered on 0.45-μm membranes (Millipore Sigma, Oakville, ON).

Genomic subsamples were taken in triplicate, while MC and nutrient samples were taken in duplicate. MC, N- Kjeldahl, and NO_2/NO_3 subsamples were frozen at −25 °C. TN, TP, and TOC samples were stored at 4 °C. Taxonomic cell count samples were stored in a dark place at ambient temperature.

4.3. Description of Analytical Methods

4.3.1. Taxonomic Cell Counts

Taxonomic cell counts were performed by an inverted microscope in a Sedgwick-Rafter chamber at magnifications of 10 and 40× according to [57–59].

4.3.2. Microcystin Analysis

Total microcystins (MCs) were analyzed by on-line solid-phase extraction ultra-high-performance liquid chromatography coupled to tandem mass spectrometry (on-line SPE-UHPLC-MS/MS). Firstly, samples were oxidized by potassium permanganate and sodium (meta) periodate (Sigma Aldrich, Oakville, ON, Canada). Secondly, samples were quenched by a 4M sodium bisulfite solution (Sigma Aldrich, Oakville, ON, Canada). Thirdly, the standard solutions of 4-phenylbutyric acid (50 ng/L) (Sigma Aldrich, Oakville, ON, Canada) and erythro-2-Methyl-3-methoxy-4-phenylbutyric acid (D3-MMPB, 10 ng/L) (Wako Pure Chemicals Industries, Ltd., Osaka, Japan) were added to the samples. Fourthly, 10 mL of solution were filtered using 0.22-µm nylon filters (Sterlitech Corporation, Kent, WA, USA). Aliquots were taken for analysis using the Thermo EQUAN™ interface (Thermo Fischer Scientific, Waltham, MA, USA). The "in-loop" injection was controlled by an HTC Thermopal autosampler (CTC analytics, Zwingen, Switzerland). Then, samples were loaded into a Thermo Hypersil Gold aQ C18 (on-line SPE) column (20 mm × 2.1 mm, 12 µm). Separation of toxins was performed on a Thermo Hypersil Gold C18 column (100 mm × 2.1 mm, 1.9 µm). MS/MS detection was performed by thermo TSQ QUANTIVA triple quadrupole mass spectrometer (Thermo Fischer Scientific) following UHPLC. Water, methanol, and acetonitrile for HPLC were purchased from Fisher Scientific (Whitby, ON, Canada) and formic acid (>95%), potassium carbonate, ammonium hydroxide (28–30% $NH3$), and ammonium acetate (≥99.0%) were obtained from Sigma Aldrich (Oakville, ON, Canada). More details are provided by Munoz et al. [60] and Roy-Lachapelle et al. [61].

4.3.3. Nutrient Analysis

Nitrogen, nitrite, nitrate, and ammonium nitrogen were analyzed by the colorimetric technique based on EPA 350.1 and 353.2 standard methods [62,63]. Phosphorous and phosphate were measured by the colorimetric technique based on EPA 365.1 and 365.3 methods [64]. TOC was analyzed by Sievers InnovOX Laboratory TOC analyzer (USA) based on USEPA 415.1 method [65].

4.4. DNA Extraction and Metagenomics Preparation

The extraction of total nucleic acid from frozen filters was performed with the RNeasy PowerWater Kit (Qiagen, Toronto, ON, USA) with slight modifications. To avoid formation of disulfide bonds protein residuals, dithiothreitol (DTT) was spiked into the pre-warmed (55 °C) PM1 buffer. Since substantial biomass remained on the surface of the membrane after the bead-beating step, the filters were transferred alongside the supernatant and were incubated with the IRS solution. Total nucleic acids were eluted with 60 µL of nuclease-free water. Half of the sample was treated with RNase If (New England Biolabs, Whitby, ON) to remove the RNA. The resulting DNA was purified using the Genomic DNA Clean & Concentrator TM-10 (Zyimo Research Corporation, Irvine, CA, USA), following the instructions of the manufacturer.

For the sludge samples, the RNeasy PowerSoil Total RNA Kit (Qiagen, Toronto, ON, Canada) was used with two modifications to improve the recovery of DNA. The modifications consisted of (1) a centrifugation step for 5 min at 13,000 rpm at 4 °C to separate the water from the sludge; and (2) incubation at −20 °C for 60 min after the addition of solution SR4 to precipitate nucleic acids. DNA was eluted using the PowerSoil DNA Elution Kit (Qiagen, Toronto, ON, Canada), following the instructions of the manufacturer. DNA quantification was done by Qubit V2.0 fluorometer (Life Technologies, Burlington, ON, Canada). The metagenomic libraries (Roche 454 FLX instrumentation with Titanium chemistry) were sent to McGill University and Genome Quebec Innovation Centre for sequencing. Ninety-six libraries were then pooled together and sequenced on a NovaSeq 6000 S4 with paired end of 150 bp and an insert of 360 bp.

Metagenomic analysis were performed on all RW, ST, and SST samples collected on 5 August and 5 September. Community dynamics was assessed using shotgun metagenomics levels of phylum, order, genus and species. The number of reads for taxonomic data was normalized by relative abundance.

4.5. Bioinformatics Analysis

DNA libraries were sequenced on the Illumina NovaSeq 6000 platform using S4 flow cells. Paired-end raw reads of 150 base pairs (bp) were further analyzed using a home-made bioinformatics pipeline. Firstly, quality trimming of raw reads was performed by the SolexaQA v3.1.7.1 program with default settings [66]. Trimmed reads shorter than 75 nt were removed for further analysis. Artificial duplicate removal was performed using an in-house script based on the screening of identical leading 20 bp. From the trimmed high-quality reads, gene fragments were predicted using FragGeneScan-Plus v3.0 [67]. Cd-hit v4.8.1 was applied to cluster predicted protein fragments at a 90% similarity [68]. One representative of each cluster was used for a similarity search on the M5nr database (https://github.com/MG-RAST/myM5NR) using the Diamond engine [69]. For assessment of taxonomic affiliation of gene fragments encoding proteins, we took into account best hits (minimal e-value of 1×10^{-5}) combined with a last common ancestor approach.

4.6. Statistical Analysis

Statistical analysis was performed by R (3.6.2). Bacterial communities at the phylum, order, and genus level were analyzed by phyloseq (1.28.0) [70]. Taxonomic data was normalized by centered log-ratio transformation using easyCODA (0.31.1) [71]. Then, the cyanobacteria species community was analyzed based on the first 25 most frequent species by pheatmap (1.0.12) (https://CRAN.R-project.org/package=pheatmap). The richness index was analyzed by phyloseq's estimate_richness function. For visualization of the species community and diversity variation, heat trees were illustrated using the metacoder (0.3.3) [72]. Beta-diversity was performed by vegan package (2.5–6) (https://CRAN.R-project.org/package=vegan). Similarity matrices were calculated according to Euclidean distance. A redundancy analysis (RDA) was performed to evaluate the impact of constrained variables on sampling points at >95% significance. The homogeneity of variances was validated before the model implementation. A model was defined by the ordistep function [73] to illustrate the impact of nutrient parameters on the distribution of cyanobacterial communities in the RW, ST, and SST at the order level. The Envfit function was used to find similar scores and to scale the fitted vectors of variables based on the correlations. The permutation test (>95% significance) was applied to select significant variables.

Supplementary Materials: The following are available online at https://www.mdpi.com/2072-6651/13/1/25/s1. Figure S1. Principal component analysis (PCA) of nutrient parameters' impact on (**a**) sampling dates, and (**b**) bacterial diversity (phylum-level) in raw water (RW) on 27 July, 15, 25, and 30 August, 1 and 5 September, and 27 October 2017. Only the significant parameters are shown ($p < 0.05$). Figure S2. Relative abundance of the top 25 major abundant genera in RW. Samples taken on 27 July, 15, 25, and 30 August, 1 and 5 September, and 27 October 2017. Figure S3. Relative abundance of the top 25 major abundant genera in the sludge holding tank (ST) and sludge holding tank supernatant (SST). Samples taken on 30 August and 5 September 2017. Figure S4. Evaluation of the cyanobacterial diversity in raw water (RW) on 27 July, 15, 25, and 30 August, 1 and 5 September, and 27 October 2017 using the Shannon index. Figure S5. Bacterial diversity in the raw water (RW) on 25 and 30 August and 5 September 2017, and in the sludge holding tank (ST) and sludge holding tank supernatant (SST) on 30 August and 5 September 2017. Table S1. Concentration of nutrients in raw water (RW), in the sludge holding tank (ST), and in sludge tank supernatant (SST) on 27 July, 15, 25, and 30 August, 1 and 5 September, and 27 October 2017. Table S2. Concentrations of cell-bound and dissolved microcystins (MCs) in the RW (raw water), sludge holding tank (ST), and sludge holding tank supernatant (SST). Table S3. Water characteristics of the studied plant in Missisquoi Bay during the sampling campaign from July to October 2017.

Author Contributions: Conceptualization, F.J., H.T., S.D., and M.P.; Methodology, F.J., H.T., J.F.G.M., N.F., B.J.S., Y.T., A.Z., S.D., S.S., and M.P.; Software, F.J., J.F.G.M., and Y.T.; Validation, F.J., H.T., J.F.G.M., N.F., Y.T., S.D., and M.P.; Formal analysis, F.J., H.T., J.F.G.M., and Y.T.; Investigation, F.J., H.T., N.F., J.F.G.M., S.D., A.Z., and M.P.; Resources: F.J., H.T., and N.F.; Data curation, F.J. and J.F.G.M.; Writing—original draft preparation, F.J., H.T., and J.F.G.M.; Writing—review and editing, F.J., H.T., N.F., J.F.G.M., S.S., S.D., and M.P.; Visualization, F.J., H.T., J.F.G.M., N.F., S.D., and M.P.; Supervision, A.Z., S.D., and M.P.; Project administration, A.Z., S.D., B.J.S., S.S., and M.P.; Funding acquisition, S.D., S.S., B.J.S., and M.P. All authors have read and agreed to the published version of the manuscript.

Funding: This research was funded by Genome Canada and Génome Québec (Algal Blooms, Treatment, Risk Assessment, Prediction and Prevention through Genomics (ATRAPP) projects).

Institutional Review Board Statement: Not applicable.

Informed Consent Statement: Not applicable.

Data Availability Statement: Data is contained within the article and supplementary material.

Acknowledgments: The authors acknowledge support from Algal Blooms, Treatment, Risk Assessment, Prediction and Prevention through Genomics (ATRAPP) projects, Génome Québec and Genome Canada. The authors thank the staff at NSERC Industrial Chair on Drinking Water at Polytechnique Montréal, Shapiro lab, GRIL lab, Dana F. Simon, Audrey Roy-Lachapelle and Sung Vo Duy (University of Montréal), Irina Moukhina (Université du Québec à Montréal), Stephanie Messina Pacheco (National Research Council Canada), and operators at the studied plant for their valuable contributions to this research.

Conflicts of Interest: The authors declare no conflict of interest.

References

1. Westrick, J.A.; Szlag, D.C.; Southwell, B.J.; Sinclair, J. A review of cyanobacteria and cyanotoxins removal/inactivation in drinking water treatment. *Anal. Bioanal. Chem.* **2010**, *397*, 1705–1714. [CrossRef] [PubMed]
2. Shang, L.; Feng, M.; Xu, X.; Liu, F.; Ke, F.; Li, W. Co-occurrence of microcystins and taste-and-odor compounds in drinking water source and their removal in a full-scale drinking water treatment plant. *Toxins* **2018**, *10*, 26. [CrossRef]
3. Zamyadi, A.; Dorner, S.; Ellis, D.; Bolduc, A.; Bastien, C.; Prévost, M. Species-dependence of *cyanobacteria* removal efficiency by different drinking water treatment processes. *Water Res.* **2013**, *47*, 2689–2700. [CrossRef] [PubMed]
4. Drikas, M.; Chow, C.W.K.; House, J.; Burch, M.D. Using coagulation, flocculation, and settling to remove toxic cyanobacteria. *J. Am. Water Works Assoc.* **2001**, *93*, 100–111. [CrossRef]
5. Newcombe, G.; Nicholson, B. Water treatment options for dissolved cyanotoxins. *Water Supply Res. Technol. Aqua* **2004**, *53*, 227–239. [CrossRef]
6. Pietsch, J.; Bornmann, K.; Schmidt, W. Relevance of intra- and extracellular cyanotoxins for drinking water treatment. *Acta Hydrochim. Hydrobiol.* **2002**, *30*, 7–15. [CrossRef]
7. Pestana, C.J.; Capelo-Neto, J.; Lawton, L.; Oliveira, S.; Carloto, I.; Linhares, H.P. The effect of water treatment unit processes on cyanobacterial trichome integrity. *Sci. Total Environ.* **2019**, *659*, 1403–1414. [CrossRef]
8. Zamyadi, A.; MacLeod, S.; Fan, Y.; McQuaid, N.; Dorner, S.; Sauvé, S.; Prévost, M. Toxic cyanobacterial breakthrough and accumulation in a drinking water plant: A monitoring and treatment challenge. *Water Res.* **2012**, *46*, 1511–1523. [CrossRef]
9. Ho, L.; Dreyfus, J.; Boyer, J.E.; Lowe, T.; Bustamante, H.; Duker, P.; Meli, T.; Newcombe, G. Fate of cyanobacteria and their metabolites during water treatment sludge management processes. *Sci. Total Environ.* **2012**, *424*, 232–238. [CrossRef]
10. Almuhtaram, H.; Cui, Y.; Zamyadi, A.; Hofmann, R. Cyanotoxins and cyanobacteria cell accumulations in drinking water treatment plants with a low risk of bloom formation at the source. *Toxins* **2018**, *10*, 430. [CrossRef]
11. Zamyadi, A.; Dorner, S.; Ndong, M.; Ellis, D.; Bolduc, A.; Bastien, C.; Prévost, M. Low-risk cyanobacterial bloom sources: Cell accumulation within full-scale treatment plants. *J. Am. Water Works Assoc.* **2013**, *102*, E651–E663. [CrossRef]
12. Li, X.; Pei, H.; Hu, W.; Meng, P.; Sun, F.; Ma, G.; Xu, X.; Li, Y. The fate of *Microcystis aeruginosa* cells during the ferric chloride coagulation and flocs storage processes. *Environ. Technol.* **2015**, *36*, 920–928. [CrossRef] [PubMed]
13. Sun, F.; Pei, H.-Y.; Hu, W.-R.; Li, X.-Q.; Ma, C.-X.; Pei, R.-T. The cell damage of *Microcystis aeruginosa* in PACl coagulation and floc storage processes. *Sep. Purif. Technol.* **2013**, *115*, 123–128. [CrossRef]
14. Sun, F.; Pei, H.-Y.; Hu, W.-R.; Ma, C.-X. The lysis of Microcystis aeruginosa in AlCl3 coagulation and sedimentation processes. *Chem. Eng. J.* **2012**, *193–194*, 196–202. [CrossRef]
15. Water Research Foundation (WRF); Water Research Australia. *Optimizing Conventional Treatment for the Removal of Cyanobacteria and Toxins*; Water Research Foundation: Denver, CO, USA, 2015; p. 185. ISBN 978-1-60573-216-9.
16. Pestana, C.J.; Reeve, P.J.; Sawade, E.; Voldoire, C.F.; Newton, K.; Praptiwi, R.; Collingnon, L.; Dreyfus, J.; Hobson, P.; Gaget, V.; et al. Fate of cyanobacteria in drinking water treatment plant lagoon supernatant and sludge. *Sci. Total Environ.* **2016**, *565*, 1192–1200. [CrossRef] [PubMed]

17. Zamyadi, A.; Romanis, C.; Mills, T.; Neilan, B.; Choo, F.; Coral, L.A.; Gale, D.; Newcombe, G.; Crosbie, N.; Stuetz, R.; et al. Diagnosing water treatment critical control points for cyanobacterial removal: Exploring benefits of combined microscopy, next-generation sequencing, and cell integrity methods. *Water Res.* **2019**, *152*, 96–105. [CrossRef]
18. Water Research Foundation (WRF); United States Environmental Protection Agency (US EPA); Veolia Water Indianapolis. *Strategies for Controlling and Mitigating Algal Growth within Water Treatment Plants*; Water Research Foundation: Denver, CO, USA, 2009; p. 312.
19. Dreyfus, J.; Monrolin, Y.; Pestana, C.J.; Reeve, P.J.; Sawade, E.; Newton, K.; Ho, L.; Chow, C.W.K.; Newcombe, G. Identification and assessment of water quality risks associated with sludge supernatant recycling in the presence of cyanobacteria. *J. Water Supply Res. Technol. Aqua* **2016**, *65*, 441–452. [CrossRef]
20. Merel, S.; Walker, D.; Chicana, R.; Snyder, S.; Baures, E.; Thomas, O. State of knowledge and concerns on cyanobacterial blooms and cyanotoxins. *Environ. Int.* **2013**, *59*, 303–327. [CrossRef]
21. Cook, D.; Newcombe, G. Removal of microcystin variants with powdered activated carbon. *Water Sci. Technol. Water Supply* **2002**, *2*, 201–207. [CrossRef]
22. Ho, L.; Lambling, P.; Bustamante, H.; Duker, P.; Newcombe, G. Application of powdered activated carbon for the adsorption of cylindrospermopsin and microcystin toxins from drinking water supplies. *Water Res.* **2011**, *45*, 2954–2964. [CrossRef]
23. Newcombe, G.; Cook, D.; Brooke, S.; Ho, L.; Slyman, N. Treatment options for microcystin toxins: Similarities and differences between variants. *Environ. Technol.* **2003**, *24*, 299–308. [CrossRef] [PubMed]
24. Tromas, N.; Fortin, N.; Bedrani, L.; Terrat, Y.; Cardoso, P.; Bird, D.; Greer, C.W.; Shapiro, B.J. Characterising and predicting cyanobacterial blooms in an 8-year amplicon sequencing time course. *ISME J.* **2017**, *11*, 1746–1763. [CrossRef] [PubMed]
25. Berry, M.A.; Davis, T.W.; Cory, R.M.; Duhaime, M.B.; Johengen, T.H.; Kling, G.W.; Marino, J.A.; Den Uyl, P.A.; Gossiaux, D.; Dick, G.J.; et al. Cyanobacterial harmful algal blooms are a biological disturbance to Western Lake Erie bacterial communities. *Environ. Microbiol.* **2017**, *19*, 1149–1162. [CrossRef] [PubMed]
26. Li, Q.; Yu, S.; Li, L.; Liu, G.; Gu, Z.; Liu, M.; Liu, Z.; Ye, Y.; Xia, Q.; Ren, L. Microbial Communities Shaped by Treatment Processes in a Drinking Water Treatment Plant and Their Contribution and Threat to Drinking Water Safety. *Front. Microbiol.* **2017**, *8*, 2465. [CrossRef] [PubMed]
27. Pinto, A.J.; Xi, C.; Raskin, L. Bacterial community structure in the drinking water microbiome is governed by filtration processes. *Environ. Sci. Technol.* **2012**, *46*, 8851–8859. [CrossRef]
28. Zhang, Y.; Oh, S.; Liu, W.T. Impact of drinking water treatment and distribution on the microbiome continuum: An ecological disturbance's perspective. *Environ. Microbiol.* **2017**, *19*, 3163–3174. [CrossRef]
29. Ma, X.; Li, G.; Chen, R.; Yu, Y.; Tao, H.; Zhang, G.; Shi, B. Revealing the changes of bacterial community from water source to consumers tap: A full-scale investigation in eastern city of China. *J. Environ. Sci.* **2020**, *87*, 331–340. [CrossRef]
30. Chao, Y.; Ma, L.; Yang, Y.; Ju, F.; Zhang, X.X.; Wu, W.M.; Zhang, T. Metagenomic analysis reveals significant changes of microbial compositions and protective functions during drinking water treatment. *Sci. Rep.* **2013**, *3*, 3550. [CrossRef]
31. Lautenschlager, K.; Hwang, C.; Ling, F.; Liu, W.-T.; Boon, N.; Köster, O.; Egli, T.; Hammes, F. Abundance and composition of indigenous bacterial communities in a multi-step biofiltration-based drinking water treatment plant. *Water Res.* **2014**, *62*, 40–52. [CrossRef]
32. Lin, W.; Yu, Z.; Zhang, H.; Thompson, I.P. Diversity and dynamics of microbial communities at each step of treatment plant for potable water generation. *Water Res.* **2014**, *52*, 218–230. [CrossRef]
33. Xu, H.; Pei, H.; Jin, Y.; Ma, C.; Wang, Y.; Sun, J.; Li, H. High-throughput sequencing reveals microbial communities in drinking water treatment sludge from six geographically distributed plants, including potentially toxic cyanobacteria and pathogens. *Sci. Total Environ.* **2018**, *634*, 769–779. [CrossRef] [PubMed]
34. Pei, H.; Xu, H.; Wang, J.; Jin, Y.; Xiao, H.; Ma, C.; Sun, J.; Li, H. 16S rRNA Gene amplicon sequencing reveals significant changes in microbial compositions during cyanobacteria-laden drinking water sludge storage. *Environ. Sci. Technol.* **2017**, *51*, 12774–12783. [CrossRef] [PubMed]
35. Teixeira, M.R.; Rosa, M.J. Comparing dissolved air flotation and conventional sedimentation to remove cyanobacterial cells of Microcystis aeruginosa. Part II. The effect of water background organics. *Sep. Purif. Technol.* **2007**, *53*, 126–134. [CrossRef]
36. Chorus, I.; Bartram, J. Chapter 6. Situation assessment, planning and management. In *Toxic Cyanobacteria in Water: A Guide to Their Public Health Consequences, Monitoring and Management*; WHO: Geneva, Switzerland, 1999; p. 28.
37. Maghsoudi, E.; Fortin, N.; Greer, C.; Vo Duy, S.; Fayad, P.; Sauvé, S.; Prévost, M.; Dorner, S. Biodegradation of multiple microcystins and cylindrospermopsin in clarifier sludge and drinking water source: Effects of particulate attached bacteria and phycocyanin. *Ecotoxicol. Environ. Saf.* **2015**, *120*, 409–417. [CrossRef] [PubMed]
38. Guedes, I.A.; Rachid, C.; Rangel, L.M.; Silva, L.H.S.; Bisch, P.M.; Azevedo, S.; Pacheco, A.B.F. Close Link Between Harmful Cyanobacterial Dominance and Associated Bacterioplankton in a Tropical Eutrophic Reservoir. *Front. Microbiol.* **2018**, *9*, 424. [CrossRef] [PubMed]
39. Kim, M.; Lee, J.; Yang, D.; Park, H.Y.; Park, W. Seasonal dynamics of the bacterial communities associated with cyanobacterial blooms in the Han River. *Environ. Pollut.* **2020**, *266*, 115198. [CrossRef]
40. Cai, H.; Jiang, H.; Krumholz, L.R.; Yang, Z. Bacterial community composition of size-fractioned aggregates within the phycosphere of cyanobacterial blooms in a eutrophic freshwater lake. *PLoS ONE* **2014**, *9*, e102879. [CrossRef]

41. Aktas, T.S.; Takeda, F.; Maruo, C.; Chiba, N.; Nishimura, O. A comparison of zeta potentials and coagulation behaviors of cyanobacteria and algae. *Desalin. Water Treat.* **2012**, *48*, 294–301. [CrossRef]
42. Arii, S.; Tsuji, K.; Tomita, K.; Hasegawa, M.; Bober, B.; Harada, K. Cyanobacterial blue color formation during lysis under natural conditions. *Appl. Environ. Microbiol.* **2015**, *81*, 2667–2675. [CrossRef]
43. Reynolds, C.S.; Oliver, R.L.; Walsby, A.E. Cyanobacterial dominance: The role of buoyancy regulation in dynamic lake environments. *N. Z. J. Mar. Freshw. Res.* **1987**, *21*, 379–390. [CrossRef]
44. Sun, F.; Hu, W.; Pei, H.; Li, X.; Xu, X.; Ma, C. Evaluation on the dewatering process of cyanobacteria-containing AlCl3 and PACl drinking water sludge. *Sep. Purif. Technol.* **2015**, *150*, 52–62. [CrossRef]
45. Lu, J.; Zhu, B.; Struewing, I.; Xu, N.; Duan, S. Nitrogen-phosphorus-associated metabolic activities during the development of a cyanobacterial bloom revealed by metatranscriptomics. *Sci. Rep.* **2019**, *9*, 2480. [CrossRef] [PubMed]
46. Jankowiak, J.; Hattenrath-Lehmann, T.; Kramer, B.J.; Ladds, M.; Gobler, C.J. Deciphering the effects of nitrogen, phosphorus, and temperature on cyanobacterial bloom intensification, diversity, and toxicity in western Lake Erie. *Limnol. Oceanogr.* **2019**, *64*, 1347–1370. [CrossRef]
47. Lopez, J.S.; Garcia, N.S.; Talmy, D.; Martiny, A.C. Diel variability in the elemental composition of the marine cyanobacterium-Synechococcus. *J. Plankton Res.* **2016**, *38*, 1052–1061. [CrossRef]
48. Moradinejad, S.; Trigui, H.; Guerra Maldonado, J.F.; Shapiro, J.; Terrat, Y.; Zamyadi, A.; Dorner, S.; Prevost, M. Diversity Assessment of Toxic Cyanobacterial Blooms during Oxidation. *Toxins* **2020**, *12*, 728. [CrossRef] [PubMed]
49. Ellegaard, M.; Clokie, M.R.J.; Czypionka, T.; Frisch, D.; Godhe, A.; Kremp, A.; Letarov, A.; McGenity, T.J.; Ribeiro, S.; John Anderson, N. Dead or alive: Sediment DNA archives as tools for tracking aquatic evolution and adaptation. *Commun. Biol.* **2020**, *3*, 169. [CrossRef]
50. Park, J.; Kim, Y.; Kim, M.; Lee, W.H. A novel method for cell counting of Microcystis colonies in water resources using a digital imaging flow cytometer and microscope. *Environ. Eng. Res.* **2018**, *24*, 397–403. [CrossRef]
51. America Water Works Association (AWWA). *Algae Source to Treatment. Manual of Water Supply Practices—M57*, 1st ed.; America Water Works Association: Washington, DC, USA, 2010; p. 481.
52. Hawkins, P.R.; Holliday, J.; Kathuria, A.; Bowling, L. Change in cyanobacterial biovolume due to preservation by Lugol's Iodine. *Harmful Algae* **2005**, *4*, 1033–1043. [CrossRef]
53. Bag, S.; Saha, B.; Mehta, O.; Anbumani, D.; Kumar, N.; Dayal, M.; Pant, A.; Kumar, P.; Saxena, S.; Allin, K.H.; et al. An improved method for high quality metagenomics DNA extraction from human and environmental samples. *Sci. Rep.* **2016**, *6*, 26775. [CrossRef]
54. Kuczynski, J.; Lauber, C.L.; Walters, W.A.; Parfrey, L.W.; Clemente, J.C.; Gevers, D.; Knight, R. Experimental and analytical tools for studying the human microbiome. *Nat. Rev. Genet.* **2011**, *13*, 47–58. [CrossRef]
55. Gevers, D.; Pop, M.; Schloss, P.D.; Huttenhower, C. Bioinformatics for the Human Microbiome Project. *PLoS Comput. Biol.* **2012**, *8*, e1002779. [CrossRef] [PubMed]
56. Teeling, H.; Glockner, F.O. Current opportunities and challenges in microbial metagenome analysis—A bioinformatic perspective. *Brief. Bioinform.* **2012**, *13*, 728–742. [CrossRef] [PubMed]
57. Lund, J.W.G.; Kipling, C.; Le Cren, E.D. The inverted microscope method of estimating algal number and the statistical basis of estimations by counting. *Hydrobiologia* **1958**, *11*, 143–170. [CrossRef]
58. Lund, J.W.G. A simple counting chamber for Nannoplankton. *Limnol. Oceanogr.* **1959**, *4*, 57–65. [CrossRef]
59. Planas, D.; Desrosiers, M.; Groulx, S.R.; Paquet, S.; Carignan, R. Pelagic and benthic algal responses in eastern Canadian Boreal Shield lakes following harvesting and wildfires. *Can. J. Fish. Aquat. Sci.* **2000**, *57*, 136–145. [CrossRef]
60. Munoz, G.; Vo Duy, S.; Roy-Lachapelle, A.; Husk, B.; Sauve, S. Analysis of individual and total microcystins in surface water by on-line preconcentration and desalting coupled to liquid chromatography tandem mass spectrometry. *J. Chromatogr.* **2017**, *1516*, 9–20. [CrossRef]
61. Roy-Lachapelle, A.; Vo Duy, S.; Munoz, G.; Dinh, Q.T.; Bahl, E.; Simon, D.F.; Sauvé, S. Analysis of multiclass cyanotoxins (microcystins, anabaenopeptins, cylindrospermopsin and anatoxins) in lake waters using on-line SPE liquid chromatography high-resolution Orbitrap mass spectrometry. *Anal. Methods* **2019**. [CrossRef]
62. United States Environmental Protection Agency (USEPA). *Method 350.1: Determination of Ammonia Nitrogen by Semi-Automated Colorimetry*; United States Environmental Protection Agency: Washington, DC, USA, 1993; pp. 1–15.
63. United States Environmental Protection Agency (USEPA). *Method 353.2, Revision 2.0: Determination of Nitrate-Nitrite Nitrogen by Automated Colorimetry*; United States Environmental Protection Agency: Washington, DC, USA, 1993; pp. 1–15.
64. United States Environmental Protection Agency (USEPA). *Method 365.3: Phosphorous, All Forms (Colorimetric, Ascorbic Acid, Two Reagent)*; United States Environmental Protection Agency: Washington, DC, USA, 1978; pp. 1–5.
65. United States Environmental Protection Agency (USEPA). *Method 415.1. Organic Carbon, Total (Combustion or Oxidation)*; United States Environmental Protection Agency: Washington, DC, USA, 1974; pp. 1–3.
66. Cox, M.P.; Peterson, D.A.; Biggs, P.J. SolexaQA: At-a-glance quality assessment of Illumina second-generation sequencing data. *BMC Bioinform.* **2010**, *11*, 485. [CrossRef]
67. Kim, D.; Hahn, A.S.; Wu, S.-J.; Hanson, N.W.; Konwar, K.M.; Hallam, S.J. FragGeneScan-plus for scalable high-throughput short-read open reading frame prediction. In Proceedings of the 2015 IEEE Conference on Computational Intelligence in Bioinformatics and Computational Biology (CIBCB), Niagara Falls, ON, Canada, 12–15 August 2015; pp. 1–8.

68. Fu, L.; Niu, B.; Zhu, Z.; Wu, S.; Li, W. CD-HIT: Accelerated for clustering the next-generation sequencing data. *Bioinformatics* **2012**, *28*, 3150–3152. [CrossRef]
69. Buchfink, B.; Xie, C.; Huson, D.H. Fast and sensitive protein alignment using DIAMOND. *Nat. Methods* **2015**, *12*, 59–60. [CrossRef]
70. McMurdie, P.J.; Holmes, S. phyloseq: An R package for reproducible interactive analysis and graphics of microbiome census data. *PLoS ONE* **2013**, *8*, e61217. [CrossRef] [PubMed]
71. Graffelman, J. Compositional data analysis in practice. Michael J.Greenacre. (2018). London: CRC Press. 136 pages, ISBN: 978-1-138-31661-4. *Biom. J.* **2019**. [CrossRef]
72. Foster, Z.S.; Sharpton, T.J.; Grunwald, N.J. Metacoder: An R package for visualization and manipulation of community taxonomic diversity data. *PLoS Comput. Biol.* **2017**, *13*, e1005404. [CrossRef] [PubMed]
73. Blanchet, F.G.; Legendre, P.; Borcard, D. Forward selection of explanatory variables. *Ecology* **2008**, *89*, 2623–2632. [CrossRef]

toxins

MDPI

Article

Oxidation of Cylindrosmpermopsin by Fenton Process: A Bench-Scale Study of the Effects of Dose and Ratio of H_2O_2 and Fe(II) and Kinetics

Matheus Almeida Ferreira, Cristina Celia Silveira Brandão * and Yovanka Pérez Ginoris

Environmental Technology and Water Resources Postgraduation Program, Department of Civil and Environmental Engineering, University of Brasília, Brasilia 70910-900, Brazil; ferreira.m.a@outlook.com (M.A.F.); yovanka@unb.br (Y.P.G.)

* Correspondence: cbrandao@unb.br

Abstract: The cyanotoxin cylindrospermopsin (CYN) has become a significant environmental and human health concern due to its high toxicological potential and widespread distribution. High concentrations of cyanotoxins may be produced during cyanobacterial blooms. Special attention is required when these blooms occur in sources of water intended for human consumption since extracellular cyanotoxins are not effectively removed by conventional water treatments, leading to the need for advanced water treatment technologies such as the Fenton process to produce safe water. Thus, the present study aimed to investigate the application of the Fenton process for the degradation of CYN at bench-scale. The oxidation of CYN was evaluated by Fenton reaction at H_2O_2/Fe(II) molar ratio in a range of 0.4 to 4.0, with the highest degradation of about 81% at molar ratio of 0.4. Doubling the concentrations of reactants for the optimized H_2O_2/Fe(II) molar ratio, the CYN degradation efficiency reached 91%. Under the conditions studied, CYN degradation by the Fenton process followed a pseudo-first-order kinetic model with an apparent constant rate ranging from 0.813×10^{-3} to 1.879×10^{-3} s^{-1}.

Keywords: cylindrospermopsin removal; advanced oxidation processes; Fenton process

Key Contribution: Information on CYN degradation by the Fenton process is minimal. To our knowledge, this is the first published study evaluating the removal of CYN from water using the Fenton process. This study provides an overall assessment of parameter optimization of Fenton process for CYN degradation.

check for updates

Citation: Ferreira, M.A.; Brandão, C.C.S.; Ginoris, Y.P. Oxidation of Cylindrosmpermopsin by Fenton Process: A Bench-Scale Study of the Effects of Dose and Ratio of H_2O_2 and Fe(II) and Kinetics. *Toxins* **2021**, *13*, 604. https://doi.org/10.3390/toxins13090604

Received: 16 July 2021
Accepted: 20 August 2021
Published: 29 August 2021

1. Introduction

Although natural aquatic ecosystems can be eutrophic as water bodies age and are filled in with sediments [1], human activities, such as agriculture, industry, and sewage disposal, can speed up the natural eutrophication of lentic systems by increasing the load of nutrients, resulting in anthropogenic eutrophication. The nutrients enrichment of an aquatic environment, mainly due to nitrogen and phosphorus, leads to a rapid growth of cyanobacteria, algae, and aquatic plants, which results in a shift in the biodiversity and ecosystem balance, compromising the water quality and various uses of water [2–7].

The fast growth of cyanobacteria in eutrophic waters, most frequently referred to as cyanobacterial bloom, can be harmful when the toxins produced by these organisms (cyanotoxins) reach dangerous concentrations to humans and animals. Dozens of genera and species of cyanobacteria are capable of producing toxins [8–10], which are classified as secondary metabolites.

The cylindrospermopsins (CYNs) are among the cyanotoxins of most significant health concern. These cyanotoxins are produced by species of various genera, including *Raphidiopsis* (previously *Cylindrospermopsis*), *Aphanizomenon*, *Dolichospermum* (previously

Anabaena), *Lyngbya*, and *Umezakia*. Several CYN producing species are found in lakes, rivers, and drinking water reservoirs all over the globe [11], and it is known that CYN is harmful to both animal and human health, with hepatotoxic, nephrotoxic, immunotoxic, cytotoxic, and genotoxic effects [12–15].

Most cyanobacterial toxins such as microcystins (MCs), nodularins, and saxitoxins are primarily intracellular, and the dissolved fraction (extracellular toxin) is detected in the water body when cell lysis occurs. However, CYNs can also be released from viable cells into the aquatic environment during its life cycle [10]. While extracellular MCs represent less than 30% of total MCs (intracellular + extracellular toxin) [16], extracellular CYN is reported ranging from 50 to 90% of the total CYN [17–19]. In addition, extracellular CYN tends to be reasonably stable in surface water under sunlight irradiation, with a half-life of 11–15 days, although the presence of cell pigments can speed up the CYN oxidation [20].

As cyanotoxins are a threat to human health and their dissolved fractions are not effectively removed by conventional water treatment processes, special attention is necessary when cyanobacterial blooms occur in drinking water reservoirs [14,21–24].

In 1996 in Caruaru, Pernambuco state, Brazil, following the use of inadequately treated water from local eutrophic reservoirs, dozens of patients died after intravenous exposure to MCs during renal dialysis treatment [25,26]. After this tragic event, in 2000, the Brazilian Ministry of Health established the guideline values of 1 $\mu g\ L^{-1}$ for MCs (mandatory) and 15 $\mu g\ L^{-1}$ for CYN (recommended) [27]. Based on the studies by Humpage and Falconer [28], the CYN recommended guideline value was reduced to 1 $\mu g\ L^{-1}$ in 2011 in Brazil [29], becoming mandatory only in 2021 [30] due to the increasing number of CYN occurrence reports.

Conventional drinking water treatments, in general, focus on cyanobacteria removal without compromising cell integrity to remove intracellular toxins, preventing lysis of cyanobacterial cells, thereby releasing toxins into the water. However, as high concentrations of extracellular CYN (usually ranging from <1 to 10 $\mu g\ L^{-1}$ but occasionally up to 800 $\mu g\ L^{-1}$ [31]) can be found in water bodies throughout the bloom development, even without cell lysis, the use of advanced water treatment process is a requirement to face the challenge of removing CYN, or any other extracellular cyanotoxin, in drinking water treatment.

In this context, advanced oxidative processes (AOPs) emerged as an important alternative to the effective removal and degradation of cyanotoxins in water treatment. These AOPS are based on the in situ generation of powerful and non-selective chemical oxidants such as hydroxyl and sulfate radicals, capable of reacting with a wide range of organic and inorganic pollutants.

Various AOPs can be applied to cyanotoxins removal from water, including O_3, O_3/UV, H_2O_2/UV, Fenton, photo-Fenton, electro-Fenton, and Fenton-like processes. Among them, the Fenton process is one of the most cost-effective [32–35] and has been gaining some attention due to its high performance, simplicity, short reaction times, and the lack of toxicity of the easy-to-handle reagents H_2O_2 and Fe(II) [36–39]. The advantages of the Fenton process compared to other AOPs can make this process more suitable for scale-up, especially in developing countries such as Brazil. In pre-existing water treatment plants, the Fenton process and iron-based coagulation can be easily combined in a single rapid-mix unit by adding H_2O_2.

In the Fenton process, the formation of hydroxyl radical involves several parallel and series reactions (Equations (1) to (7)) [40,41] that can be represented by a global reaction (Equation (8)) [42]. The kinetic rate constants for Equations (1) to (7) were reported elsewhere [40,41,43–46].

$$\mathrm{Fe(II)} + \mathrm{H_2O_2} \rightarrow \mathrm{Fe\,(III)} + {}^{\bullet}\mathrm{OH} + \mathrm{OH^-}\ k \approx 70\ \mathrm{M^{-1}s^{-1}} \tag{1}$$

$$\mathrm{Fe(III)} + \mathrm{H_2O_2} \rightarrow \mathrm{Fe(II)} + \mathrm{HO_2^{\bullet}} + \mathrm{H^+}\ k = 0.001\text{–}0.01\ \mathrm{M^{-1}s^{-1}} \tag{2}$$

$${}^{\bullet}\mathrm{OH} + \mathrm{H_2O_2} \rightarrow \mathrm{HO_2^{\bullet}} + \mathrm{H_2O}\ k = 3.3 \times 10^{7}\ \mathrm{M^{-1}s^{-1}} \tag{3}$$

$$^{\bullet}OH + Fe(II) \rightarrow Fe(III) + OH^{-}\ k = 3.2 \times 10^{8}\ M^{-1}s^{-1} \quad (4)$$

$$Fe(III) + HO_2^{\bullet} \rightarrow Fe(II) + O_2H^{+}\ k \leq 2 \times 10^{3}\ M^{-1}s^{-1} \quad (5)$$

$$Fe(II) + HO_2^{\bullet} + H^{+} \rightarrow Fe(III) + H_2O_2\ k = 1.2 \times 10^{6}\ M^{-1}s^{-1} \quad (6)$$

$$2HO_2^{\bullet} \rightarrow H_2O_2 + O_2\ k = 8.3 \times 10^{5}\ M^{-1}s^{-1} \quad (7)$$

$$2Fe(II) + H_2O_2 + 2H^{+} \rightarrow 2Fe(III) + 2H_2O \quad (8)$$

The degradation efficiency of the Fenton process depends on several parameters: pH, reaction time, temperature, initial concentration of pollutant, as well as reagents dosage and H_2O_2/Fe(II) molar ratio [47,48] since H_2O_2 and Fe(II) can also scavenge hydroxyl radicals (Equations (3) and (4)) if their concentrations are not optimized. The reagents dosage also reflects on the cost of the process and on the solids concentration, which can increase the iron sludge production and impair further treatment steps or discharge when high concentrations of Fe(II) are used [49].

Due to the potential to completely mineralize a variety of organic compounds, the Fenton process has been extensively studied over the past few decades. Numerous studies have been carried out applying the Fenton process for the removal of a diversity of pollutants, including phenol [50–52], bisphenol A [53,54], persistent organic pollutants [55], landfill leachate [56–58]. Concerning the removal of cyanotoxins, the studies have focused on MCs achieving degradation efficiency ranging from 18 to 100% [36,59–63].

To the best of our knowledge, no previously published study focused on CYN degradation by the traditional Fenton process. Only recently, however, few studies concerning the CYN oxidation by Fenton-like processes are available and reported a CYN degradation efficiency of around 90% [64,65]. The high CYN degradation efficiency obtained by using Fenton-like processes (Equation (2)), which are generally slower than Fenton process (Equation (1)), shows the promising potential for the CYN degradation by Fenton process. Additionally, the uracil ring, which is critical to CYN toxicity [66], is very susceptible to oxidation by hydroxyl radical [67,68]. Thus, the present study aimed to evaluate the oxidation (degradation) of CYN using Fenton process, at bench-scale, with emphasis on the effects of H_2O_2/Fe(II) molar ratio and the initial concentrations of H_2O_2 and Fe (II) on this cyanotoxin oxidation efficiency as well as the oxidation kinetics.

2. Results and Discussion

2.1. The Effect of H_2O_2/Fe(II) Molar Ration on CYN Degradation

As pointed out in the Materials and Methods (Section 4.1), two sets of experiments were carried out to evaluate the effect of H_2O_2/Fe(II) molar ratio on the CYN degradation. In the first set, the initial H_2O_2 concentration was kept constant at 25 μM; and, in the second one the initial Fe(II) concentration was kept constant at 25 μM. Figure 1 presents the average results of the two sets.

The highest CYN degradation efficiency of 81% was achieved with an H_2O_2/Fe(II) molar ratio of 0.4 in the first set of experiments (Figure 1a) and 65% with H_2O_2/Fe(II) molar ratio of 1.0 in the second set of experiments. The effect of H_2O_2/Fe(II) molar ratio on the CYN oxidation presents a similar behavior in both sets of experiments, although the difference in the optimum H_2O_2/Fe(II) molar ratio. Regarding the blank synthetic water used to evaluate the degradation of CYN over the reaction time in the absence of Fenton reagents, less than 5% of CYN degradation was observed as expected because CYN standard is relatively stable in ultrapure water [20].

Comparing the CYN degradation efficiency at H_2O_2/Fe(II) molar ratio of 0.4 in the two sets of experiments, the lower degradation of 58% observed in the second set (Figure 1b) can be explained by the initial H_2O_2 and Fe(II) concentrations, which were 2.5 times lower in the second set than in the first set of experiments (see Figure 5). Additionally, comparing the relative residual Fenton reagents, while the relative residual H_2O_2 was approximately similar in both sets of experiments, the relative residual Fe(II) was at least two times lower in the first set (Figure 1b) than in the second set (Figure 1a), indicating that the increase in the Fenton reagents, especially Fe(II), increased the hydroxyl radical scavenging activity (see Equation (4)) and consequently the Fe(II) consumption.

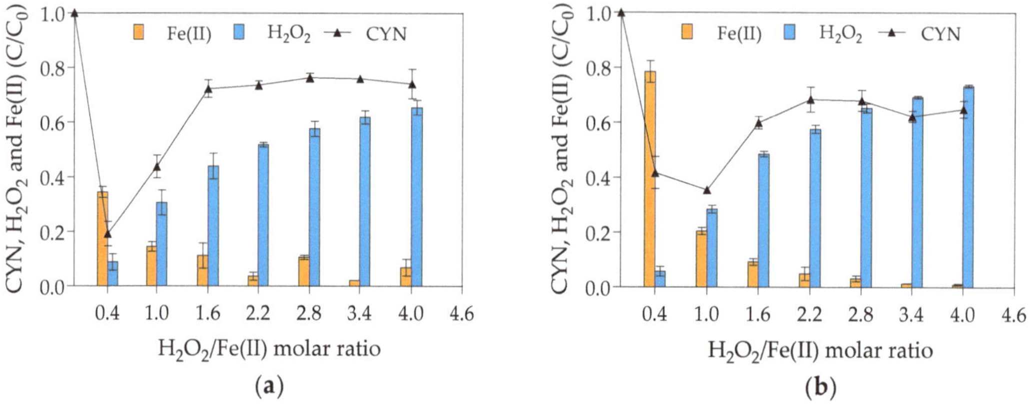

Figure 1. Relative residual concentration (C/C_0) of CYN, H_2O_2 and Fe(II) for various H_2O_2/Fe(II) molar ratios. (**a**) First set of experiments conducted with initial H_2O_2 concentration of 25 μM and initial Fe(II) concentrations from 6.25 to 62.5 μM. (**b**) Second set of experiments conducted with initial Fe(II) concentration of 25 μM and initial H_2O_2 concentrations from 10 to 100 μM. (Initial CYN concentration ≈ 0.05 μM; initial pH = 5.0 ± 0.1; reaction time = 30 min. The values are averages of three replicates, and error bars indicate the standard deviation).

The increase of the initial Fe(II) concentration may have led to a reduction in final pH as the hydrolysis of Fe(III) yielded in the Fenton reaction (Equation (1)) contributes to water acidification [69] (Figure 2a). The final pH at H_2O_2/Fe(II) molar ratio of 0.4 was lower in the first set of experiments (Figure 2a) than in the second set (Figure 2b). As higher degradation efficiencies of the Fenton process were reported at acidic conditions (pH values between 3 and 4) [61,70,71], the lowest final pH observed in the first set of experiments might have increased hydroxyl radical generation and consequently the CYN oxidation.

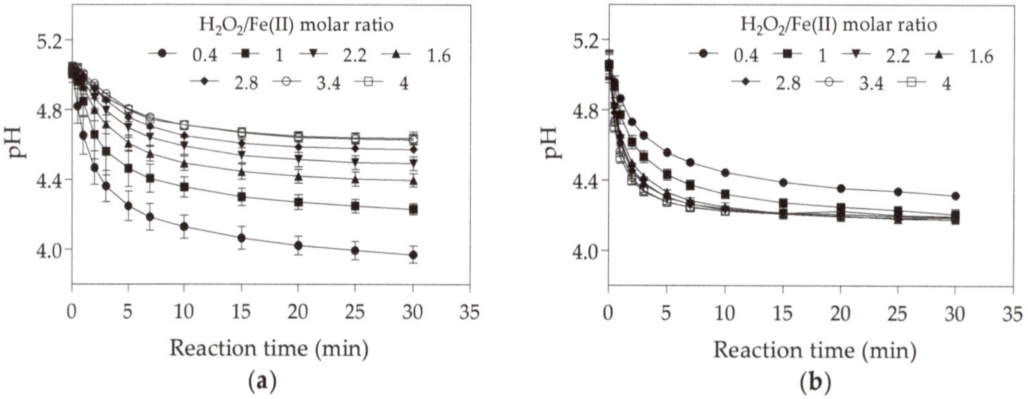

Figure 2. pH-time profile during Fenton oxidation under different H_2O_2/Fe(II) molar ratios. (**a**) The first set of experiments was conducted with initial H_2O_2 concentration of 25 μM and initial Fe(II) concentrations from 6.25 to 62.5 μM. (**b**) The second set of experiments was conducted with initial Fe(II) concentration of 25 μM and initial H_2O_2 concentrations from 10 to 100 μM. (The values presented are averages of three replicates, and error bars indicate the standard deviation).

As can be seen in Figure 1, the degradation efficiency of CYN decreased when H_2O_2/Fe(II) molar ratio increased from 0.4 to 1.6 and remained constant at higher H_2O_2/Fe(II) ratios. The observed reduction in the degradation efficiency of CYN may be explained by the scavenging of hydroxyl radicals by the excess H_2O_2 (see Equation (3)) since the residual CYN concentrations showed a similar trend as the residual H_2O_2. Furthermore, the scavenging of hydroxyl radicals by H_2O_2 leads to the generation of hydroperoxyl radical that has a lower oxidation potential, 1.7 V (SHE), than the hydroxyl radical, 2.8 V (SHE) [72].

Concerning the high percentage of residual H_2O_2 (from 50 to 69%) and the low percentage of residual Fe(II) (from 1 to 9%) at H_2O_2/Fe(II) molar ratios above 1.0, higher degradation efficiency could be achieved over a long time since the regeneration of Fe(II) can be accomplished by residual H_2O_2 by the Fenton-like reaction (Equation (2)). However, the Fenton-like reaction is very slow (second-order rate constant of 0.001 to 0.01 M^{-1} s^{-1} [44]) in comparison with the Fenton main reaction (second-order rate constant of 70 M^{-1} s^{-1} [43]), Equation (1).

Despite the high reactivity of hydroxyl radical with Fe(II) in comparison with H_2O_2, with second-order rate constants respectively of 3.2×10^8 and 3.3×10^7 M^{-1} s^{-1} [45], low H_2O_2/Fe(II) molar ratios resulted in higher CYN degradation. However, the optimal H_2O_2/Fe(II) molar ratios observed in this study are in accordance with the molar ratios of the Fenton main (Equation (1)) and global (Equation (8)) reactions, which suggests that H_2O_2/Fe(II) molar ratios less than or equal to 1 might favor the hydroxyl radical generation. It must be emphasized that there is no consensus regarding the optimum H_2O_2/Fe(II) molar ratio since it also depends on specific conditions such as pH and type and concentration of pollutant. For MCs degradation by the Fenton process, the optimum H_2O_2/Fe(II) molar ratio reported is in a range of 0.7 to 15 [36,59–61].

Based on the CYN oxidation efficiencies, the value of 0.4 obtained from the first set of experiments was adopted as the optimum H_2O_2/Fe(II) molar ratio (25 μM H_2O_2 and 62.5 μM Fe(II)) for CYN degradation by the Fenton process for the conditions evaluated in this study.

2.2. The Effect of Initial H_2O_2 and Fe(II) Concentration on CYN Degradation

The effect of initial Fenton reagents concentrations on CYN degradation was evaluated at the optimal H_2O_2/Fe(II) molar ratio of 0.4 obtained from the first set of experiments in Phase 1 with initial H_2O_2 and Fe(II) concentrations of 0.4, 1.0, and 2.0 times their optimal concentrations (25 μM H_2O_2 and 62.5 μM Fe(II)). Figure 3 shows the relative residual concentration (C/C_0) of CYN, H_2O_2, and Fe(II) and the pH-time profile for various initial concentrations of H_2O_2 and Fe(II).

The appropriate concentrations of H_2O_2 and Fe(II) are a key factor in enhancing the efficiency of the Fenton process. The analysis of reagent blank synthetic water showed that no significant CYN degradation (less than 10%) was observed at the highest concentrations of Fenton reagents when tested alone, that is, 100 μM H_2O_2 alone and 125 μM Fe(II) alone. Munoz and co-workers [64] also found no significant CYN degradation by heterogeneous Fenton-like reagents (H_2O_2 and Fe_3O_4-R400) when tested independently.

On the other hand, the Fe(II) and H_2O_2 interaction can lead to higher CYN degradation, even at low initial concentrations. As can be seen in Figure 3a, the CYN degradation was 49% with 10 μM H_2O_2 and 25 μM Fe(II), 81% with 25 μM H_2O_2 and 62.5 μM Fe(II), and 91% with 50 μM H_2O_2 and 125 μM Fe(II).

In addition, the relative residual H_2O_2 has remained almost constant, around 6%, while the relative residual Fe(II) decreased from 76% to 21% when the initial Fe(II) concentration increased from 25 to 125 μM (Figure 3a). As previously mentioned, this reduction in the relative residual Fe(II) was probably caused by the oxidation of Fe(II) to Fe(III) by hydroxyl radicals (Equation (4)).

Such behaviour suggests that, even at a fixed H_2O_2/Fe(II) molar ratio, the increase in the Fenton reagents led to an increase in the hydroxyl radical scavenging activity that can slow the rise in CYN degradation, as observed in Figure 3a. This explains why the increase from 25 μM H_2O_2 and 62.5 μM Fe(II) to 50 μM H_2O_2 and 125 μM Fe(II) only resulted in a rise of 10% in the CYN degradation, suggesting an asymptotic trend. This trend indicates that the CYN degradation by the Fenton process follows a reaction with an order greater than zero as the different initial concentrations of Fenton reagents resulted in different relative residual CYN after 30 min reaction (Figure 3).

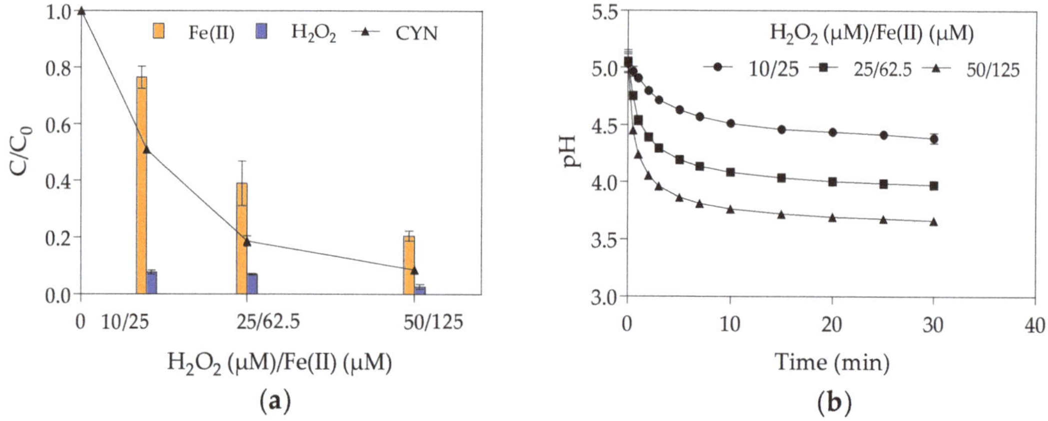

Figure 3. The effect of the Fenton reagent on CYN degradation by Fenton process at H_2O_2/Fe(II) molar ratio of 0.4. (**a**) relative residual concentration (C/C_0) of CYN, H_2O_2, and Fe(II) and (**b**) pH-time profile. (Initial CYN concentration ≈ 0.05 μM. The values are averages of three replicates, and error bars indicate the standard deviation).

Similarly, Park and co-workers [59], concerning the degradation of MC-LR by the Fenton process, also observed that increasing the H_2O_2 and Fe(II) concentration at a fixed H_2O_2/Fe(II) ratio increased the degradation efficiency until they reach a certain concentration above which all degradation efficiency increases appear to be insignificant.

The increase of the initial Fe(II) concentration diminished the final pH of the solution (Figure 3b), as also observed in the first set of experiments in Phase 1 (Figure 2a). As previously mentioned, acidic conditions may favour hydroxyl radical generation and consequently CYN oxidation. Although the positive effect of the high initial Fe(II) concentration on decreasing pH, the negative effect on hydroxyl radicals scavenging (Equation (4)) seems more significant.

2.3. Kinetic Assessment

During the Fenton process, H_2O_2 reacts with Fe(II) to produce mainly hydroxyl radical, hydroperoxyl radical and high-valent iron complexes, which can oxidize organic and inorganic compounds [48,73].

In this study, the kinetic experiments were performed with excess Fenton reagents. The H_2O_2 and Fe(II) concentrations at the optimal H_2O_2/Fe(II) molar ratio obtained from Phase 1 were respectively 500 and 1250 times greater than the initial CYN concentration. Due to the excess Fenton reagents, reaction order and rate constant were estimated from a pseudo reaction. The fitting of the kinetic models to the experimental data of one replicate is shown in Figure 4.

The kinetics parameters of the oxidation of CYN for each replicate are shown in Table 1. As shown in Figure 4 and based on R^2 (Table 1), CYN oxidation by the Fenton process was best described by the pseudo-first-order kinetic model. The results obtained from all three replicates showed a similar trend concerning fitting this kinetic model to the experimental data.

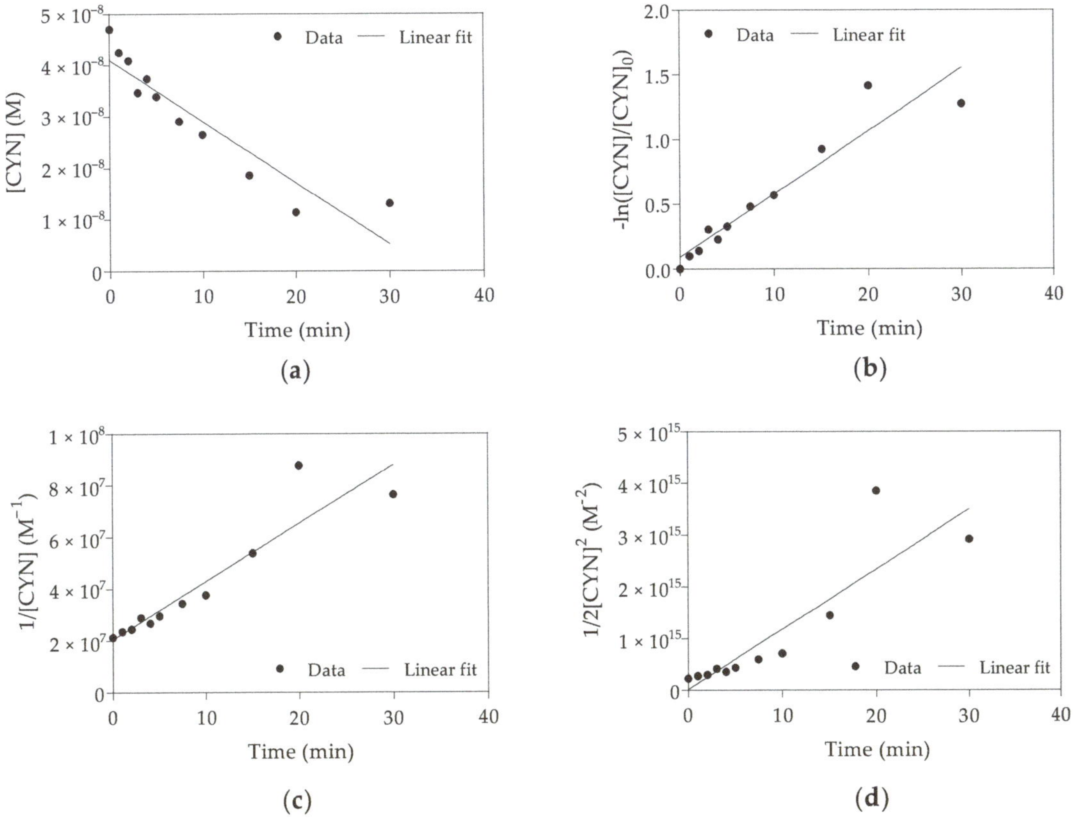

Figure 4. Linear plots and fitting of pseudo-zero-order (**a**), pseudo-first-order (**b**), pseudo-second-order (**c**) and pseudo-third-order (**d**) kinetic models to the experimental data obtained from Replicate 2. (Initial CYN concentration ≈ 0.05 μM; H_2O_2 = 25 μM; Fe(II) = 62.5 μM; initial pH 5.0 ± 0.1; 30 min reaction).

Table 1. Kinetics parameters of CYN oxidation by Fenton process. (Initial CYN concentration ≈ 0.05 μM; H_2O_2 = 25 μM; Fe(II) = 62.5 μM; initial pH 5.0 ± 0.1; 30 min reaction).

Kinetic Model	Replicate 1			Replicate 2 [b]			Replicate 3			Average		
	k [a]	$T_{1/2}$ (min)	R^2	k [a]	$T_{1/2}$ (min)	R^2	k [a]	$T_{1/2}$ (min)	R^2	k [a]	$T_{1/2}$ (min)	R^2
Zero	1.925×10^{-11}	19.1	0.76	1.987×10^{-11}	19.7	0.87	3.799×10^{-11}	11.9	0.90	2.570×10^{-11}	16.9	0.84
First	1.007×10^{-3}	11.5	0.80	0.813×10^{-3}	14.2	0.90	1.879×10^{-3}	6.1	0.98	1.233×10^{-3}	10.6	0.89
Second	6.435×10^{4}	5.9	0.73	3.748×10^{4}	9.5	0.87	13.390×10^{4}	2.3	0.87	7.858×10^{4}	5.9	0.82
Third	4.978×10^{12}	2.6	0.60	1.934×10^{12}	5.9	0.79	13.460×10^{12}	0.6	0.72	9.219×10^{12}	3.0	0.70

[a] k is the apparent pseudo-zero-order rate constant (M s^{-1}), pseudo-first-order rate constant (s^{-1}), pseudo-second-order rate constant (M^{-1} s^{-1}), pseudo-third-order rate constant (M^{-2} s^{-1}); [b] data showed in Figure 4.

As previously mentioned, to the best of our knowledge, no studies evaluating the CYN oxidation by Fenton process were published until the present. However, the oxidation of CYN and also MC-RR, MC-LR, anatoxin-a, and saxitoxin by heterogeneous Fenton-like process (H_2O_2/Fe_3O_4-R400) was reported by Munoz and co-workers [64]. The authors studied the oxidation of 1.2 μM of CYN, and 0.5 μM of MC-RR diluted in deionized water at pH 5 and with excess Fenton reagents (58.8 μM of H_2O_2 for CYN, 75.0 μM of H_2O_2 for MC-RR and fixed Fe_3O_4-R400 concentration of 863.8 μM). Under these conditions, the apparent pseudo-first-order rate constants of 7.4167 s^{-1} for CYN and 10.0167 s^{-1} for MC-RR were obtained.

The rate constant for CYN obtained by Munoz and co-workers [64] is significantly higher than that obtained herein (1.233×10^{-3} s^{-1}, Table 1). Likewise, the rate constant for MC-RR found by the same authors [64] is also considerably greater than that reported by Zhong and co-workers (2.165×10^{-3} s^{-1}) [61], who evaluated the degradation of 0.7 μM of MC-RR diluted in deionized-distilled water at pH 3 and using an excess of Fenton reagents (1500 μM of H_2O_2 and 100 μM of Fe(II)).

The results obtained by Munoz and co-workers [64] may be attributed to the nanocatalyst itself, which was specially designed and boosted for the heterogeneous Fenton-like oxidation [74]. As nanocatalysts have high surface areas and low diffusional resistance, they are more efficient than conventional heterogeneous catalysts [75].

The observed differences in the apparent pseudo-first-order rate constants reflect the different radical generations in each process. It should be pointed out that higher apparent rate constants may indicate higher hydroxyl radical concentration since the hydroxyl radical concentration is incorporated in the apparent rate constant and/or higher susceptibility to hydroxyl radical oxidation.

Despite these differences, the apparent rate constant for CYN degradation found in the present study is in the same order of magnitude as that (4×10^{-3} s^{-1}) reported by Chen and co-workers [67] obtained by using UV-TiO_2 photocatalysis under the following conditions: 2.4 μM of initial CYN, 313 μM of TiO_2, O_2 saturation, and 350 nm irradiation with an intensity of about 1.12×10^{16} photons s^{-1} cm^{-3}.

3. Summary and Conclusions

The degradation of CYN standard in ultrapure water was investigated by means of the Fenton process. The results showed that the CYN removal increased as the H_2O_2/Fe(II) molar ratio decreased. Within the range of H_2O_2/Fe(II) molar ratio tested (0.4 to 4.0), the highest CYN degradation of 81% was obtained when H_2O_2/Fe(II) molar ratio was 0.4 (25 μM H_2O_2 and 62.5 μM Fe(II)).

The increase of the dosage of Fenton reagents (50 μM H_2O_2 and 125 μM Fe(II)) at the optimal H_2O_2/Fe(II) molar ratio of 0.4 resulted in an increase of the CYN oxidation efficiency to 91%. The CYN oxidation by the Fenton process followed a pseudo-first-order kinetic model with an apparent rate constant of 1.233×10^{-3} s^{-1}.

Based on the results herein obtained, the Fenton process was effective for the removal of CYN from ultrapure water. Thus, the Fenton process seems to be a promising alternative for the CYN removal in drinking water treatment that could be easily implemented in full-scale worldwide since Fenton process is quite simple and highly cost-effective. However, further studies are necessary to evaluate matrix effects and analyze the feasibility and applicability of the Fenton process to treat natural water with high concentrations of CYN and much higher concentrations of hydroxyl radical scavenging compounds such as natural organic matter.

4. Materials and Methods

4.1. Chemicals

Cylindrospermopsin standard (purity > 95%) was purchased from Eurofins/Abraxis (Eurofins/Abraxis, Warminster, PA, USA). Methanol (99.9% HPLC grade), Ferrozine (97%), and peroxidase from horseradish (type II) were obtained from Sigma-Aldrich (Sigma-Aldrich, São Paulo, SP, Brazil). Acetic acid glacial (99.7% HPLC grade) was purchased from J.T Baker (J. T. Baker, Brazil). Hydroxylamine hydrochloride (96%) and ammonium hydroxide (27% *v*/*v*) were purchased from Synth (Synth, Diadema, SP, Brazil). Ammonium acetate (97%), N,N-Diethyl-p-phenylenediamine sulfate salt (98%), sodium phosphate dibasic (98%), sodium phosphate monobasic (98%), and hydrogen peroxide (35% *v*/*v*) were obtained from Neon (Neon, Suzano, SP, Brazil). Sodium sulfite (98%), sulfuric acid (98% *v*/*v*), hydrochloric acid (36.5% *v*/*v*), iron (II) sulfate heptahydrate (99%) and iron

(III) chloride hexahydrate (97%) were purchased from Dinâmica (Dinâmica, Indaiatuba, SP, Brazil).

All solutions used in the experiments were prepared using ultrapure water (Milli-Q Reference water purification system, C79625, Merck Millipore, Darmstadt, Hesse, Germany).

4.2. Experimental Setup

Fenton experiments were performed in 250 or 500 mL borosilicate glass beaker batch reactors at room temperature (23 to 25 °C). The oxidation experiments were carried out with a CYN solution with initial concentration (C_0) of 0.05 μM prepared by diluting a CYN stock solution (0.12 μM) with ultrapure water. This solution will be referred to as "synthetic water". Oxidation reactions were started by adding to the synthetic water predetermined amounts of Fe(II), immediately followed by the addition of H_2O_2, under vigorous magnetic stirring. After the desired reaction time, a sodium sulfite solution (2 times stoichiometric excess of Na_2SO_3 to H_2O_2 [76]) was added to quench the residual H_2O_2, stopping the generation of hydroxyl radicals. The Fe(II), H_2O_2, and sodium sulfite stock solutions were always prepared fresh.

All of the experiments were carried out in triplicates. The synthetic water initial pH was adjusted to 5.0 ± 0.1 with 50 mM H_2SO_4 and measured with a pH probe (Scientific Orion 3 Star portable pH meter, Thermo Fisher Scientific, Waltham, MA, USA).

The experimental setup can be divided into three phases (Figure 5).

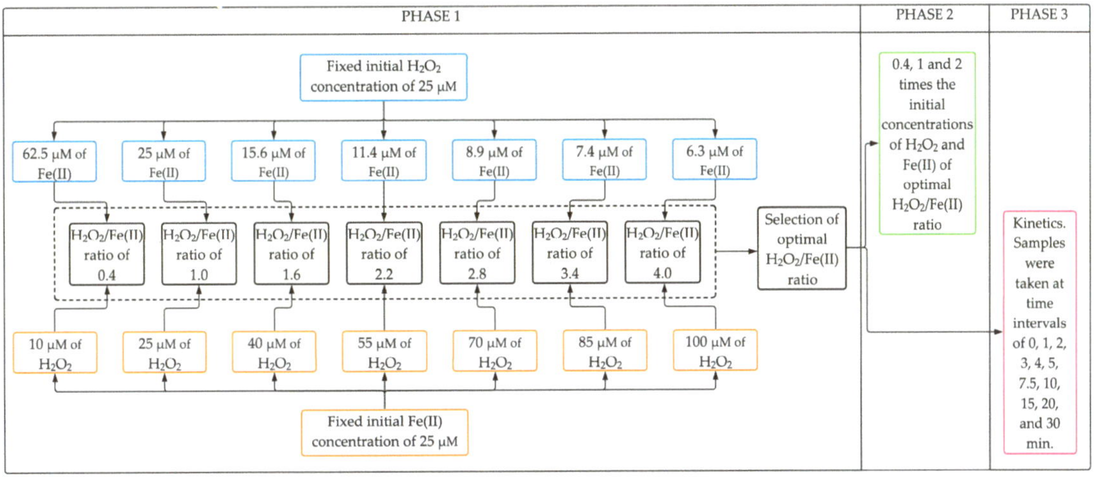

Figure 5. Flowchart of the experimental setup.

Phase 1 aimed to evaluate the effect of H_2O_2/Fe(II) molar ratio on the CYN degradation after 30 min Fenton reaction. Seven H_2O_2/Fe(II) molar ratios (0.4, 1.0, 1.6, 2.2, 2.8, 3.4, and 4.0) were evaluated in two sets of experiments. In the first set, experiments were conducted with a fixed initial H_2O_2 concentration of 25 μM with initial Fe(II) concentrations ranging from 6.25 to 62.5 μM (see Phase 1 blue boxes in Figure 5). After that, similar experiments were performed in the second set, keeping fixed the initial Fe(II) concentration at 25 μM with initial H_2O_2 concentrations ranging from 10 to 100 μM (see Phase 1 orange boxes in Figure 5). The optimal H_2O_2/Fe(II) molar ratio was chosen based on the highest CYN degradation obtained from both sets of experiments. Additionally, In Phase 1, blank synthetic water was used to evaluate the degradation of CYN over the reaction time in the absence of Fenton reagents.

In Phase 2, experiments were conducted to evaluate the effect of the initial concentrations of H_2O_2 and Fe(II) on CYN degradation after a 30 min reaction time. These oxidation experiments were performed at initial concentrations of H_2O_2 and Fe(II) 0.4, 1, and 2 times the initial concentrations of the optimal $H_2O_2/Fe(II)$ molar ratio obtained from Phase 1. Additionally, in Phase 2, the degradation of CYN over the reaction time was also evaluated using Fenton reagents (H_2O_2 and Fe(II)) separately at the highest concentrations tested in this study.

In Phase 3, oxidation kinetics of CYN was investigated in Fenton experiments conducted at the optimal $H_2O_2/Fe(II)$ molar ratio as determined in Phase 1. Samples were taken at time intervals of 0, 1, 2, 3, 4, 5, 7.5, 10, 15, 20, and 30 min.

4.3. Detection and Quantification of H_2O_2

The H_2O_2 concentration was measured using the horseradish peroxidase (POD)-N,N-Diethyl-p-phenylenediamine (DPD) photometric method as described by Hahn and co-workers [77]. Hydrogen peroxide was measured immediately after the reaction time, withdrawing a sample from the reactor before adding the sodium sulfite solution.

In the analytical routine, a 13.5 mL aliquot of the oxidized synthetic water was transferred to a 50 mL beaker and, under magnetic stirring, 1.5 mL of a solution of 0.5 M Na_2HPO_4 plus 0.5 M NaH_2PO_4 was added. Right after that, 25 μL of 38.12 mM DPD solution (prepared in 50 mM H_2SO_4) and 25 μL of 100 units mL^{-1} POD were added. The mixture was allowed to react for 40 s and then was transferred to a 5 cm path length quartz cuvette cell. The absorbance was measured at a wavelength of 551 nm using a UV-Vis spectrophotometer (DR 5000, Hach, Loveland, CO, USA).

For H_2O_2 quantification, a 6-point calibration curve encompassing H_2O_2 concentrations over the range of 0 to 10.29 μM was used. Samples were diluted using ultrapure water when their concentration was higher than the range of the calibration curve. The limit of detection (LoD) of 0.06 μM was determined according to Eurachem guidelines [78].

4.4. Detection and Quantification of Fe(II) and Total Iron

The concentrations of Fe(II) and total iron were measured by the ferrozine photometric method [79]. Similar to the hydrogen peroxide, residual Fe(II) and total iron were measured immediately after the reaction time.

Initially, to quantify residual Fe(II), a 0.3 mL of a 10 mM ferrozine solution (prepared in 0.1 M ammonium acetate) was added to a 2.7 mL sample of the oxidized synthetic water to form the Fe(II)-ferrozine complex whose absorbance can be measured at a wavelength of 562 nm using a UV-Vis spectrophotometer (DR 5000, Hach) employing 1 cm path length quartz cuvette cell.

Following the analytical routine, to quantify the total iron present, 0.45 mL of 1.4 M hydroxylamine hydrochloride (prepared in 2 M HCl) was added to a 2.4 mL aliquot of Fe(II)-ferrozine complex solution in order to reduce the Fe(III) to Fe(II) species. This mixture was allowed to react for ten minutes, and then 0.15 mL of 10 M ammonium acetate was added, and the absorbance of the resultant solution was also measured at a wavelength of 562 nm. The Fe(III) concentration was calculated as the difference between total iron and Fe(II) concentrations.

For Fe(II) and total iron quantifications, a 6-point calibration curve encompassing Fe(III) concentrations over the range of 0 to 6.72 μM was used. Samples were diluted using ultrapure water when their concentration was higher than the range of the calibration curve. The LoD of 0.61 μM for Fe(II) and 0.64 μM for total iron were determined according to Eurachem guidelines [78].

4.5. Detection and Quantification of CYN

Cylindrospermopsin was determined by high-performance liquid chromatography (Agilent 1200 Series, Agilent Technologies, Palo Alto, CA, USA) coupled to mass spectrometry (3200 QTRAP, Sciex, Toronto, ON, Canada) with an electrospray ion source operating in the positive mode, using N_2 as curtain (20 psi) and source gas (40 psi) under a capillary spray voltage of 5 kV at 450 °C.

Separation from matrix interferents was performed using a Kromasil 100-5 C18 column (100 × 4.6 mm, 5 μm, Akzo Nobel, Bohus, Sweden), coupled to its corresponding guard column (3.0 × 4.6 mm, 5 μm) at room temperature (19 to 21 °C), using 0.15% (*v*/*v*) acetic acid solutions prepared in ultrapure water (A) and methanol (B) as mobile phase, at a flow rate of 0.55 mL min^{-1}. Gradient elution was achieved by increasing B from 10% (initial condition) to 30% in 0.5 min, to 90% in 7.5 min, held B constant for 2 min, and returning to the initial condition in 2 min. Under these conditions, CYN eluted at approximately 4.6 min.

For MS acquisition, a declustering potential of 56 V was applied to the orifice to prevent the ions from clustering together. Multiple reaction monitoring (MRM) was used for CYN detection and quantification through the monitoring of three precursor-to-product ion transitions. The most intense one, at m/z 416.1 to 194.3 (43 eV CE), was used for quantification, while transitions at m/z 416.1 to 336.1 (29 eV CE) and 416.1 to 176.2 (45 eV CE) were used for confirmation purposes.

Quantification was performed by external calibration using a 6-point analytical curve encompassing CYN concentrations over the range of 2.41 nM to 0.12 μM. The LoD of 0.24 nM was determined according to Eurachem guidelines [78]. Since the LoD was quite low, sample extraction and extract concentration were not necessary for quantification of CYN.

Author Contributions: Conceptualization, M.A.F., C.C.S.B. and Y.P.G.; methodology, M.A.F. and C.C.S.B.; validation, M.A.F.; formal analysis, M.A.F. and C.C.S.B.; investigation, M.A.F.; resources, C.C.S.B.; writing—original draft preparation, M.A.F.; writing—review and editing, C.C.S.B. and Y.P.G.; supervision, C.C.S.B. All authors have read and agreed to the published version of the manuscript.

Funding: This research was partially funded by the Post-Graduate Provost Board (DPG) of the University of Brasília grant numbers 001/2019, 011/2019 and 002/2021. The authors Scholarship granted the National Council for Scientific and Technological Development (CNPq), grant number 132427/2019-2.

Institutional Review Board Statement: Not applicable.

Informed Consent Statement: Not applicable.

Acknowledgments: The authors are grateful to the Professors Raquel M. Soares and Fernando F. Sodré for their help and suggestions. The authors acknowledge the important assistance of Katyeny Manuela da Silva and Daniel V. Cárdenas in the development of the CYN LC-MS/MS method and also wishes to thank the staff at the Laboratory of Environmental Sanitation at University of Brasília.

Conflicts of Interest: The authors declare no conflict of interest.

References

1. Carpenter, S.R. Submersed Vegetation: An Internal Factor in Lake Ecosystem Succession. *Am. Nat.* **1981**, *118*, 372–383. [CrossRef]
2. Thomas, E.A. The process of eutrophication in central European lakes. In *Eutrophication: Causes, Consequences, and Correctives*; National Academy of Sciences: Washington, DC, USA, 1969; pp. 29–49.
3. Colby, P.J.; Spangler, G.R.; Hurley, D.A.; McCombie, A.M. Effects of Eutrophication on Salmonid Communities in Oligotrophic Lakes. *J. Fish. Res. Board Can.* **1972**, *29*, 975–983. [CrossRef]
4. Nixon, S.W. Coastal marine eutrophication: A definition, social causes, and future concerns. *Ophelia* **1995**, *41*, 199–219. [CrossRef]
5. Dolman, A.M.; Rücker, J.; Pick, F.; Fastner, J.; Rohrlack, T.; Mischke, U.; Wiedner, C. Cyanobacteria and Cyanotoxins: The Influence of Nitrogen versus Phosphorus. *PLoS ONE* **2012**, *7*, e38757. [CrossRef]
6. Smith, V.H. Eutrophication of freshwater and coastal marine ecosystems a global problem. *Environ. Sci. Pollut. Res.* **2003**, *10*, 126–139. [CrossRef]
7. Smith, V.H.; Joye, S.; Howarth, R. Eutrophication of freshwater and marine ecosystems. *Limnol. Oceanogr.* **2006**, *51*, 351–355. [CrossRef]
8. Codd, G.A. Cyanobacterial toxins: Occurrence, properties and biological significance. *Water Sci. Technol.* **1995**, *32*, 149–156. [CrossRef]
9. Carmichael, W.W. The cyanotoxins. *Adv. Bot. Res.* **1997**, *27*, 211–256.
10. Antoniou, M.G.; De La Cruz, A.A.; Dionysiou, D.D. Cyanotoxins: New Generation of Water Contaminants. *J. Environ. Eng.* **2005**, *131*, 1239–1243. [CrossRef]

11. Buratti, F.M.; Manganelli, M.; Vichi, S.; Stefanelli, M.; Scardala, S.; Testai, E.; Funari, E. Cyanotoxins: Producing organisms, occurrence, toxicity, mechanism of action and human health toxicological risk evaluation. *Arch. Toxicol.* **2017**, *91*, 1049–1130. [CrossRef] [PubMed]
12. Humpage, A.R.; Fontaine, F.; Froscio, S.; Burcham, P.; Falconer, I. Cylindrospermopsin Genotoxicity and Cytotoxicity: Role of Cytochrome P-450 and Oxidative Stress. *J. Toxicol. Environ. Health Part A* **2005**, *68*, 739–753. [CrossRef] [PubMed]
13. Falconer, I.R.; Hardy, S.J.; Humpage, A.R.; Froscio, S.M.; Tozer, G.J.; Hawkins, P.R. Hepatic and renal toxicity of the blue-green alga (cyanobacterium) Cylindrospermopsis raciborskii in male Swiss Albino mice. *Environ. Toxicol.* **1999**, *14*, 143–150. [CrossRef]
14. Falconer, I.R. *Cyanobacterial Toxins of Drinking Water Supplies*; CRC Press: Boca Raton, FL, USA, 2004.
15. Poniedziałek, B.; Rzymski, P.; Wiktorowicz, K. Experimental immunology First report of cylindrospermopsin effect on human peripheral blood lymphocytes proliferation in vitro. *Cent. Eur. J. Immunol.* **2012**, *4*, 314–317. [CrossRef]
16. Graham, J.L.; Loftin, K.A.; Meyer, M.; Ziegler, A.C. Cyanotoxin Mixtures and Taste-and-Odor Compounds in Cyanobacterial Blooms from the Midwestern United States. *Environ. Sci. Technol.* **2010**, *44*, 7361–7368. [CrossRef] [PubMed]
17. Preußel, K.; Wessel, G.; Fastner, J.; Chorus, I. Response of cylindrospermopsin production and release in Aphanizomenon flos-aquae (Cyanobacteria) to varying light and temperature conditions. *Harmful Algae* **2009**, *8*, 645–650. [CrossRef]
18. US EPA. *Cyanobacteria and Cyanotoxins: Information for Drinking Water Systems*; US EPA: Washington, DC, USA, 2014.
19. Kokociński, M.; Cameán, A.M.; Carmeli, S.; Guzmán-Guillén, R.; Jos, Á.; Mankiewicz-Boczek, J.; Metcalf, J.S.; Moreno, I.M.; Prieto, A.I.; Sukenik, A. Cylindrospermopsin and Congeners. In *Handbook of Cyanobacterial Monitoring and Cyanotoxin Analysis*; Meriluoto, J., Spoof, L., Codd, G.A., Eds.; John Wiley & Sons, Ltd.: Hoboken, NJ, USA, 2017; pp. 127–137.
20. Chiswell, R.K.; Shaw, G.R.; Eaglesham, G.; Smith, M.J.; Norris, R.L.; Seawright, A.A.; Moore, M.R. Stability of cylindrospermopsin, the toxin from the cyanobacterium,Cylindrospermopsis raciborskii: Effect of pH, temperature, and sunlight on decomposition. *Environ. Toxicol.* **1999**, *14*, 155–161. [CrossRef]
21. Keijola, A.M.; Himberg, K.; Esala, A.L.; Sivonen, K.; Hiis-Virta, L. Removal of cyanobacterial toxins in water treatment processes: Laboratory and pilot-scale experiments. *Environ. Toxicol. Water Qual.* **1988**, *3*, 643–656. [CrossRef]
22. Himberg, K.; Keijola, A.-M.; Hiisvirta, L.; Pyysalo, H.; Sivonen, K. The effect of water treatment processes on the removal of hepatotoxins fromMicrocystis andOscillatoria cyanobacteria: A laboratory study. *Water Res.* **1989**, *23*, 979–984. [CrossRef]
23. Teixeira, M.R.; Rosa, M.J. Comparing dissolved air flotation and conventional sedimentation to remove cyanobacterial cells of Microcystis aeruginosa: Part I: The key operating conditions. *Sep. Purif. Technol.* **2006**, *52*, 84–94. [CrossRef]
24. Van Apeldoorn, M.E.; Van Egmond, H.P.; Speijers, G.J.A.; Bakker, G.J.I. Toxins of cyanobacteria. *Mol. Nutr. Food Res.* **2007**, *51*, 7–60. [CrossRef]
25. Azevedo, S.M.; Carmichael, W.W.; Jochimsen, E.M.; Rinehart, K.L.; Lau, S.; Shaw, G.R.; Eaglesham, G.K. Human intoxication by microcystins during renal dialysis treatment in Caruaru—Brazil. *Toxicology* **2002**, *181–182*, 441–446. [CrossRef]
26. Pouria, S.; de Andrade, A.; Barbosa, J.; Cavalcanti, R.; Barreto, V.; Ward, C.; Preiser, W.; Poon, G.K.; Neild, G.; Codd, G. Fatal microcystin intoxication in haemodialysis unit in Caruaru, Brazil. *Lancet* **1998**, *352*, 21–26. [CrossRef]
27. Ministério da Saúde Brasil. *Portaria no 1469, de 29 de Dezembro de 2000. Procedimentos de Controle e Vigilância da Qualidade da Água para Consumo Humano e Seu Padrão de Potabilidade*; Diário Oficial da União; Federal Government of Brazil: Brasília, Brazil, 2000.
28. Humpage, A.R.; Falconer, I. Oral toxicity of the cyanobacterial toxin cylindrospermopsin in male Swiss albino mice: Determination of no observed adverse effect level for deriving a drinking water guideline value. *Environ. Toxicol.* **2003**, *18*, 94–103. [CrossRef]
29. Ministério da Saúde Brasil. *Portaria no 2914, de 12 de Dezembro de 2011. Procedimentos de Controle e Vigilância da Qualidade da Água para Consumo Humano e Seu Padrão de Potabilidade*; Diário Oficial da União; Federal Government of Brazil: Brasília, Brazil, 2011.
30. Ministério da Saúde Brasil. *Portaria de Consolidação GM/MS n° 888, de 4 de Maio de 2021. Procedimentos de Controle e de Vigilância da Qualidade da Água para Consumo Humano e Seu Padrão de Potabilidade*; Diário Oficial da União; Federal Government of Brazil: Brasília, Brazil, 2021.
31. Humpage, A.; Fastner, J. Cylindrospermopsins. In *Toxic Cyanobacteria in Water*; Chorus, I., Welker, M., Eds.; CRC Press: Abingdon, UK, 2021; pp. 53–71.
32. Azbar, N.; Yonar, T.; Kestioglu, K. Comparison of various advanced oxidation processes and chemical treatment methods for COD and color removal from a polyester and acetate fiber dyeing effluent. *Chemosphere* **2004**, *55*, 35–43. [CrossRef]
33. Cañizares, P.; Paz, R.; Saez, C.; Rodrigo, M.A. Costs of the electrochemical oxidation of wastewaters: A comparison with ozonation and Fenton oxidation processes. *J. Environ. Manag.* **2009**, *90*, 410–420. [CrossRef] [PubMed]
34. Gadipelly, C.; Pérez-González, A.; Yadav, G.D.; Ortiz, I.; Ibañez, R.; Rathod, V.K.; Marathe, K. Pharmaceutical Industry Wastewater: Review of the Technologies for Water Treatment and Reuse. *Ind. Eng. Chem. Res.* **2014**, *53*, 11571–11592. [CrossRef]
35. Xu, M.; Wu, C.; Zhou, Y. Advancements in the Fenton Process for Wastewater Treatment. *Adv. Oxid. Process.* **2020**, 61. [CrossRef]
36. Al Momani, F.; Smith, D.W.; El-Din, M.G. Degradation of cyanobacteria toxin by advanced oxidation processes. *J. Hazard. Mater.* **2008**, *150*, 238–249. [CrossRef] [PubMed]
37. Bautista, P.; Mohedano, A.F.; Casas, J.A.; Zazo, J.A.; Rodriguez, J.J. An overview of the application of Fenton oxidation to industrial wastewaters treatment. *J. Chem. Technol. Biotechnol.* **2008**, *83*, 1323–1338. [CrossRef]
38. Jiang, F.; Cao, G.; Zhuang, Y.; Wu, Z. Kinetic fluorimetry for determination of bisphenol S in plastics based on its promoting effect on the Fenton process. *React. Kinet. Mech. Catal.* **2020**, *130*, 1093–1108. [CrossRef]
39. Schneider, M.; Bláha, L. Advanced oxidation processes for the removal of cyanobacterial toxins from drinking water. *Environ. Sci. Eur.* **2020**, *32*, 1–24. [CrossRef]

40. De Laat, J.; Gallard, H. Catalytic Decomposition of Hydrogen Peroxide by Fe(III) in Homogeneous Aqueous Solution: Mechanism and Kinetic Modeling. *Environ. Sci. Technol.* **1999**, *33*, 2726–2732. [CrossRef]
41. Bielski, B.H.J.; Cabelli, D.E.; Arudi, R.L.; Ross, A.B. Reactivity of HO2/O−2 Radicals in Aqueous Solution. *J. Phys. Chem. Ref. Data* **1985**, *14*, 1041–1100. [CrossRef]
42. Tang, W.Z.; Huang, C.P. 2,4-Dichlorophenol oxidation kinetics by Fenton's reagent. *Environ. Technol.* **1996**, *17*, 1371–1378. [CrossRef]
43. Rigg, T.; Taylor, W.; Weiss, J. The Rate Constant of the Reaction between Hydrogen Peroxide and Ferrous Ions. *J. Chem. Phys.* **1954**, *22*, 575–577. [CrossRef]
44. Walling, C.; Goosen, A. Mechanism of the ferric ion catalyzed decomposition of hydrogen peroxide. Effect of organic substrates. *J. Am. Chem. Soc.* **1973**, *95*, 2987–2991. [CrossRef]
45. Buxton, G.V.; Greenstock, C.L.; Helman, W.P.; Ross, A.B. Critical Review of rate constants for reactions of hydrated electrons, hydrogen atoms and hydroxyl radicals (OH/O−) in Aqueous Solution. *J. Phys. Chem. Ref. Data* **1988**, *17*, 513–886. [CrossRef]
46. Rush, J.D.; Bielski, B.H.J. Pulse radiolysis studies of alkaline iron(III) and iron(VI) solutions. Observation of transient iron complexes with intermediate oxidation states. *J. Am. Chem. Soc.* **1986**, *108*, 523–525. [CrossRef]
47. Roudi, A.M.; Chelliapan, S.; Mohtar, W.H.M.W.; Kamyab, H. Prediction and Optimization of the Fenton Process for the Treatment of Landfill Leachate Using an Artificial Neural Network. *Water* **2018**, *10*, 595. [CrossRef]
48. Miller, C.J.; Wadley, S.; Waite, T.D. Fenton, photo-Fenton and Fenton-like processes. In *Advanced Oxidation Processes for Water Treatment: Fundamentals and Applications*; Stefan, M.I., Ed.; IWA Publishing: London, UK, 2017; pp. 297–332.
49. Vasquez-Medrano, R.; Prato-Garcia, D.; Vedrenne, M. Ferrioxalate-mediated processes. In *Advanced Oxidation Processes for Waste Water Treatment: Emerging Green Chemical Technology*; Ameta, S.C., Ameta, R., Eds.; Academic Press: Cambridge, MA, USA, 2018; pp. 89–113.
50. Babuponnusami, A.; Muthukumar, K. Degradation of Phenol in Aqueous Solution by Fenton, Sono-Fenton and Sono-photo-Fenton Methods. *CLEAN Soil Air Water* **2011**, *39*, 142–147. [CrossRef]
51. Kavitha, V.; Palanivelu, K. The role of ferrous ion in Fenton and photo-Fenton processes for the degradation of phenol. *Chemosphere* **2004**, *55*, 1235–1243. [CrossRef] [PubMed]
52. Vione, D.; Merlo, F.; Maurino, V.; Minero, C. Effect of humic acids on the Fenton degradation of phenol. *Environ. Chem. Lett.* **2004**, *2*, 129–133. [CrossRef]
53. Ioan, I.; Wilson, S.R.; Lundanes, E.; Neculai, A. Comparison of Fenton and sono-Fenton bisphenol A degradation. *J. Hazard. Mater.* **2007**, *142*, 559–563. [CrossRef] [PubMed]
54. Chen, W.; Zou, C.; Liu, Y.; Li, X. The experimental investigation of bisphenol A degradation by Fenton process with different types of cyclodextrins. *J. Ind. Eng. Chem.* **2017**, *56*, 428–434. [CrossRef]
55. Cravotto, G.; Di Carlo, S.; Tumiatti, V.; Roggero, C.; Bremner, H.D. Degradation of Persistent Organic Pollutants by Fenton's Reagent Facilitated by Microwave or High-intensity Ultrasound. *Environ. Technol.* **2005**, *26*, 721–724. [CrossRef] [PubMed]
56. Göde, J.N.; Souza, D.H.; Trevisan, V.; Skoronski, E. Application of the Fenton and Fenton-like processes in the landfill leachate tertiary treatment. *J. Environ. Chem. Eng.* **2019**, *7*, 103352. [CrossRef]
57. Zhang, H.; Choi, H.J.; Huang, C.-P. Optimization of Fenton process for the treatment of landfill leachate. *J. Hazard. Mater.* **2005**, *125*, 166–174. [CrossRef]
58. Hermosilla, D.; Cortijo, M.; Huang, C.P. Optimizing the treatment of landfill leachate by conventional Fenton and photo-Fenton processes. *Sci. Total. Environ.* **2009**, *407*, 3473–3481. [CrossRef] [PubMed]
59. Park, J.-A.; Yang, B.; Park, C.; Choi, J.-W.; van Genuchten, C.; Lee, S.-H. Oxidation of microcystin-LR by the Fenton process: Kinetics, degradation intermediates, water quality and toxicity assessment. *Chem. Eng. J.* **2017**, *309*, 339–348. [CrossRef]
60. Bandala, E.R.; Martínez, D.; Martínez, E.; Dionysiou, D.D. Degradation of microcystin-LR toxin by Fenton and Photo-Fenton processes. *Toxicon* **2004**, *43*, 829–832. [CrossRef]
61. Zhong, Y.; Jin, X.; Qiao, R.; Qi, X.; Zhuang, Y. Destruction of microcystin-RR by Fenton oxidation. *J. Hazard. Mater.* **2009**, *167*, 1114–1118. [CrossRef]
62. Gajdek, P.; Lechowski, Z.; Bochnia, T.; Kępczyński, M. Decomposition of microcystin-LR by Fenton oxidation. *Toxicon* **2001**, *39*, 1575–1578. [CrossRef]
63. Bober, B.; Pudas, K.; Lechowski, Z.; Bialczyk, J. Degradation of microcystin-LR by ozone in the presence of Fenton reagent. *J. Environ. Sci. Health Part A* **2008**, *43*, 186–190. [CrossRef]
64. Munoz, M.; Nieto-Sandoval, J.; Cirés, S.; de Pedro, Z.M.; Quesada, A.; Casas, J.A. Degradation of widespread cyanotoxins with high impact in drinking water (microcystins, cylindrospermopsin, anatoxin-a and saxitoxin) by CWPO. *Water Res.* **2019**, *163*, 114853. [CrossRef]
65. Henz, S.K.F.; De sousa, D.S.; Ginoris, Y.P.; Brandão, C.C.S. Remoção de cilindrospermopsinas por meio do processo Fenton no tratamento de água. In *Anais do XXIII Simpósio Brasileiro de Recursos Hídricos*; ABRHidro: Foz do Iguaçu, Brasil, 2019; pp. 1–10.
66. Banker, R.; Carmeli, S.; Werman, M.; Teltsch, B.; Porat, R.; Sukenik, A. Uracil Moiety is Required for Toxicity of the Cyanobacterial Hepatotoxin Cylindrospermopsin. *J. Toxicol. Environ. Health Part A* **2001**, *62*, 281–288. [CrossRef] [PubMed]
67. Chen, L.; Zhao, C.; Dionysiou, D.D.; O'Shea, K.E. TiO_2 photocatalytic degradation and detoxification of cylindrospermopsin. *J. Photochem. Photobiol. A Chem.* **2015**, *307–308*, 115–122. [CrossRef]

68. Zhang, G.; Wurtzler, E.; He, X.; Nadagouda, M.; O'Shea, K.; El-Sheikh, S.M.; Ismail, A.A.; Wendell, D.; Dionysiou, D. Identification of TiO_2 photocatalytic destruction byproducts and reaction pathway of cylindrospermopsin. *Appl. Catal. B Environ.* **2015**, *163*, 591–598. [CrossRef]
69. Hurowitz, J.A.; Tosca, N.J.; Dyar, M.D. Acid production by $FeSO_4{\cdot}nH_2O$ dissolution and implications for terrestrial and martian aquatic systems. *Am. Mineral.* **2009**, *94*, 409–414. [CrossRef]
70. Tekin, H.; Bilkay, O.; Ataberk, S.S.; Balta, T.H.; Ceribasi, I.H.; Sanin, F.D.; Dilek, F.B.; Yetis, U. Use of Fenton oxidation to improve the biodegradability of a pharmaceutical wastewater. *J. Hazard. Mater.* **2006**, *136*, 258–265. [CrossRef] [PubMed]
71. Pignatello, J.J.; Oliveros, E.; Mackay, A. Advanced Oxidation Processes for Organic Contaminant Destruction Based on the Fenton Reaction and Related Chemistry. *Crit. Rev. Environ. Sci. Technol.* **2006**, *36*, 1–84. [CrossRef]
72. Lawton, L.; Robertson, P. Physico-chemical treatment methods for the removal of microcystins (cyanobacterial hepatotoxins) from potable waters. *Chem. Soc. Rev.* **1999**, *28*, 217–224. [CrossRef]
73. Lee, H.; Lee, H.J.; Sedlak, D.L.; Lee, C. pH-Dependent reactivity of oxidants formed by iron and copper-catalyzed decomposition of hydrogen peroxide. *Chemosphere* **2013**, *92*, 652–658. [CrossRef] [PubMed]
74. Álvarez-Torrellas, S.; Munoz, M.; Mondejar, V.; De Pedro, Z.M.; Casas, J.A. Boosting the catalytic activity of natural magnetite for wet peroxide oxidation. *Environ. Sci. Pollut. Res.* **2018**, *27*, 1176–1185. [CrossRef] [PubMed]
75. Garrido-Ramírez, E.; Theng, B.; Mora, M.L. Clays and oxide minerals as catalysts and nanocatalysts in Fenton-like reactions—A review. *Appl. Clay Sci.* **2010**, *47*, 182–192. [CrossRef]
76. Liu, W.; Andrews, S.A.; Stefan, M.I.; Bolton, J.R. Optimal methods for quenching H_2O_2 residuals prior to UFC testing. *Water Res.* **2003**, *37*, 3697–3703. [CrossRef]
77. Bader, H.; Sturzenegger, V.; Hoigné, J. Photometric method for the determination of low concentrations of hydrogen peroxide by the peroxidase catalyzed oxidation of N,N-diethyl-p-phenylenediamine (DPD). *Water Res.* **1988**, *22*, 1109–1115. [CrossRef]
78. Magnusson, B. *Örnemark, U. Eurachem Guide: The Fitness for Purpose of Analytical Methods—A Laboratory Guide to Method Validation and Related Topics*, 2nd ed.; LGC: London, UK, 2014.
79. Viollier, E.; Inglett, P.; Hunter, K.; Roychoudhury, A.; Van Cappellen, P. The ferrozine method revisited: Fe(II)/Fe(III) determination in natural waters. *Appl. Geochem.* **2000**, *15*, 785–790. [CrossRef]

toxins

MDPI

Article

Enhanced Photocatalytic Removal of Cyanotoxins by Al-Doped ZnO Nanoparticles with Visible-LED Irradiation

Majdi Benamara [1], Elvira Gómez [2,3], Ramzi Dhahri [4,5,*] and Albert Serrà [2,3,*]

1. Laboratory of Physics of Materials and Nanomaterials Applied at Environment (LaPhyMNE), Faculty of Sciences, University of Gabes, Erriadh Manara Zrig, 6072 Gabes, Tunisia; majdibenamara1@gmail.com
2. Thin Films and Nanostructures Electrodeposition Group (GE-CPN), Department of Materials Science and Physical Chemistry, University of Barcelona, Martí i Franquès 1, E-08028 Barcelona, Spain; e.gomez@ub.edu
3. Institute of Nanoscience and Nanotechnology (IN2UB), University of Barcelona, E-08028 Barcelona, Spain
4. Department of Engineering, University of Messina, 98166 Messina, Italy
5. ITE Sarl Laboratories for Analysis Water, Processing and Packaging Industry (ITE), Moukondo, Brazzaville, Congo

* Correspondence: r_dhahri@yahoo.fr (R.D.); a.serra@ub.edu (A.S.)

Abstract: The ZnO-based visible-LED photocatalytic degradation and mineralization of two typical cyanotoxins, microcystin-LR (MC-LR), and anatoxin-A were examined. Al-doped ZnO nanoparticle photocatalysts, in Al:Zn ratios between 0 and 5 at.%, were prepared via sol-gel method and exhaustively characterized by X-ray diffraction, transmission electron microscopy, UV-vis diffuse reflectance spectroscopy, photoluminescence spectroscopy, and nitrogen adsorption-desorption isotherms. With both cyanotoxins, increasing the Al content enhances the degradation kinetics, hence the use of nanoparticles with 5 at.% Al content (A5ZO). The dosage affected both cyanotoxins similarly, and the photocatalytic degradation kinetics improved with photocatalyst concentrations between 0.5 and 1.0 g L^{-1}. Nevertheless, the pH study revealed that the chemical state of a species decisively facilitates the mutual interaction of cyanotoxin and photocatalysts. A5ZO nanoparticles achieved better outcomes than other photocatalysts to date, and after 180 min, the mineralization of anatoxin-A was virtually complete in weak alkaline medium, whereas only 45% of MC-LR was in neutral conditions. Moreover, photocatalyst reusability is clear for anatoxin-A, but it is adversely affected for MC-LR.

Keywords: cyanotoxins; microcystin-LR; anatoxin-(a); photocatalysis; ZnO-doped nanoparticles; visible light; LEDs

Key Contribution: The efficient mineralization of cyanotoxins is proposed, using Al-doped ZnO nanoparticles under visible-LED irradiation. Concerning photocatalytic performance, the Al-doped ZnO nanoparticles are equal to, if not better than, the most competitive state-of-the-art photocatalysts for cyanotoxin mineralization.

Citation: Benamara, M.; Gómez, E.; Dhahri, R.; Serrà, A. Enhanced Photocatalytic Removal of Cyanotoxins by Al-Doped ZnO Nanoparticles with Visible-LED Irradiation. *Toxins* **2021**, *13*, 66. https://doi.org/10.3390/toxins13010066

Received: 11 November 2020
Accepted: 13 January 2021
Published: 17 January 2021

1. Introduction

Cyanobacteria are common prokaryotic photosynthetic inhabitants of bodies of water worldwide that can survive in extreme environmental conditions and through dramatic changes in salinity, temperature, desiccation, and irradiance [1,2]. Under favorable conditions, cyanobacteria can achieve high rates of population density in water and thus facilitate the formation of dense blooms that can significantly compromise water quality (e.g., increase the turbidity of freshwater and change its odor and taste) [3–5]. Even so, the greater threat to freshwater resources posed by cyanobacterial blooms is the formation of cyanotoxins, which, in naturally occurring concentrations, are highly toxic to plants, invertebrates, and vertebrates [6,7]. Because the ingestion of cyanotoxins generally causes liver, digestive, and neurological diseases, their presence could make water too dangerous for recreational activities, drinking-water purification, fisheries, and, more broadly, human health [1,2,8,9].

Cyanotoxins are typically classified as hepatotoxins (e.g., microcystin, nodularin, and cylindrospermopsin), neurotoxins (e.g., anatoxin and saxitoxin), or dermatoxins (e.g., aplysiatoxin), depending on their toxicological effects [1,2,8,10]. However, they can also be classified according to their chemical composition—that is, as cyclic peptides (e.g., microcystin and nodularin), alkaloids (e.g., cylindrospermopsin, anatoxin, and saxitoxin), or lipopolysaccharides (e.g., endotoxin). No exhaustive list of cyanotoxins exists, however, and far more could be discovered, pending the identification of cyanobacterial secondary metabolites. Regardless of classification, cyanotoxins generally present a high risk of illness and mortality and rank among the most lethal groups of known biotoxins [8–11].

In recent years, mounting evidence of increased cyanobacterial blooms and cyanotoxins worldwide has raised public awareness of their toxic effects on water use and water quality. Research has shown that the increased prevalence, occurrence, and abundance of cyanobacterial blooms, and, in turn, concentration of cyanotoxins in bodies of water on a global scale, are driven by eutrophication, rising concentrations of atmospheric CO_2, and global warming, all of which stem from anthropogenic activities, especially climate change [2,9,11,12]. Because more frequent and prolonged blooms are likely to occur worldwide in the coming decades, the global context of the threat urges the development of simple, low-cost strategies and technologies for detoxicating and degrading cyanotoxins that can be readily implemented worldwide [2,9].

Although conventional water-treatment processes, such as coagulation, flocculation, rapid sand filtration, sedimentation, and chlorination, can efficiently remove cyanobacteria and low levels of various cyanotoxins, they cannot completely degrade some cyanotoxins (e.g., anatoxin-A). In the past two decades, the relatively low efficiency of those methods gave way to the use of advanced oxidation processes (AOPs) [9,13–16]. In general, AOPs are based on the mineralization of organic pollutants by highly reactive and non-selective species generated in situ (i.e., H_2O_2, $\cdot OH$, $O_2\cdot^-$, and O_3). Among the different species generated, hydroxyl radicals are often considered to be the most important contributors to water decontamination during AOPs [17,18]. As a testament to their versatility, AOPs can generate such reactive species by accommodating various processes, especially radiation, oxidation, and catalysis, and even their combination [9,13,14].

Of the various AOPs, heterogeneous photocatalysis, particularly when using TiO_2-based photocatalysts, has emerged as an efficient approach to degrading, and eventually mineralizing, microcystins and cylindrospermopsin by using UV-light irradiation [9,12,13]. However, despite considerable efforts to improve TiO_2's photocatalytic performance in mineralizing cyanotoxins under visible light or sunlight irradiation, usually via doping or surface modification, such laboratory-scale research has rarely assessed the integration of the photocatalysts into photocatalytic reactors [19]. On top of that, few studies have involved investigating the photocatalytic degradation of other cyanotoxins, such as anatoxin-A, or the use of alternative, inexpensive sources of light [9,13,20,21]. Importantly, it is clear that photocatalysis does not seem to be the most appropriate way to remove cyanotoxins from a body of water (e.g., reservoirs, lakes, and ponds), but it may be especially relevant for treating water that contains cyanotoxins (and other organic contaminants), which is destined for human, animal, or industrial use.

Research in the future may face formidable challenges with improving the economic, practical, engineering, and environmental quality of full-scale applications that can promote the use of photocatalysis in water-treatment processes [9]. Whereas solar light, despite its advantages, entails the complexity of designing efficient solar photocatalytic reactors and, beyond that, using solar light, a smarter way to facilitate photocatalytic reactor design for full-scale plants could be by using light-emitting diodes (LEDs) as a light source. LEDs are characterized by low energy consumption, which, in turn, reduces the energy consumption of UV or visible lamps; they can also simplify photoreactor designs [9,13,22].

In our study, a visible-light active Al-doped ZnO nanoparticle photocatalyst, synthesized by following a facile, scalable sol–gel new method, was examined for how the Al:Zn ratio affected its morphology, structure, and optoelectronic properties. The photocatalytic activity of the Al-doped ZnO nanoparticle photocatalyst was evaluated according to its photocatalytic degradation of microcystin-LR (MC-LR) and anatoxin-A, under visible-LED irradiation, compared with the photocatalytic activity of ZnO and Al-doped ZnO nanoparticles. The effects of the dosage of photocatalyst and the solution's pH were also investigated, as were its reusability and chemical and photochemical stability.

2. Results and Discussion

Al-doped ZnO photocatalysts, all synthesized via the sol-gel method, were labeled A0ZO, A1ZO, A3ZO, and A5ZO for Al:Zn atomic ratios of 0%, 1%, 3%, and 5%, respectively. The synthesized materials were characterized by X-ray diffraction (XRD), transmission electron microscopy (TEM), UV-vis diffuse reflectance spectroscopy, photoluminescence (PL) spectroscopy, and nitrogen adsorption-desorption isotherms.

2.1. Characterization of Photocatalysts

2.1.1. Microstructural and Morphological Characterization

The diffractogram in Figure 1 illustrates the XRD patterns of Al-doped ZnO (AZO) nanoparticles annealed at 400 °C and their numerous diffraction peaks. As shown, all peaks had different intensities, which could indicate anisotropic growth. Notably, the peaks corresponded to a reticular plane of (100), (002), (101), (102), (110), (103), (200), (112), and (201) in ZnO's hexagonal structure, with space group P63mc, according to JCPDS no. 01-073-8765. The calculated lattice parameters values (a,c), the concluded unit cell volume (V), the average sizes of the crystallites (D), the micro-deformations (ε), and the dislocation densities (δ) are presented in Table 1 [23–26]. Among other results, the smaller average size (D) and higher dislocation density (δ) of A1ZO indicate its relatively high concentration of structural defects. Structural defects can improve the photocatalytic performance of ZnO-based photocatalysts.

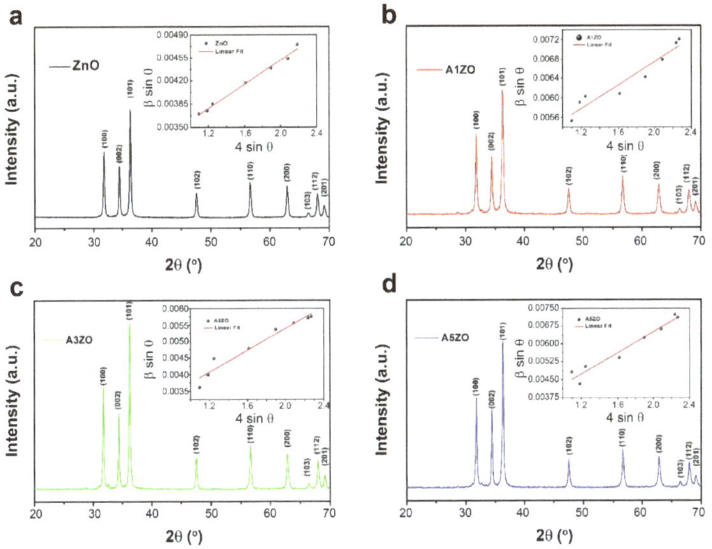

Figure 1. X-ray diffraction patterns of (**a**) A0ZO, (**b**) A1ZO, (**c**) A3ZO, and (**d**) A5ZO nanopowders with Williamson-Hall plots (inset).

Table 1. Structural parameters of the AZO samples.

	a (Å)	*c* (Å)	*V* (Å^3)	$D_{W\text{-}H}$ (nm)	D_{TEM} (nm)	ε (10^{-4})	δ (10^{-3} Line × nm^{-2})
A0ZO	3.244	5.195	47.338	52	39	9.1	81
A1ZO	3.246	5.201	47.465	32	33	12.0	174
A3ZO	3.245	5.200	47.431	66	43	16.5	116
A5ZO	3.245	5.199	47.423	69	48	22.2	148

The distribution of the AZO samples prepared by the sol-gel method by shape and particle size appears in multiple TEM images in Figure 2. Pure ZnO nanoparticles were widely distributed in terms of size and irregularly in shape. Doping ZnO with Al, by comparison, produced smaller nanoparticles. Low doping with Al appeared to reduce the size of particles, while higher loadings generated not only more of the small nanoparticles but also nanoparticles greater than 60 nm in size. The small particles were likely derived from the presence of secondary phases formed during heterogeneous nucleation and the accompanying spontaneous increase in the grain area [27]. For comparison's sake, Table 1 lists the average particle sizes of AZO samples estimated by TEM images. On average, particles in the sample, with low doping (i.e., A1ZO), were smaller than the ones in the pure ZnO samples and the samples with high doping (i.e., A3ZO and A5ZO).

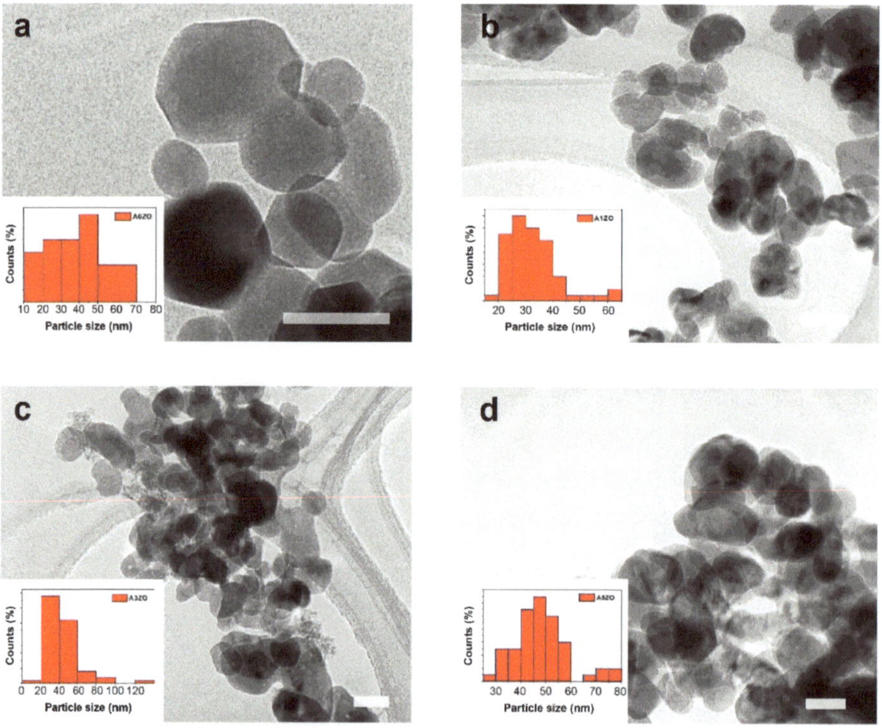

Figure 2. Transmission electron microscopic images and grain sizes distribution of (**a**) A0ZO, (**b**) A1ZO, (**c**) A3ZO, and (**d**) A5ZO nanopowders. Scale bar: 50 nm.

The surface area of pure ZnO and Al-doped ZnO nanopowders was investigated by the BET measurements. The surface areas of pure ZnO was found to be 11.8 m^2/g. The surface area increased to 16.6, 19.2, and 17.8 m^2/g when ZnO was doped with 1, 3, and 5 at.% of Al, respectively. The increase in surface area could be accounted for by the change of size of the particles. A similar increase in surface areas was observed in other studies involving doping on ZnO with aluminum.

2.1.2. Optoelectronic Characterization

The UV-vis absorption spectra of ZnO and AZO are depicted in Figure 3a. Not only were the edge values of AZO nanoparticles blue-shifted relative to ZnO nanoparticles, but absorbance in the visible region had improved primarily due to deep levels in ZnO's bandgap created by the increased concentration of defects. The bandgaps of ZnO and AZO, determined with an $(\alpha h \nu)^2$ plot as a function of photon energy, revealed that higher Al concentrations coincided with lower bandgap energies, which fell from 3.22 to 2.99 eV.

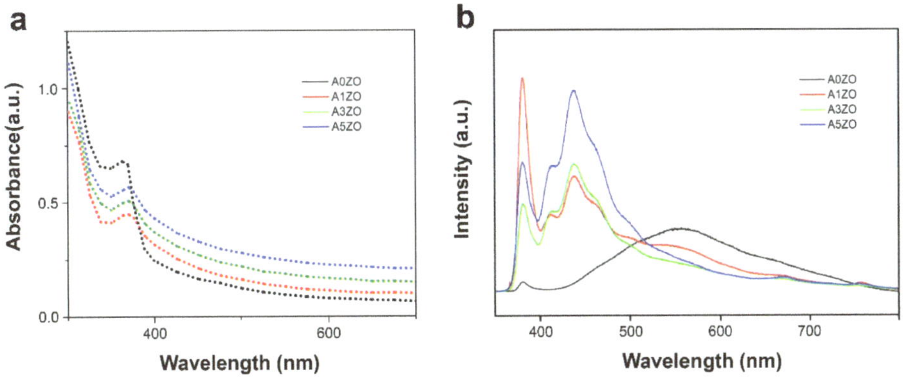

Figure 3. (**a**) UV-Vis DRS absorption spectra in diffuse reflectance spectroscopy and (**b**) photoluminescence spectra of ZnO and Al-doped ZnO nanoparticles at room temperature.

The PL spectra of the pure and Al-doped ZnO nanopowders appear in Figure 3b. The spectrum of the reference sample (i.e., pure ZnO) showed two bands: a narrow one at 380 nm of UV emission and a wide one of visible emission at 552 nm. UV emission was used to refer to band-to-band emissions, while PL emissions in the visible-light range were based on emissions attributed to inter-band defects. As such, the wide visible band was attributed to several intrinsic defects, including interstitial zinc (Zn_i), oxygen vacancies (V_o), and oxygen antisites (O_{Zn}), present in the ZnO's structure [28,29].

Beyond that, the PL emission spectra of the Al-doped ZnO samples and pure ZnO samples differed in appearance. As the intensity of the green emission band gradually disappeared, the 380 nm's emission band increased in intensity. In addition, a new band centered at 437 nm appeared, which we attributed to the blue emission, possibly from intrinsic defects and/or from the recombination of the donor-acceptor pair linked to the acceptor Al. Altogether, our results suggest that, as Al content increased, so did the intensity of the blue band, whereas the green band's intensity dropped [30].

2.2. Photocatalytic Degradation of Cyanotoxins

The photocatalytic activity of ZnO and AZO nanoparticle photocatalysts was evaluated in terms of the photocatalytic degradation of MC-LR and anatoxin-A under visible-LED irradiation at pH 7.0. As shown in Figure 4a,b, the photolytic degradation of both cyanotoxins was negligible during 80 min of visible-LED irradiation. Prior to investigating the photocatalytic degradation of the cyanotoxins, the time required to attain the adsorption-desorption equilibrium was determined in dark conditions for both the ZnO and AZO nanoparticle photocatalysts. After 60 min, the adsorption-desorption equilibrium was reached for all of the photocatalysts. Thus, only 5 ± 1% and 3 ± 1% of MC-LR and anatoxin-A, respectively, could be adsorbed by ZnO and AZO nanoparticles in dark conditions.

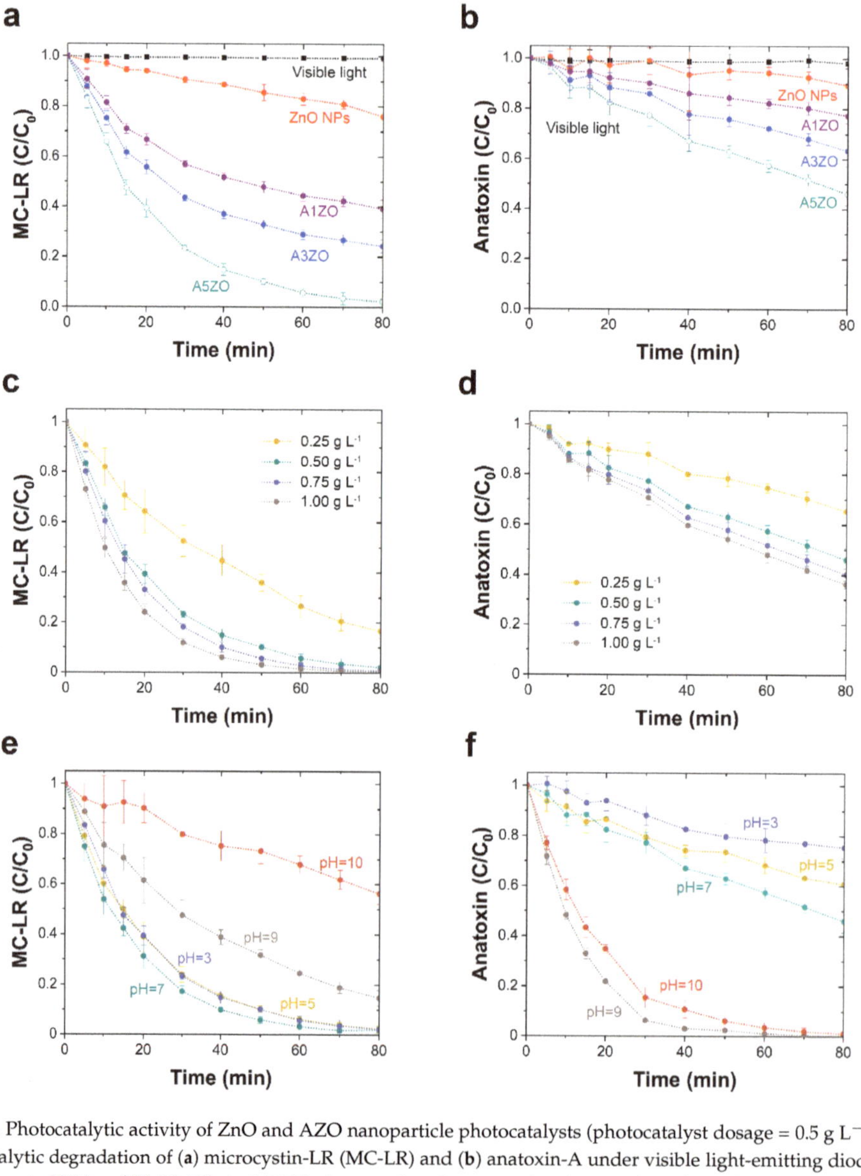

Figure 4. Photocatalytic activity of ZnO and AZO nanoparticle photocatalysts (photocatalyst dosage = 0.5 g L^{-1}) in the photocatalytic degradation of (**a**) microcystin-LR (MC-LR) and (**b**) anatoxin-A under visible light-emitting diode (LED) irradiation at pH 7.0 and 20 °C. Effect of A5ZO photocatalyst dosage on the photocatalytic degradation of (**c**) MC-LR and (**d**) anatoxin-A under visible LED irradiation at pH 7.0 and 20 °C. Effect of pH of A5ZO photocatalyst (0.5 g L^{-1}) on the photocatalytic degradation of (**e**) MC-LR and (**f**) anatoxin-A under visible LED irradiation at 20 °C. Error bars indicate standard deviations of the four replicated experiments.

As shown in Figure 4a,b, pure ZnO nanoparticles exhibited the least efficiency in removing MC-LR and anatoxin-A—only approximately 24 ± 3% and 10 ± 2%, respectively—within 80 min. Moreover, as also shown in Figure 4, as Al doping increased, the photocatalytic degradation achieved by AZO nanoparticles with MC-LR and anatoxin-A increased from 61 ± 2% to 98 ± 2% and from 23 ± 2% to 54 ± 1%, respectively, probably due to the improved surface area and enhanced visible-light absorption. Al doping enhanced the photocatalytic performance in the samples, and, for that reason, A5ZO displayed the highest degradation efficiency (i.e., approximately 98 $\pm$ 2% and 54 $\pm$ 1% for MC-LR and anatoxin-A, respectively) under the same conditions. The effects of the dosage of photocatalyst and the solution's pH were investigated for the A5ZO nanoparticle photocatalyst only.

The effects the dosage of photocatalyst on the photocatalytic degradation of MC-LR and anatoxin-A under visible LED illumination were systematically assessed by altering the catalyst loading from 0.25 to 1.0 g L^{-1}, with all other parameters unchanged. As shown in Figure 4c,d, the photocatalytic degradation of both cyanotoxins increased as the dosage of photocatalyst rose from 0.25 to 0.5 g L^{-1} but less so from 0.5 to 1.0 g L^{-1}. In turn, increases in dosage further decreased the degradation rate of both cyanotoxins. In that dynamic, increasing the dosage of the photocatalyst creates more sites for actively adsorbing both cyanotoxins and generates highly reactive radicals, which together translate into an increase in photocatalytic degradation. At the same time, high dosages of photocatalysts (>1.0 g L^{-1}) hinder the penetration of visible light and can induce the agglomeration of nanoparticles, which together translate into a reduced rate of photocatalytic degradation. Thus, the optimal dosage for enhanced photocatalytic activity was found to range from 0.5 to 1.0 g $L^{-1.}$

The effect of pH on the photocatalytic degradation of MC-LR and anatoxin-A was assessed in the pH ranges of 3.0 to 10.0 (Figure 4e,f). As is well-known, because the pH of reaction media controls the surface-charge properties of photocatalysts and the speciation of cyanotoxins, it plays a pivotal role in the adsorption and photocatalytic degradation of cyanotoxins. In our study, the isoelectric points of ZnO and A5ZO were identified to be pH 8.8 and pH 9.0, respectively. Such findings suggest that, when the pH is less than the isoelectric point, the surface is positively charged; when the pH equals the isoelectric point, the surface is neutral; and when the pH is greater than the isoelectric point, the surface is negatively charged. By comparison, MC-LR was positively charged when the pH was less than 2.09 neutral when the pH ranged between 2.09 and 2.19, and negatively charged when the pH was higher than 2.19 [9]. However, anatoxin-A, the dominant species of this cyanotoxin, was neutral at a pH less than 9.36 [9].

As shown in Figure 4e, the photocatalytic degradation of MC-LR was higher and virtually identical after 80 min of visible-LED irradiation at pH 3.0 to 7.0, when the photocatalyst surface was positively charged and the dominant species of MC-LR in the reactant media was negatively charged. As expected, further increases in pH translated into less photocatalytic degradation when both the photocatalyst's surface and the cyanotoxins were negatively charged, which promoted the electrostatic repulsion, while hindering the interaction between both entities. By contrast, the maximum photocatalytic degradation of anatoxin-A occurred at pH 9.0 (i.e., 99.8 $\pm$ 0.2%) to 10.0 (i.e., 99.6 $\pm$ 0.4%), when A5ZO was negatively charged and anatoxin-A was neutral (Figure 4f). At a pH lower than 9.0, because both A5ZO and anatoxin-A were negatively charged, the interaction between the entities was unfavorable, which effectively reduced the efficiency of photocatalytic degradation. Taken together, it is well-known that acidic media are not optimal for ZnO, because they affect the photocatalyst's stability as a result of ZnO's dissolution. However, in the presence of cyanobacterial blooms, the pH of bodies of water increases and can exceed 8.0, due to the consumption of dissolved CO_2. Thus, the photocatalytic degradation of cyanotoxins is necessary and has to be efficient at pH levels close to neutral or slightly alkaline [9].

The aim of treating water photocatalytically is to completely mineralize pollutants—in our case, cyanotoxins. The photocatalytic degradation of MC-LR and anatoxin-A produces numerous organic intermediates as they mineralize [9,31–34]. As shown in Figure 5, the mineralization of MC-LR after 180 min of visible-LED irradiation was less than 50% of the theoretical amount expected for complete mineralization at pH $\leq$ 7.0. In the case of anatoxin-A, practically complete mineralization occurred at pH $\geq$ 9.0 after 180 min of visible-LED irradiation. Such results imply that the mechanism of MC-LR's degradation is more complex than the one proposed for anatoxin-A. The complete mineralization of both cyanotoxins was confirmed after 380 min of visible-LED irradiation. At the same time, the reported values are highly competitive with outcomes achieved with state-of-the-art photocatalytic techniques for mineralizing cyanotoxins [9,12,13,31,35–37]. By contrast, most studies on the photocatalytic degradation of cyanotoxins have not involved analyzing mineralization but only the degradation of the initial cyanotoxins, often without considering that some by-products and/or intermediates could be as harmful as the cyanotoxins [9,12,13,31,35–37].

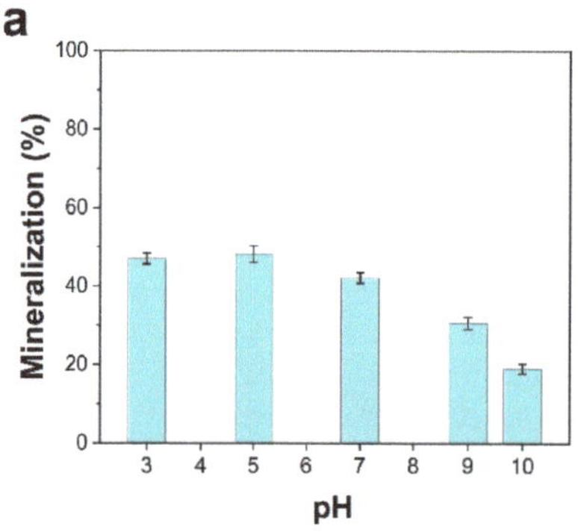

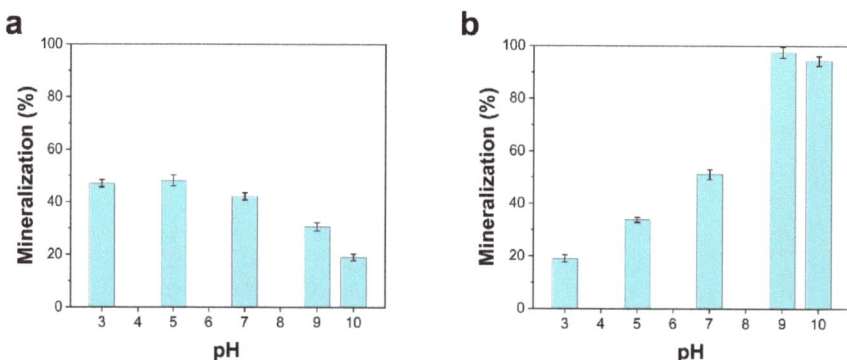

Figure 5. Mineralization (180 min) as a function of pH for AZO nanoparticle photocatalysts (photocatalyst dosage = 0.5 g L^{-1}) of (**a**) microcystin-LR and (**b**) anatoxin-A under visible-light-emitting-diode irradiation at 20 °C. Error bars indicate standard deviations of the four replicated experiments.

It is significant that the cost associated with the photocatalytic treatment of cyanotoxins (or organic pollutants in general) is typically translated directly to the energy consumption, which is closely related to the wattage of the lamp. Therefore, the use of visible-LED irradiation should reduce the energy consumption versus conventional light sources. However, large-scale experiments, incorporation of photocatalysts in reactors, photocatalytic reactor design, and water-matrix effects, among others, must be necessarily considered to predict costs [9,38–41].

2.3. Photocatalyst Reusability

The reusability and stability of A5ZO nanoparticles in the degradation of MC-LR at pH 7.0 and anatoxin-A at pH 9.0 were also investigated. Figure 6a shows the five consecutive cyclic photocatalytic degradations of both cyanotoxins after 80 min irradiation under visible LED. For MC-LR, the efficiency of photocatalytic degradation dropped significantly after five cycles, possibly due to the adsorption of by-products formed during MC-LR's degradation, which results in photocatalyst poisoning. At pH 7.0, MC-LR's mineralization after 80 min of irradiation was less than 20 ± 2%. Conversely, the photocatalytic degradation of anatoxin-A was virtually constant during the 80 min of five consecutive cycles of irradiation, as well as especially efficient (i.e., 98 ± 2%) with A5ZO. By contrast, at pH 9.0, the mineralization of anatoxin-A after 80 min of visible-LED irradiation was approximately 67 ± 2%. The different mineralization rate of MC-LR at pH 7.0 and anatoxin-A at pH 9.0 seemed to affect the reusability of A5ZO, which was higher when the rate of mineralization was greater.

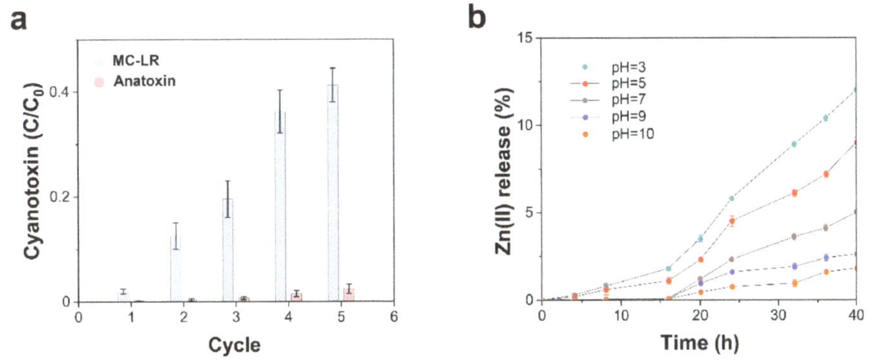

Figure 6. (**a**) Degradation of MC-LR at pH 7.0 and anatoxin-A at pH 9.0, in five consecutive reusability experiments with A5ZO nanoparticle photocatalysts (photocatalyst dosage = 0.5 g L^{-1}), after 80 min of visible light-emitting diode (LED) irradiation at 20 °C. (**b**) Time-dependent concentration of Zn(II) ions from A5ZO at different pH levels during 40 h of visible-LED irradiation at 20 °C. Error bars indicate standard deviations of the four replicated experiments.

The chemical and photochemical stability of A5ZO was also investigated by analyzing the Zn(II) and Al(III) released in aqueous media at different pH levels for 40 h of visible-LED irradiation. ZnO's well-known photocorrosive behavior occurs under UV and visible-light irradiation, and at extreme pH values (i.e., 2.0 < pH > 10.0), ZnO even dissolves, which of course defeats its purpose in photocatalysis [39]. However, corrosion and photocorrosion can be suppressed via surface modification or doping strategies. As shown in Figure 6b, the corrosion and photocorrosion of A5ZO nanoparticles were minimal in weak alkaline conditions and moderate at neutral pH. Importantly, no Al(III) species in the aqueous medium were detected during the 40 h of visible-LED irradiation. Such findings matter because cyanotoxins are usually generated in neutral or slightly alkaline bodies of water. The findings also confirm the stability of synthesized A5ZO photocatalysts in the photocatalytic degradation of cyanotoxins from natural bodies of water.

3. Conclusions

In search of a complementary economic gain relative to energy consumption, we examined the photocatalytic activity of Al-doped ZnO nanoparticles in degrading and mineralizing two cyanotoxins under visible-LED irradiation. The sol-gel method was found to be as reproducible method of synthetizing well-defined Al-doped ZnO nanoparticles with controlled percentages of Al, namely ranging from 0 to 5 at.%. Doping has a twofold effect: It reducing bandgaps, thus widening the visible activity; and it minimizes deleterious photocorrosion, that is, the easy photochemical dissolution of zinc oxide.

Increasing the dosage of photocatalyst in the solution did not exert a linear effect on the photocatalytic activity. Upon having a certain amount of photocatalyst, photocatalytic activity increased; however, similar behavior was observed from a threshold value as well (0.5 g L^{-1}). Considering that the easy agglomeration of the nanoparticles could have been responsible, we pragmatically decided that a dose of 0.5 g L^{-1} could maximize efficacy and save material.

Among other findings, the pH of the solution for mineralizing cyanotoxins was critical. The optimal pH is the one that favors the interaction of the substrate and the photocatalyst, which itself relates to the surface charge of the species. Because the pH value is responsible for the molecule's chemical form, and thus its electrical charge, the value of the isoelectric point of the species proved critical as well. According to our results, the best pH value for solutions for mineralizing anatoxin-A is 9.0, under which conditions mineralization completed, and for MC-LR is pH 7.0, under which conditions the reaction yield became balanced with the photocatalyst's chemical stability. The reusability of the photocatalysts also depended on the entities and reaction conditions. In any case, neutral or weak alkaline media are best suited to avoiding the risk of the photocorrosion or dissolution of the photocatalyst, as has been proven by controlling the release of Zn(II).

The results additionally show that the photocatalytic activity of the synthesized photocatalysts was greater than that previously reported for degradation of cyanotoxins. Even when mineralization remains incomplete, as was the case with MC-LR, degrading a high percentage of any pollutant benefits the environment, because the intermediate-reaction products exert lower harmful effects on water.

4. Materials and Methods

4.1. Synthesis of Nanoparticles

Al-doped ZnO nanopowders were elaborated via the sol–gel method, as shown in Scheme 1. We dissolved 16 g of zinc acetate dihydrate ($Zn(CH_3COO)_2{\cdot}2H_2O$; ≥99.0%, Sigma-Aldrich) in 112 mL of methanol (99.8%, Sigma-Aldrich), which acted as solvent. The solution was stirred for 10 min, at room temperature, after which we added an adequate quantity of aluminum nitrate-9-hydrate (≥98%, Sigma-Aldrich) as a dopant precursor, according to Al:Zn atomic ratios of 0%, 1%, 3%, and 5%. The obtained solutions were stirred for 15 min and placed in an autoclave, to dry in the supercritical condition of ethyl alcohol. Last, the synthesized powders were calcined for 2 h, using an air muffle furnace at 400 °C.

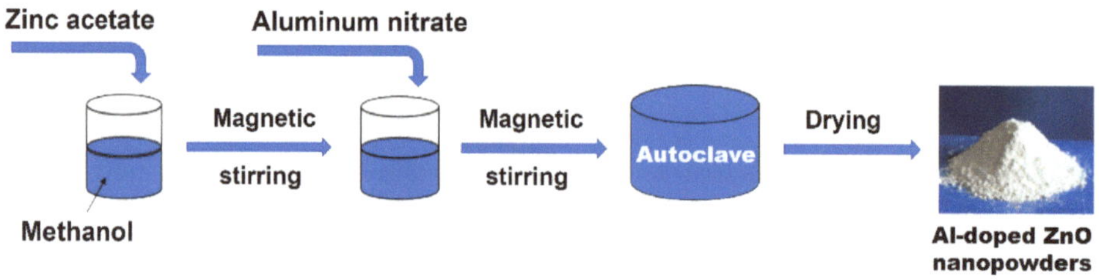

Scheme 1. Synthesis of AZO nanoparticles by the sol-gel method.

4.2. Materials Characterization

The surface morphology, crystalline structure, and optical properties of ZnO-based photocatalysts were investigated by TEM, XRD, and PL measurements. First, we studied the microstructure of the samples by using an XRD machine (Bruker AXS D8 Advance, MA, USA), at the wavelength of $CuK\alpha$ (i.e., 1.5405 Å). The network parameters (*a*,*c*) and the unit cell volume (*V*) were determined by using the following equations [23,24]:

$$a = \frac{\lambda}{\sqrt{3}\sin\theta_{(100)}} \tag{1}$$

$$c = \frac{\lambda}{\sin\theta_{(002)}} \tag{2}$$

$$V = \frac{\sqrt{3}}{2}a^2{\cdot}c \tag{3}$$

where λ is the wavelength of the radiation used (i.e., 1.54060 Å for $CuK\alpha$) and θ is the Bragg diffraction angle. The calculated lattice parameters values (*a*,*c*) and the concluded unit cell volume (*V*) are presented in Table 1. The micro-deformations (ε) were estimated according to the Williamson-Hall model [25]:

$$\beta\cos\theta = \frac{k\lambda}{D} + 4\,\varepsilon\sin\theta \tag{4}$$

The plot of ($\beta\cos\theta$) versus ($4\sin\theta$) for the ZnO nanopowder, as shown in Figure 1, was fitted according to Equation (4). The D value was estimated from the slope of the extrapolation of the linear fit, while the ε value was given by the fit slope. In addition, the dislocation density (δ) was calculated by using the following relationship [26]:

$$\delta = \frac{15\,\varepsilon}{a\,D} \quad (5)$$

Next, the shapes and sizes of the prepared nanoparticles were examined with a transmission electron microscope (JEOL JEM 2010, Tokyo, Japan), an LaB6 electron gun that operates at 200, which was equipped with a Gatan 794 MultiScan CCD camera used to view the digital images. The Brunauer-Emmett-Teller (BET) surface area of each photocatalyst was determined by using the N_2 adsorption-desorption isotherms obtained at 77 K, using a Micrometrics Tristar-II (Micrometrics, Canada). The UV-vis diffuse reflectance spectra (DRS) were recorded by using a Lambda 900 UV spectrophotometer (PerkinElmer, Waltham, MA, USA). Last, a NanoLog Horiba modular spectrofluorometer (Horiba, Kyoto, Japan) used for PL measurements was equipped with an Xe lamp, as the excitation light source, with a wavelength of 325 nm, at room temperature. Emissions were studied between 350 and 800 nm.

4.3. Photocatalytic Activity

To examine the photocatalytic activity of the Al-doped ZnO nanoparticles, the photocatalytic degradation of single-pollutant aqueous solutions of 1.25 mg L^{-1} MC-LR (Sigma-Aldrich) and 2.5 mg L^{-1} (±)-anatoxin-A fumarate (Hello Bio Ltd., >99%) was observed under irradiation from three 6.2 W LEDs (irradiance = 65 mW cm^{-2}; energy consumption = 7 kWh or 1000 h) for 80 min. The photocatalytic experiments were conducted in a quartz dish reactor (volume = 40 mL, diameter = 10 cm) sealed with a quartz cover and cooled with fans, to prevent the evaporation of the solution. All the experiments were performed in quadruplicate. Different dosages of Al-doped ZnO nanoparticles (i.e., 0.25, 0.5, 0.75, and 1.0 g L^{-1}) were ultrasonically (power = 5 W L^{-1}; time = 60 s) suspended in the pollutant solution, to determine the effect of the dosage of the photocatalyst. To examine the effect of pH on the photocatalytic degradation of the cyanotoxins, the initial solution's pH was adjusted to 3.0, 5.0, 7.0, 9.0, and 10.0 with 0.1 M HCl or NaOH solutions. Before irradiation commenced, the suspensions were magnetically stirred at 200 rpm, in dark conditions, in the presence of each cyanotoxin, for 1 h, in order to achieve adsorption-desorption equilibrium. Once achieved, the suspensions were irradiated with visible light by way of continuous magnetic stirring. With the temperature maintained at 20 °C, samples (500 μL) were taken at specified times and filtered through a syringe filter (0.02 μm, Whatman), to separate the photocatalyst. The temporal photocatalytic degradation of cyanotoxins was monitored by quantifying the reduction of the concentration of MC-LR and anatoxin-A, using a high-performance liquid chromatograph (Agilent; Santa Clara, CA, USA) with a photodiode array detector set at 238 nm and ultrahigh-performance liquid chromatography (Thermo Fisher Scientific; Waltham, MA, USA), respectively.

To quantify MC-LR, a C18 Discovery HS column 150 mm by 2.1 mm i.d. for particles 5 μm in size (Supelco, Inc.; Bellefonte, PA, USA) was used for the stationary and mobile phases; it consisted of 0.05% (*v/v*) trifluoroacetic acid in Milli-Q water and 0.05% trifluoroacetic acid in acetonitrile in a ratio of 60:40. Analysis was performed under equilibrium conditions at a column temperature of 40 °C, a flow rate of 0.2 mL min^{-1}, and an injection volume of 20 μL [40]. By contrast, to quantify anatoxin-A, a Kinetex HILIC column measuring 100 mm by 2.1 mm i.d. for particles 2.6 μm in size (Phenomenex, Inc.; Basel, Switzerland) was employed for the stationary phase, while the mobile phase consisted of 95% (*v/v*) acetonitrile +5% (*v/v*) aqueous solution (i.e., 2 mM ammonium formate +3.6 mM formic acid) and an aqueous solution of 2 mM ammonium formate and 3.6 mM formic acid in a ratio of 85:15 [12]. The analysis was performed under equilibrium conditions, with a column temperature of 25 °C, a flow rate of 0.2 mL min^{-1}, and an injection volume of 20 μL.

Last, the mineralization of cyanotoxins was determined by comparing the reduction of total organic content (TOC) after 180 min of visible LED irradiation, using a TOC analyzer (Shimadzu; Nakagyo-ku, Kyoto, Japan). The photocatalytic experiments for each catalyst were performed in quadruplicate.

4.4. Photocatalyst Reusability

To evaluate the reusability of the Al-doped ZnO nanoparticles, the photocatalytic degradation of MC-LR (Sigma-Aldrich) and (±)-anatoxin-A fumarate was analyzed by reusing the same photocatalyst sample, under the same reaction conditions, during five consecutive 80 min irradiation cycles. For comparison, the photocatalytic degradation of cyanotoxins was also determined by liquid chromatography after 80 min of visible LED irradiation. After each experiment, the Al-doped nanoparticles were recuperated by centrifugation, washed, and reused. Meanwhile, photostability was analyzed by quantifying the time-dependent evolution concentration of Zn(II) ions in 40 mL Milli-Q water, under visible LED irradiation for 8 h. The initial solution's pH was adjusted to 3.0, 5.0, 7.0, 9.0, and 10.0 with 0.1 M HCl or NaOH solutions. At different times during the 40 h period, 0.5 mL of irradiated solution was filtered through a syringe filter (0.02 µm, Whatman), to separate the photocatalyst. The concentration of Zn(II) ions was determined spectrophotometrically by using Zincon monosodium salt (Sigma-Aldrich) in borate buffer (50 mM, pH = 9) and measuring the absorbance associated with the Zn(II)-bound Zincon complex at 620 nm [42,43]. For those measurements, a UV-1800 Shimadzu UV-vis spectrophotometer (Shimadzu; Japan), accompanied by a quartz cuvette with an optical length of 1 cm, was used. The Al(III) concentration was determined by ICP–OES analysis. The dosage of the photocatalyst for all experiments described in this section was set at 0.5 g L^{-1}. The reusability experiments were performed in quadruplicate.

Author Contributions: Conceptualization, M.B., E.G., R.D., and A.S.; methodology, M.B., and A.S.; validation, E.G., R.D., and A.S.; formal analysis, M.B., E.G., R.D., and A.S.; investigation, M.B., and A.S.; resources, E.G., R.D., and A.S.; data curation, M.B. and A.S.; writing—original draft preparation, M.B. and A.S.; writing—review and editing, E.G., R.D., and A.S.; supervision, E.G., R.D., and A.S.; project administration, E.G., R.D., and A.S.; funding acquisition, E.G., and R.D. All authors have read and agreed to the published version of the manuscript.

Funding: This work was partially supported by the Metrohm foundation. Partial funding from the TEC2017-85059-C3-2-R project (co-financed by the *Fondo Europeo de Desarrollo Regional*, FEDER) from the Spanish *Ministerio de Economía y Competitividad* (MINECO) is also acknowledged. Albert Serrà would like to acknowledge funding from the EMPAPOSTDOCS-II program. The EMPAPOSTDOCS-II program has received funding from the European Union's Horizon 2020 research and innovation program under the Marie Skłodowska-Curie grant agreement, number 754364.

Data Availability Statement: Data is contained within the article.

Conflicts of Interest: The authors declare no conflict of interest.

References

1. Huisman, J.; Codd, G.A.; Paerl, H.W.; Ibelings, B.W.; Verspagen, J.M.H.; Visser, P.M. Cyanobacterial blooms. *Nat. Rev. Microbiol.* **2018**, *16*, 471–483. [CrossRef] [PubMed]
2. Buratti, F.M.; Manganelli, M.; Vichi, S.; Stefanelli, M.; Scardala, S.; Testai, E.; Funari, E. Cyanotoxins: Producing organisms, occurrence, toxicity, mechanism of action and human health toxicological risk evaluation. *Arch. Toxicol.* **2017**, *91*, 1049–1130. [CrossRef] [PubMed]
3. Dittmann, E.; Wiegand, C. Cyanobacterial toxins—Occurrence, biosynthesis and impact on human affairs. *Mol. Nutr. Food Res.* **2006**, *50*, 7–17. [CrossRef] [PubMed]
4. Meriluoto, J.; Blaha, L.; Bojadzija, G.; Bormans, M.; Brient, L.; Codd, G.A.; Drobac, D.; Faassen, E.J.; Fastner, J.; Hiskia, A.; et al. Toxic cyanobacteria and cyanotoxins in European waters—Recent progress achieved through the CYANOCOST action and challenges for further research. *Adv. Oceanogr. Limnol.* **2017**, *8*, 161–178. [CrossRef]
5. Liu, Y.; Chen, W.; Li, D.; Huang, Z.; Shen, Y.; Liu, Y. Cyanobacteria-/cyanotoxin-contaminations and eutrophication status before Wuxi Drinking Water Crisis in Lake Taihu, China. *J. Environ. Sci.* **2011**, *23*, 575–581. [CrossRef]

6. Gkelis, S.; Zaoutsos, N. Cyanotoxin occurrence and potentially toxin producing cyanobacteria in freshwaters of Greece: A multi-disciplinary approach. *Toxicon* **2014**, *78*, 1–9. [CrossRef]
7. Ibelings, B.W.; Backer, L.C.; Kardinaal, W.E.A.; Chorus, I. Current approaches to cyanotoxin risk assessment and risk management around the globe. *Harmful Algae* **2014**, *40*, 63–74. [CrossRef]
8. Du, X.; Liu, H.; Yuan, L.; Wang, Y.; Ma, Y.; Wang, R.; Chen, X.; Losiewicz, M.D.; Guo, H.; Zhang, H. The diversity of cyanobacterial toxins on structural characterization, distribution and identification: A systematic review. *Toxins* **2019**, *11*, 530. [CrossRef]
9. Serrà, A.; Philippe, L.; Perreault, F.; Garcia-Segura, S. Photocatalytic treatment of natural waters. Reality or hype? The case of cyanotoxins remediation. *Water Res.* **2020**, *188*, 116543. [CrossRef]
10. Christoffersen, K.; Kaas, H. *Toxic Cyanobacteria in Water. A Guide to Their Public Health Consequences, Monitoring, and Management*; CRC Press: London, UK, 2000; Volume 45, ISBN 0419239308.
11. Berry, M.A.; Davis, T.W.; Cory, R.M.; Duhaime, M.B.; Johengen, T.H.; Kling, G.W.; Marino, J.A.; Den Uyl, P.A.; Gossiaux, D.; Dick, G.J.; et al. Cyanobacterial harmful algal blooms are a biological disturbance to Western Lake Erie bacterial communities. *Environ. Microbiol.* **2017**, *19*, 1149–1162. [CrossRef]
12. Serrà, A.; Pip, P.; Gómez, E.; Philippe, L. Efficient magnetic hybrid ZnO-based photocatalysts for visible-light-driven removal of toxic cyanobacteria blooms and cyanotoxins. *Appl. Catal. B Environ.* **2020**, *268*, 118745. [CrossRef]
13. Schneider, M.; Bláha, L. Advanced oxidation processes for the removal of cyanobacterial toxins from drinking water. *Environ. Sci. Eur.* **2020**, *32*. [CrossRef]
14. Zhang, G.; Nadagouda, M.N.; O'Shea, K.; El-Sheikh, S.M.; Ismail, A.A.; Likodimos, V.; Falaras, P.; Dionysiou, D.D. Degradation of cylindrospermopsin by using polymorphic titanium dioxide under UV-Vis irradiation. *Catal. Today* **2014**, *224*, 49–55. [CrossRef]
15. Antoniou, M.G.; Zhao, C.; O'Shea, K.E.; Zhang, G.; Dionysiou, D.D.; Zhao, C.; Han, C.; Nadagouda, M.N.; Choi, H.; Fotiou, T.; et al. *Photocatalytic Degradation of Organic Contaminants in Water: Process Optimization and Degradation Pathways*; The Royal Society of Chemistry: London, UK, 2016; Volume 2016, ISBN 9781782620419.
16. Mauter, M.S.; Zucker, I.; Perreault, F.; Werber, J.R.; Kim, J.H.; Elimelech, M. The role of nanotechnology in tackling global water challenges. *Nat. Sustain.* **2018**, *1*, 166–175. [CrossRef]
17. Antoniou, M.G.; Shoemaker, J.A.; de la Cruz, A.A.; Dionysiou, D.D. LC/MS/MS structure elucidation of reaction intermediates formed during the TiO_2 photocatalysis of microcystin-LR. *Toxicon* **2008**, *51*, 1103–1118. [CrossRef]
18. Lawton, L.A.; Robertson, P.K.J.; Cornish, B.J.P.A.; Marr, I.L.; Jaspars, M. Processes influencing surface interaction and photocatalytic destruction of microcystins on titanium dioxide photocatalysts. *J. Catal.* **2003**, *213*, 109–113. [CrossRef]
19. Loeb, S.K.; Alvarez, P.J.J.; Brame, J.A.; Cates, E.L.; Choi, W.; Crittenden, J.; Dionysiou, D.D.; Li, Q.; Li-Puma, G.; Quan, X.; et al. The Technology Horizon for Photocatalytic Water Treatment: Sunrise or Sunset? *Environ. Sci. Technol.* **2019**, *53*, 2937–2947. [CrossRef]
20. Serrà, A.; Philippe, L. Simple and scalable fabrication of hairy ZnO@ZnS core@shell Cu cables for continuous sunlight-driven photocatalytic water remediation. *Chem. Eng. J.* **2020**, *401*, 126164. [CrossRef]
21. Pirhashemi, M.; Habibi-Yangjeh, A.; Rahim Pouran, S. Review on the criteria anticipated for the fabrication of highly efficient ZnO-based visible-light-driven photocatalysts. *J. Ind. Eng. Chem.* **2018**, *62*, 1–25. [CrossRef]
22. Serrà, A.; Zhang, Y.; Sepúlveda, B.; Gómez, E.; Nogués, J.; Michler, J.; Philippe, L. Highly reduced ecotoxicity of ZnO-based micro/nanostructures on aquatic biota: Influence of architecture, chemical composition, fixation, and photocatalytic efficiency. *Water Res.* **2020**, *169*, 115210. [CrossRef]
23. Karthika, K.; Ravichandran, K. Tuning the Microstructural and Magnetic Properties of ZnO Nanopowders through the Simultaneous Doping of Mn and Ni for Biomedical Applications. *J. Mater. Sci. Technol.* **2015**, *31*, 1111–1117. [CrossRef]
24. Kahouli, M.; Barhoumi, A.; Bouzid, A.; Al-Hajry, A.; Guermazi, S. Structural and optical properties of ZnO nanoparticles prepared by direct precipitation method. *Superlattices Microstruct.* **2015**, *85*, 7–23. [CrossRef]
25. Reddy, A.J.; Kokila, M.K.; Nagabhushana, H.; Chakradhar, R.P.S.; Shivakumara, C.; Rao, J.L.; Nagabhushana, B.M. Structural, optical and EPR studies on ZnO:Cu nanopowders prepared via low temperature solution combustion synthesis. *J. Alloys Compd.* **2011**, *509*, 5349–5355. [CrossRef]
26. Kahraman, S.; Çetinkara, H.A.; Bayansal, F.; Çakmak, H.M.; Güder, H.S. Characterisation of ZnO nanorod arrays grown by a low temperature hydrothermal method. *Philos. Mag.* **2012**, *92*, 2150–2163. [CrossRef]
27. Hjiri, M.; El Mir, L.; Leonardi, S.G.; Pistone, A.; Mavilia, L.; Neri, G. Al-doped ZnO for highly sensitive CO gas sensors. *Sens. Actuators B Chem.* **2014**, *196*, 413–420. [CrossRef]
28. Han, N.; Chai, L.; Wang, Q.; Tian, Y.; Deng, P.; Chen, Y. Evaluating the doping effect of Fe, Ti and Sn on gas sensing property of ZnO. *Sens. Actuators B Chem.* **2010**, *147*, 525–530. [CrossRef]
29. Han, N.; Hu, P.; Zuo, A.; Zhang, D.; Tian, Y.; Chen, Y. Photoluminescence investigation on the gas sensing property of ZnO nanorods prepared by plasma-enhanced CVD method. *Sens. Actuators B Chem.* **2010**, *145*, 114–119. [CrossRef]
30. Ding, J.J.; Chen, H.X.; Ma, S.Y. Structural and photoluminescence properties of Al-doped ZnO films deposited on Si substrate. *Phys. E Low-Dimens. Syst. Nanostruct.* **2010**, *42*, 1861–1864. [CrossRef]
31. Pinho, L.X. Photocatalytic Degradation of Cyanobacteria and Cyanotoxins Using Suspended and Immobilized TiO_2. Ph.D. Thesis, University of Porto, Porto, Portugal, 2014; p. 168.
32. Hu, X.; Hu, X.; Tang, C.; Wen, S.; Wu, X.; Long, J.; Yang, X.; Wang, H.; Zhou, L. Mechanisms underlying degradation pathways of microcystin-LR with doped TiO_2 photocatalysis. *Chem. Eng. J.* **2017**, *330*, 355–371. [CrossRef]

33. Andersen, J.; Han, C.; O'Shea, K.; Dionysiou, D.D. Revealing the degradation intermediates and pathways of visible light-induced NF-TiO_2 photocatalysis of microcystin-LR. *Appl. Catal. B Environ.* **2014**, *154–155*, 259–266. [CrossRef]
34. Antoniou, M.G.; de la Cruz, A.A.; Dionysiou, D.D. Degradation of microcystin-LR using sulfate radicals generated through photolysis, thermolysis and e- transfer mechanisms. *Appl. Catal. B Environ.* **2010**, *96*, 290–298. [CrossRef]
35. Likodimos, V.; Han, C.; Pelaez, M.; Kontos, A.G.; Liu, G.; Zhu, D.; Liao, S.; De La Cruz, A.A.; O'Shea, K.; Dunlop, P.S.M.; et al. Anion-doped TiO_2 nanocatalysts for water purification under visible light. *Ind. Eng. Chem. Res.* **2013**, *52*, 13957–13964. [CrossRef]
36. Antoniou, M.G.; Boraei, I.; Solakidou, M.; Deligiannakis, Y.; Abhishek, M.; Lawton, L.A.; Edwards, C. Enhancing photocatalytic degradation of the cyanotoxin microcystin-LR with the addition of sulfate-radical generating oxidants. *J. Hazard. Mater.* **2018**, *360*, 461–470. [CrossRef] [PubMed]
37. Zhao, C.; Li, D.; Liu, Y.; Feng, C.; Zhang, Z.; Sugiura, N.; Yang, Y. Photocatalytic removal of microcystin-LR by advanced WO3-based nanoparticles under simulated solar light. *Sci. World J.* **2015**, *2015*. [CrossRef] [PubMed]
38. Brillas, E.; Serrà, A.; Garcia-Segura, S. Biomimicry designs for photoelectrochemical systems: Strategies to improve light delivery efficiency. *Curr. Opin. Electrochem.* **2021**, *26*, 100660. [CrossRef]
39. Serrà, A.; Gómez, E.; Philippe, L. Bioinspired ZnO-based solar photocatalysts for the efficient decontamination of persistent organic pollutants and hexavalent chromium in wastewater. *Catalysts* **2019**, *9*, 974. [CrossRef]
40. Zhang, G.; Zhang, Y.C.; Nadagouda, M.; Han, C.; O'Shea, K.; El-Sheikh, S.M.; Ismail, A.A.; Dionysiou, D.D. Visible light-sensitized S, N and C co-doped polymorphic TiO_2 for photocatalytic destruction of microcystin-LR. *Appl. Catal. B Environ.* **2014**, *144*, 614–621. [CrossRef]
41. Serrà, A.; Gómez, E.; Michler, J.; Philippe, L. Facile cost-effective fabrication of Cu@Cu_2O@CuO-microalgae photocatalyst with enhanced visible light degradation of tetracycline. *Chem. Eng. J.* **2021**, 127477. [CrossRef]
42. Säbel, C.E.; Neureuther, J.M.; Siemann, S. A spectrophotometric method for the determination of zinc, copper, and cobalt ions in metalloproteins using Zincon. *Anal. Biochem.* **2010**, *397*, 218–226. [CrossRef]
43. Serrà, A.; Artal, R.; García-Amorós, J.; Sepúlveda, B.; Gómez, E.; Nogués, J.; Philippe, L. Hybrid Ni@ZnO@ZnS-Microalgae for Circular Economy: A Smart Route to the Efficient Integration of Solar Photocatalytic Water Decontamination and Bioethanol Production. *Adv. Sci.* **2020**, *7*, 1–9. [CrossRef]

toxins

MDPI

Article

Kinetics of Microcystin-LR Removal in a Real Lake Water by UV/H_2O_2 Treatment and Analysis of Specific Energy Consumption

Sabrina Sorlini [1,*], Carlo Collivignarelli [1], Marco Carnevale Miino [2], Francesca Maria Caccamo [2] and Maria Cristina Collivignarelli [2,3]

1 Department of Civil, Environmental, Architectural Engineering and Mathematics, University of Brescia, 25123 Brescia, Italy; carlo.collivignarelli@unibs.it
2 Department of Civil Engineering and Architecture, University of Pavia, 27100 Pavia, Italy; marco.carnevalemiino01@universitadipavia.it (M.C.M.); francescamaria.caccamo01@universitadipavia.it (F.M.C.); mcristina.collivignarelli@unipv.it (M.C.C.)
3 Interdepartmental Centre for Water Research, University of Pavia, 27100 Pavia, Italy
* Correspondence: sabrina.sorlini@unibs.it

Received: 24 November 2020; Accepted: 18 December 2020; Published: 21 December 2020

Abstract: The hepatotoxin microcystin-LR (MC-LR) represents one of the most toxic cyanotoxins for human health. Considering its harmful effect, the World Health Organization recommended a limit in drinking water (DW) of 1 μg L^{-1}. Due to the ineffectiveness of conventional treatments present in DW treatment plants against MC-LR, advanced oxidation processes (AOPs) are gaining interest due to the high redox potential of the $OH^{\bullet}$ radicals. In this work UV/H_2O_2 was applied to a real lake water to remove MC-LR. The kinetics of the UV/H_2O_2 were compared with those of UV and H_2O_2 showing the following result: UV/H_2O_2 > UV > H_2O_2. Within the range of H_2O_2 tested (0–0.9 mM), the results showed that H_2O_2 concentration and the removal kinetics followed an increasing quadratic relation. By increasing the initial concentration of H_2O_2, the consumption of oxidant also increased but, in terms of MC-LR degraded for H_2O_2 dosed, the removal efficiency decreased. As the initial MC-LR initial concentration increased, the removal kinetics increased up to a limit concentration (80 μg L^{-1}) in which the presence of high amounts of the toxin slowed down the process. Operating with UV fluence lower than 950 mJ cm^{-2}, UV alone minimized the specific energy consumption required. UV/H_2O_2 (0.3 mM) and UV/H_2O_2 (0.9 mM) were the most advantageous combination when operating with UV fluence of 950–1400 mJ cm^{-2} and higher than 1400 mJ cm^{-2}, respectively.

Keywords: cyanobacteria; cyanotoxins; drinking water; AOPs; hydrogen peroxide; algal bloom; microcystin-LR

Key Contribution: UV/H_2O_2 showed higher kinetics in free MC-LR removal and allowed it to minimize the specific energy consumption operating with UV fluence higher than 950 mJ cm^{-2}.

1. Introduction

Microcystin-LR (MC-LR) is a hepatotoxin produced by cyanobacteria such as *Microcystis aeruginosa*, *Planktothrix*, *Nostoc* and *Anabaea* and represents one of the most common and most toxic cyanotoxins for human health [1–3]. Cyanobacteria growth is enhanced in the presence of particular conditions such as mild temperature of water (25–35 °C), low flow rates, high concentration of nitrogen and phosphorous [4,5]. Therefore, lakes in areas with a temperate and warm climate represent a perfect habitat for their growth.

Cyanotoxins can interact and alter different parts of human metabolism with consequent effects on health of varying severity. For example, all cyanobacteria genera can produced cyanotoxins belonging to the group of Lipopolysaccharides, which have only a potential irritating effect on the tissues they have come into contact with [6]. On the other hand, the microcystins and nodularins, belonging to the group of cyclic peptides, have the liver as their main target of action being able to cross cell membranes mainly through the bile acid transporter [2,6]. Several studies highlighted the effects on liver tissue in humans exposed chronically to MC-LR [7–9].

The effect of microcystins was also studied by Zhou et al. [10]. They identified that the incidence rate of colorectal cancer was significantly higher in the population who drank water with high concentration of microcystins (e.g., river water) than those who drank tap water [10]. This harmful effect on intestinal cells was also confirmed by subsequent studies [11]. Alosman et al. [12] also pointed out that, aside the liver, MC-LR can cause also cardiogenic complications even if standardized animal models would be needed before the cardiotoxicity of the toxin can be defined with certainty.

The ingestion/inhalation of contaminated water in recreation (e.g., watersports) and, above all, the consumption of contaminated drinking water (DW) represent the main routes of exposure of humans to the toxin [2,6]. The effects due to secondary exposure, such as those due to the presence of MC-LR in plants and vegetables irrigated with water rich in toxins, are also being studied and quantified [13,14].

Considering the harmful effect of MC-LR revealed in literature results, the International Agency for Research on Cancer (IARC) classified this cyanotoxin as possible carcinogenic to humans (Group 2B) [6]. Based on this classification, the World Health Organization (WHO) included the MC-LR within the parameters to be monitored in DW, recommending a temporary limit of 1 $\mu g\ L^{-1}$ for total MC-LR (free plus cell-bound) [2]. The European Union implemented this recommendation by including the MC-LR in the revision of the Drinking Water Directive in 2018, providing for a limit of 1 $\mu g\ L^{-1}$ [15]. At the current date (15 December 2020) the proposed revision of the directive has not yet been approved so there is currently no unitary legislation at the European Union level about the presence of this toxin in DW.

However, several EU countries where the presence of cyanotoxins within surface water bodies is more widespread have already legislated on the matter providing national limits. For example, in 2012, Italy introduced a limit of 1 $\mu g\ L^{-1}$ as equivalent MC-LR referring to the sum of the concentrations of the different microcystins congeners present in DW [16]. In France, in 2001 a decree set the limit of 1 $\mu g\ L^{-1}$ for MC-LR in DW [17]. Instead, in 2007, this limit was referred to the sum of the microcystins quantified in the sample [17,18].

Even some non-European countries promoted laws or guidelines in order to minimize the risks for human health related to cyanobacteria and cyanotoxins in DW. For example, Canada established legislative limits for DW, with a seasonal maximum acceptable concentration of 1.5 $\mu g\ L^{-1}$ for total microcystins [19], and provided a draft of guidelines for recreational water quality a maximum acceptable concentration of 10 $\mu g\ L^{-1}$ for total microcystins [20]. Australia has provided non-mandatory guidelines suggesting that the concentration of total microcystins in DW should not exceed 1.3 $\mu g\ L^{-1}$ expressed as microcystin-LR toxicity equivalents [21].

There are two possible approaches to address the problem of the presence of MC-LR in waters: (i) remove the cyanobacteria that produce them or (ii) directly eliminate the free toxin [1]. The conventional treatments present in a drinking water treatment plant (DWTP; e.g., coagulation, flocculation, sedimentation and filtration) allow one to implement the first of the two approaches [22–25]. However, these treatments are not able to remove the MC-LR already secreted by cyanobacteria and present in the water in dissolved form [26]. On the contrary, adsorption on AC is confirmed to be a viable solution for the removal of low molecular weight substances such as cyanotoxins in general and specifically MC-LR [27–29].

Recently, even advanced oxidation processes (AOPs) are gaining interest in the removal of MC-LR due to the high redox potential of the hydroxyl radicals ($OH^{\bullet}$) or sulfate radicals ($SO_4^{\bullet -}$) developed in the process, which allows one to overcome some limitations given by the limited oxidizing power of other oxidizing agents towards MC-LR, such as H_2O_2 [30,31].

UV/H_2O_2 represents one of the AOPs. The main advantages are given by the absence of chemical sludge production (as opposed for instance to the Fenton process [32]), by the absence of toxic DBPs formation (unlike processes that involve chlorine and ozone [33–36]), and by the great ease of finding the oxidants used (as opposed to processes involving the use of nanostructured metals [37]).

In literature, several examples of application of this process for the removal of MC-LR are reported. UV lamps are used that emit at 254 nm of wavelength [38], close to the wavelength of maximum absorption of the MC-LR (235–238 nm [38,39]) or at 268 nm of wavelength [40]. However, most of the experiments involved the use of synthetic waters, thus only partially evaluating the combined effect of the presence of scavenger substances in the process such as the carbonates. On the contrary, this paper aims to evaluate the effectiveness of UV/H_2O_2 on a real lake water studying kinetics of free MC-LR removal to understand the influence of UV fluence, H_2O_2 dosage and initial MC-LR concentration on the process effectiveness. Moreover, the total specific energy consumption of UV/H_2O_2 for MC-LR removal has been evaluated and compared with UV alone to find the optimal operational conditions.

2. Results and Discussion

2.1. *Effect of the Oxidant*

The effectiveness of the H_2O_2, UV and UV/H_2O_2 processes for MC-LR removal was investigated. Figure 1 shows the degradation of MC-LR as a function of the UV fluence using different oxidants.

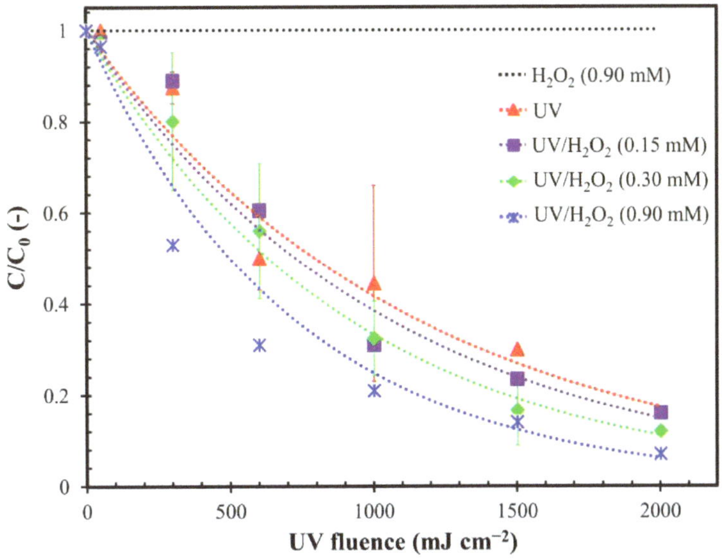

Figure 1. Degradation of MC-LR as a function of UV fluence in H_2O_2 alone, UV and UV/H_2O_2 processes. The colored curves represent exponential decay curve fitting. Error bars represent the confidence interval (n = 3). In case of the H_2O_2 alone treatment, the samples were taken at the time interval corresponding to the same UV fluence of the other tests. Conditions: MC-LR_0 = 50 μg L^{-1}; pH = 7.5 and fluence rate = 0.2 mW cm^{-2}.

H_2O_2 alone did not allow us to remove the toxin even with high contact time. This result is confirmed by Liu et al. [40] who observed an almost absent removal of dissolved MC-LR (0.1 μM) using H_2O_2 (0.1 mM) at pH nearly 7. On the contrary, the photolysis treatment with UV-C was found to be weakly effective in removing MC-LR (about 50%), with UV fluence equal to or lower than 1000 mJ cm^{-2}. The toxin removal enhanced to 80% using the maximum UV fluence tested (2000 mJ cm^{-2}).

The UV/H_2O_2 combination ensured removal yields higher than 90% with UV fluence equal to 2000 mJ cm^{-2} and H_2O_2 concentration of 0.9 mM. This result can be attributed to the production of $OH^{\bullet}$ radicals capable of almost completely oxidizing the MC-LR due to their high redox potential [41]. He et al. [38] confirmed the higher effectiveness of UV/H_2O_2 with respect with UV alone, obtaining more than 90% of MC-LR removal after 80 mJ cm^{-2} of fluence dose (H_2O_2: 1.76 mM) compared to around 20% obtained with UV alone. On the contrary to He et al. study [38], in the present work, a lower MC-LR removal was obtained using the same fluence dose, probably due to the higher initial pH that may have favored the scavenging effect on $OH^{\bullet}$ production [42].

2.2. *Influence of H_2O_2 Dosage*

The influence H_2O_2 dosage on the kinetics of MC-LR removal was studied. As shown in Figure A1 in the Appendix A, all results were well fitted by applying the first-order kinetics model to calculate the MC-LR removal kinetic constants (Figure 2a and Table A1 in Appendix A). As already shown in Figure 1, the H_2O_2 dosage generally allowed a better removal yield of the MC-LR. However, this result appeared to be dependent on the concentration of H_2O_2 dosed in the reaction. In fact, the UV/H_2O_2 combination ensured an efficacy in removing the toxin directly proportional to the quantity of chemical oxidant dosed. The half-life time (HLT) of the MC-LR was reduced from 64.2 (UV alone) to 57.8 min (−10%), 52.5 min (−18%) and 41.3 min (−36%) in the case of UV/H_2O_2 (0.15 mM), UV/H_2O_2 (0.30 mM) and UV/H_2O_2 (0.90 mM), respectively.

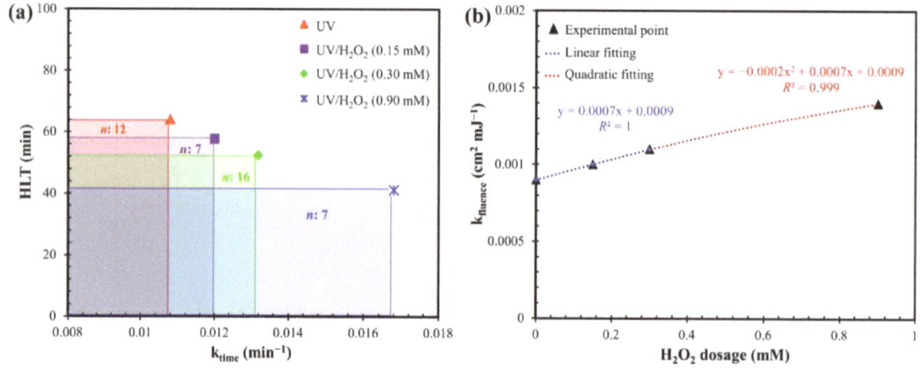

Figure 2. (**a**) First-order kinetic constant (k_{time}) and half-life time (HLT) during degradation by UV and UV/H_2O_2. (**b**) First-order kinetic constant ($k_{fluence}$) as a function of the H_2O_2 dosage. Conditions: $MC\text{-}LR_0$ = 50 µg L^{-1}; pH = 7.5; fluence rate = 0.2 mW cm^{-2} and *n*: number of data.

This relation was even more evident by comparing the apparent constant rate of the process (expressed as $k_{fluence}$) as a function of the H_2O_2 dosage. In Figure 2b it can be observed that, for low dosages of H_2O_2 (≤0.3 mM), the increase of removal kinetics can be perfectly linearly fitted (R^2 = 1). The increase in the kinetics of MC-LR removal was attributable to the increase in $OH^{\bullet}$ production due to the initial higher concentration of H_2O_2 as already seen for anatoxin-a removal [43]. This result is in agreement with He et al. [38] who studied MC-LR removal from synthetic DW using UV/H_2O_2. They highlighted that, with initial H_2O_2 concentrations below 1 mM, the MC-LR degradation rate constant seemed to increase proportionally with the chemical oxidant concentration following a linear relation [38].

However, considering also a higher H_2O_2 dosages (0.9 mM), the best fitting has been obtained with a quadratic function (R^2 > 0.99). In fact, when the H_2O_2 concentration reached high level (1 mM for [38] or more than 3 mM for [42]), the production of $OH^{\bullet}$ could be inhibited due to scavenging effect and the removal of MC-LR could remain almost constant or decrease [38,42].

Compared to results obtained by He et al. [38] and Loaiza-González [44], in the present work, lower removal kinetics were obtained using the same concentration of oxidizing agent. This could be related with the presence in the real lake water of higher concentrations of carbonates (Table A3 in Appendix A), which have a high scavenger effect, unlike chlorides and sulphates [38].

Furthermore, the amount of oxidant consumed in the UV/H_2O_2 process and the H_2O_2 efficiency in removing MC-LR were evaluated. By increasing the initial concentration of H_2O_2, the consumption of oxidant also increased (Figure 3a). This can explain the kinetics detailed in Figure 2b: for constant UV fluence, higher H_2O_2 consumption means higher $OH^{\bullet}$ production and therefore higher MC-LR removal. However, as the H_2O_2 consumed increased, the removal efficiency of the MC-LR decreased, in terms of MC-LR degraded for H_2O_2 dosed (Figure 3b). This result was also observed by m [45] in the application of the UV/H_2O_2 for the removal of organic matter and it was attributed to the scavenging effect of hydroxyl radicals that can limit the oxidative power of the process.

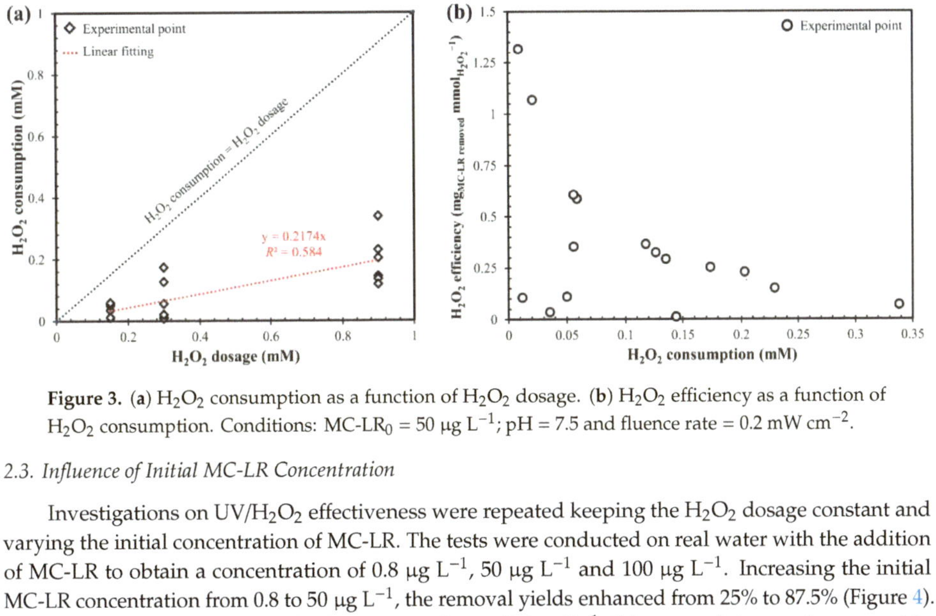

Figure 3. (**a**) H_2O_2 consumption as a function of H_2O_2 dosage. (**b**) H_2O_2 efficiency as a function of H_2O_2 consumption. Conditions: $MC\text{-}LR_0 = 50\ \mu g\ L^{-1}$; pH = 7.5 and fluence rate = 0.2 mW cm^{-2}.

2.3. *Influence of Initial MC-LR Concentration*

Investigations on UV/H_2O_2 effectiveness were repeated keeping the H_2O_2 dosage constant and varying the initial concentration of MC-LR. The tests were conducted on real water with the addition of MC-LR to obtain a concentration of 0.8 μg L^{-1}, 50 μg L^{-1} and 100 μg L^{-1}. Increasing the initial MC-LR concentration from 0.8 to 50 μg L^{-1}, the removal yields enhanced from 25% to 87.5% (Figure 4). Further increasing the initial MC-LR concentration to 100 μg L^{-1} did not result in an enhancement in toxin removal yields.

As shown in Figure A2 in Appendix A, all results were well fitted by applying the first-order kinetics model to calculate the MC-LR removal kinetic constants (Table A2 in Appendix A). By increasing the initial concentration of toxin from 0.8 to 50 μg L^{-1}, HLT decreased by about 82% (from 288.8 to 52.5 min). The further increase in the initial MC-LR concentration to 100 μg L^{-1} did not lead to a change in the removal kinetics.

Comparing the apparent rate constant of the process as a function of the initial toxin concentration (Figure 5), the experimental points were well fitted by a second degree polynomial function that predicted an increase of MC-LR degradation kinetics when the initial concentration moved from 0 to 80 μg L^{-1}. On the contrary, as the initial MC-LR increased after 80 μg L^{-1}, a lowering of the kinetics of removal was detected. This result was confirmed in the literature by several studies where the lower kinetics, obtained with high MC-LR concentration, are linked to the increase of the internal optical density, which decrease the fraction of light absorbed by H_2O_2 limiting $OH^{\bullet}$ production [38,40,42].

On the contrary, with initial MC-LR concentration lower than 80 μg L^{-1}, the obscuration of UV rays can be considered absent.

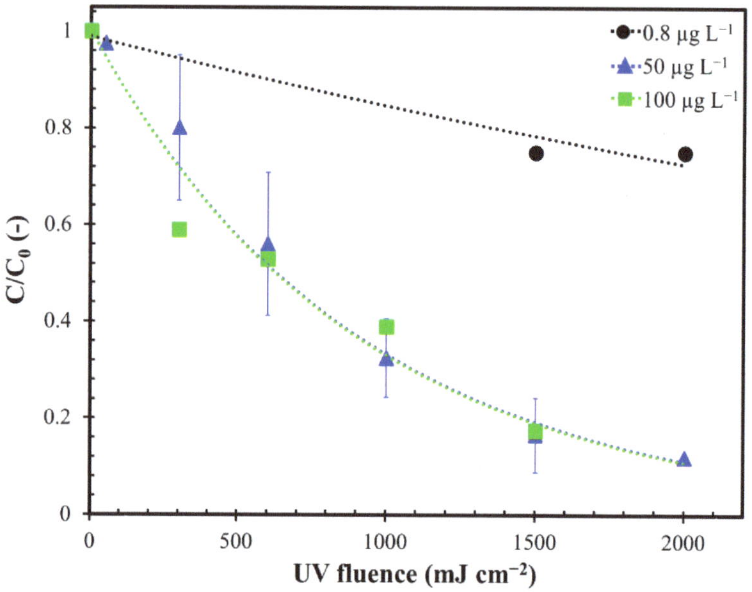

Figure 4. Degradation of MC-LR as a function of UV fluence in UV/H_2O_2 processes with different initial toxin concentration. The colored curves represent exponential decay curve fitting. Error bars represent the confidence interval (n = 3). Conditions: H_2O_2 dosage = 0.3 mM; pH = 7.5 and fluence rate = 0.2 mW cm^{-2}.

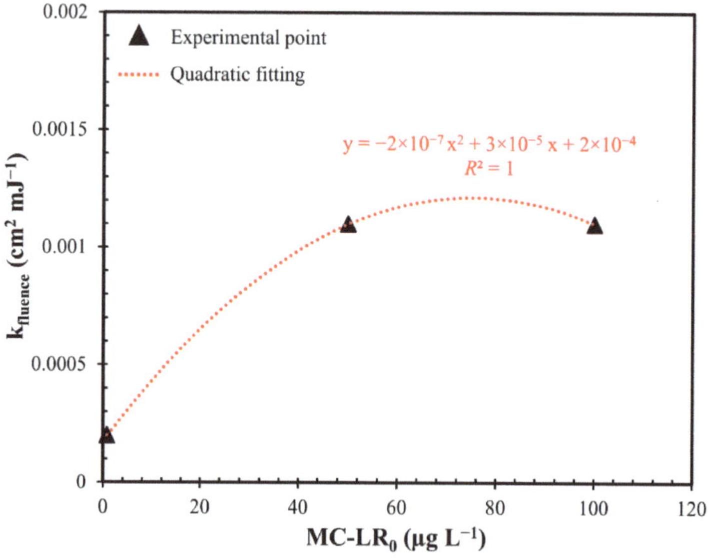

Figure 5. First-order kinetic constant ($k_{fluence}$) as a function of the initial MC-LR concentration ($MC\text{-}LR_0$). Conditions: H_2O_2 dosage = 0.3 mM; pH = 7.5 and fluence rate = 0.2 mW cm^{-2}.

2.4. Energy Consumption

The specific energy consumption required to remove MC-LR by one order of magnitude (E_{EO}) was assessed considering both the consumption given by the presence of UV ($E_{EO,\,UV}$) and the use of H_2O_2 ($E_{EO,\,oxidant}$). As shown in Figure A3 in Appendix A, electrical energy per order (E_{EO}) was reported as a function of H_2O_2 dosage with tested UV fluence.

The results show that the total specific energy consumption ($E_{EO,\,total}$) followed two different behaviors depending on the UV fluence considered and the H_2O_2 dosed (Figure A4 in the Appendix A). As the dosage of H_2O_2 increased, $E_{EO,\,total}$ decreased with high UV fluences (1000 mJ cm^{-2}, 1500 mJ cm^{-2} and 2000 mJ cm^{-2}) while, in the presence of lower UV fluences (50 mJ cm^{-2}, 300 mJ cm^{-2} and 600 mJ cm^{-2}), $E_{EO,\,total}$ increased even significantly as the concentration of the chemical oxidant dosed increased (Figure A4 in Appendix A).

In fact, significant H_2O_2 dosages in the presence of low UV radiation did not produce a significant increase in the effectiveness in removing MC-LR. On the contrary, by keeping the H_2O_2 dosage constant and increasing the fluence dose, the removal of MC-LR was more effective due to a greater production of $OH^{\bullet}$ radicals. Therefore, the specific energy consumption in relation to the MC-LR removed was lower.

By analyzing the behavior of $E_{EO,\,total}$ as a function of UV fluence (Figure 6), the optimal dosage of H_2O_2 that minimized the total specific energy consumption related with MC-LR removed was identified. Operating at low fluence rates (lower than 950 mJ cm^{-2}), UV alone allowed it to minimize the $E_{EO,\,total}$. Operating between 950 and 1400 mJ cm^{-2}, UV/H_2O_2 (0.3 µg L^{-1}) combination minimized specific consumption while, for UV fluence higher than 1400 mJ cm^{-2}, the combination UV/H_2O_2 (0.9 µg L^{-1}) was the most advantageous due to higher MC-LR removal.

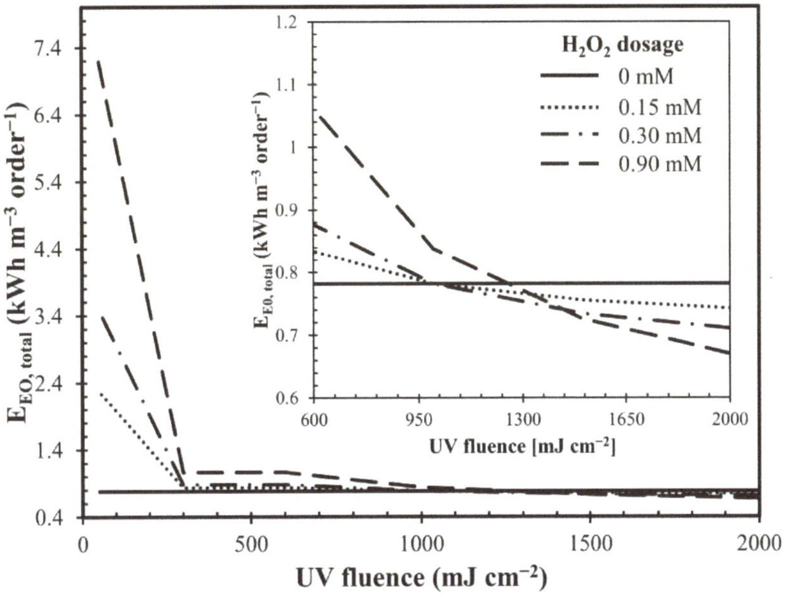

Figure 6. Total electrical energy per order (E_{EO}), by different H_2O_2 dosage, as a function of UV fluence. Conditions: MC-LR$_0$ = 50 µg L^{-1}; pH = 7.5 and fluence rate = 0.2 mW cm^{-2}.

However, E_{EO} values strongly depend on the removal yields of the toxin and therefore on the production of $OH^{\bullet}$ radicals. In addition to the concentration of oxidants used, the production of hydroxyl radicals also depends on many other aspects including the hydrodynamics of the reactor and its configuration [46]. Therefore, a direct comparison with other research is difficult to make. However, although there are not many literature data on E_{EO} related to the removal of MC-LR and most of those reported do not evaluate the energetic consumption related to the use of oxidants, the values obtained are of the same order of magnitude as those reported by Schneider and Bláha (2020) [47].

3. Conclusions

The kinetics of the UV/H_2O_2 process was compared with those of UV and H_2O_2 showing the following result: UV/H_2O_2 > UV > H_2O_2. The UV/H_2O_2 combination allowed the removal of up to 93% of MC-LR (MC-LR_0: 50 μg L^{-1}; H_2O_2: 0.9 mM; UV fluence: 2000 mJ cm^{-2} and fluence rate: 0.2 mW cm^{-2}). Within the range of H_2O_2 concentrations tested (0–0.9 mM), the results showed that H_2O_2 concentration and the removal kinetics followed a quadratic relation. By increasing the initial concentration of H_2O_2, the consumption of oxidant also increased but, in terms of MC-LR degraded for H_2O_2 dosed, the removal efficiency decreased. The initial concentration of MC-LR can significantly influence the kinetics of removal. The results showed that as the MC-LR_0 increased, the removal kinetics increased, up to a limit concentration (80 μg L^{-1}) in which the presence of high amounts of the toxin slowed down the process. About the specific energy consumption, UV alone minimized the specific energy consumption required when operating with UV fluence lower than 950 mJ cm^{-2}. Operating between 950 and 1400 mJ cm^{-2}, UV/H_2O_2 (0.3 mM) was the most advantageous combination while for UV fluence higher than 1400 mJ cm^{-2}, the use of UV/H_2O_2 (0.9 mM) was the solution that involved lower energy consumption in relation to the quantity of MC-LR removed.

4. Materials and Methods

4.1. Water Preparation

In this study, powdered MC-LR (type ALX–350–012–C500; purity ≥ 95%; Enzo Life Sciences Farmingdale, NY, USA) was stored at −20 °C and used to prepare a 50 mg L^{-1} solution by adding 10 mL ethanol (≥99.8%) to 0.5 mg powdered MC-LR.

In order to better simulate conditions of treatment of a real DW, raw water was collected from Iseo Lake, in Peschiera Maraglio of Monteisola, in Northern Italy (province of Brescia, Lombardy) at 40 m depth and 40 m from the shore. Characteristics of raw water are reported in Table A2 in Appendix A. To separate dissolved MC-LR from cells, lake water samples were filtered using a 0.45 μm (pore size) glass fiber filter [48,49] and the permeate (MC-LR = 0.1 μg L^{-1}) was spiked with the MC-LR solution to obtain toxin concentrations of 0.8 μg L^{-1}, 50 μg L^{-1} and 100 μg L^{-1}. Spiked waters were stored at 5 °C.

4.2. The Lab-Scale System

The batch system used for experimental tests was composed as described in Figure 7. A low-pressure mercury UV lamp, which emits at 254 nm of wavelength, was used. A black lampshade avoided the dispersion of the light beams to the sides and allowed it to concentrate the radiation on the reactor. The intensity of the irradiation given by the UV-C rays incident on the reactor was 0.2 mW cm^{-2}. A 50 mL Petri dish (diameter 5.45 cm, height 3.525 cm and thickness 0.25 cm), without lid, was used as a reactor. Inside the reactor, the water (depth 2.60 cm) was kept in constant stirring due to a magnetic stirring apparatus.

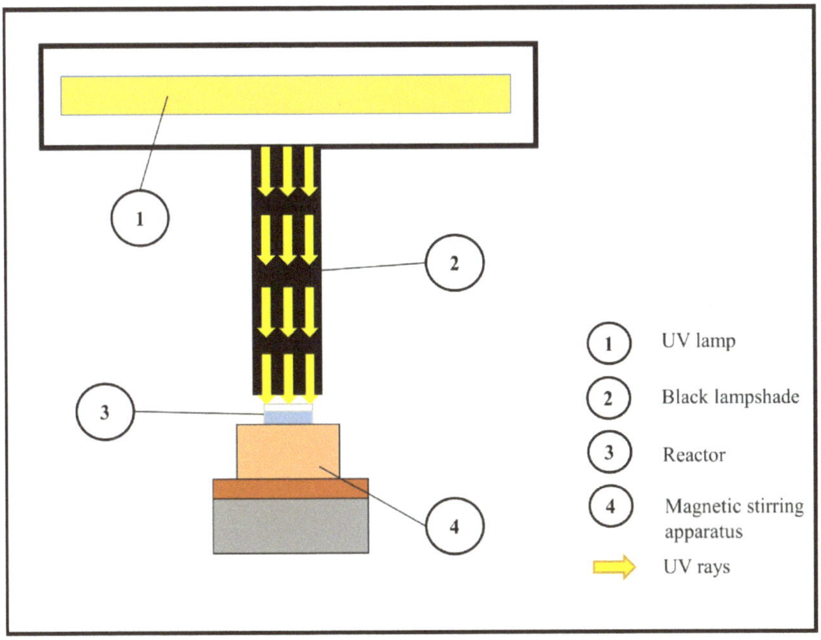

Figure 7. Scheme of the reactor used in test with UV, H_2O_2 and UV/H_2O_2.

4.3. Experimental Set-Up and Analytical Methods

This study was carried out testing the following processes:

- H_2O_2 alone;
- UV alone;
- UV/H_2O_2 combination.

Tests were conducted at room temperature (22 ± 2 °C) and aimed to study the kinetics of MC-LR removal and investigate the effects of H_2O_2 dosage and initial MC-LR concentration.

Hydrogen peroxide (30%, *w*/*v*) was purchased from Carlo Erba Reagents S.r.l (Cornaredo, Lombardy, Italy) and, during tests, the residual concentration was measured using the triiodide method [50].

Before each experiment, to ensure a stable radiation, the UV lamps were allowed to warm up for 15 min. Fluence rate was measured with iodide/iodate actinometry method [51] and was equal to 0.2 mW cm^{-2}.

pH value was monitored by means a portable multiparameter instrument (WTW 3410 SET4, Xylem Analytics Germany Sales GmbH, Weilheim, Germany).

After each fluence interval, samples were collected and catalase from *Micrococcus lysodeikticus* solution (Sigma Aldrich, St. Louis, MO, USA) was used to quench H_2O_2 reaction in samples before analysis according to Liu et al. [52]. The residual MC-LR concentration was measured with enzyme-linked immunosorbent assay (ELISA) kit, purchased from Eurofins Abraxis (Warminster, PA, USA). LOD and LOQ were equal to 0.1 µg L^{-1} and 5.0 µg L^{-1}, respectively. The treated samples were diluted with Milli-Q water in order to obtain measurable values.

4.4. MC-LR Degradation

The results were elaborated according to a pseudo-first order kinetic as reported in Equation (1) [38]:

$$C = C_0 \times e^{-k_{fluence} \times F} \quad (1)$$

where C_0 represents the initial concentration of MC-LR and C is the current i-th value. $k_{fluence}$ represents the apparent rate constant of the process ($cm^2\ mJ^{-1}$) and F is the UV fluence ($mJ\ cm^{-2}$). k_{time} (min^{-1}) was calculated considering the fluence rate of the system (0.2 $mW\ cm^{-2}$), and consequently half-life time (HLT) of MC-LR during treatments was found using the following equation [14]:

$$HLT = \ln(2) \times {k_{time}}^{-1} \quad (2)$$

4.5. Hydrogen Peroxide Consumption

Considering the amount of H_2O_2 consumed and the MC-LR removed, the H_2O_2 efficiency ($mg\ mmol^{-1}$) was calculated according to Equation (3) [45]:

$$H_2O_2\ efficiency = MC\text{-}LR_{removed} \times {H_2O_{2consumed}}^{-1} \quad (3)$$

4.6. Energy Consumption

When the concentration of the contaminant is very low, the amount of electric energy required to reduce the contaminant concentration by one order of magnitude can be considered independent of the initial concentration [46,53]. The water depth and the distance between the lamp and the water surface could affect the order of magnitude of the removal [46]. Although the lamp was not submerged into the reactor, in this work these two effects were neglected considering: (i) the low level of water inside the reactor and (ii) the presence of the black lampshade that avoided the dispersion of UV rays conveying them onto the reactor. Therefore, the energy consumption of the UV system was evaluated following the kinetic model of the electrical energy per order (E_{EO}) according to Equation (4) [46,54,55]:

$$E_{EO,UV} = (P \times t \times 10^3) \times (V \times \log_{10}(C_0\ C^{-1}))^{-1} \quad (4)$$

where P is the nominal power (kW) of the system, t (h) is the processing time and V (L) is the volume of water. The nominal power (P) was assumed equal to the energy input to the system, considering a fluence rate of 0.2 $mW\ cm^{-2}$ and assuming a UV-C production yield of the lamp equal to 35%.

In view of the application on a larger scale, it is important to know not only the energy consumption necessary to produce UV-C but also the energy consumption due to the dosage of the oxidizing reagent (H_2O_2). In this work also, the chemical energy consumption associated to H_2O_2 was evaluated according to Equation (5) [56]:

$$E_{EO,oxidant} = (C_{H2O2} \times CF) \times (\log_{10}(C_0\ C^{-1}))^{-1} \quad (5)$$

where C_{H2O2} is the concentration of H_2O_2 ($g\ m^{-3}$) and CF is a conversion factor equal to 6.67×10^{-3} $kWh\ g^{-1}$ [56–58]. Therefore, total energy consumption can be calculated as reported in Equation (6):

$$E_{EO,total} = E_{EO,UV} + E_{EO,oxidant} \quad (6)$$

Author Contributions: Conceptualization, S.S., M.C.M. and M.C.C.; Data curation, S.S., C.C. and M.C.M.; Investigation, S.S. and C.C.; Methodology, S.S., C.C. and M.C.C.; Resources, S.S. and C.C.; Supervision, S.S., C.C. and M.C.C.; Validation, S.S., C.C. and M.C.C.; Visualization, F.M.C.; Writing—original draft, S.S., M.C.M., F.M.C. and M.C.C.; Writing—review and editing, S.S., M.C.M. and M.C.C. All authors have read and agreed to the published version of the manuscript.

Funding: This research received no external funding.

Acknowledgments: The authors wish to thank Trojan Technologies for providing the lab-scale system used in the experimentation and Acque Ovest Bresciano 2 S.r.l. (Acque Bresciane S.r.l.) for having made their laboratories available for the analysis of MC-LR concentration. The Author thank also Eng. Francesca Gialdini and Eng. Michela Biasibetti for their technical support to the research.

Conflicts of Interest: The authors declare no conflict of interest.

Abbreviations

AOPs	Advanced oxidation processes
DW	Drinking water
DWTP	Drinking water treatment plant
HLT	Half-life time
IARC	International Agency for Research on Cancer
LOD	Limit of detection
LOQ	Limit of quantification
MC-LR	Microcystin-LR
UV	Ultraviolet
WHO	World Health Organization

Appendix A

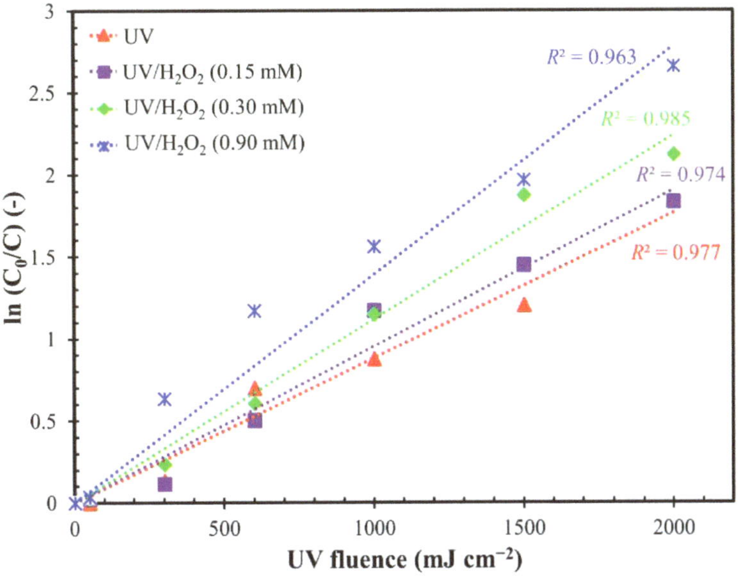

Figure A1. Normalized MC-LR concentration decay following a first-order kinetics model in UV alone and UV/H_2O_2 processes. The colored lines represented curve fitting. Conditions: $[MC\text{-}LR]_0 = 50\ \mu g\ L^{-1}$; pH = 7.5 and fluence rate = 0.2 mW cm^{-2}.

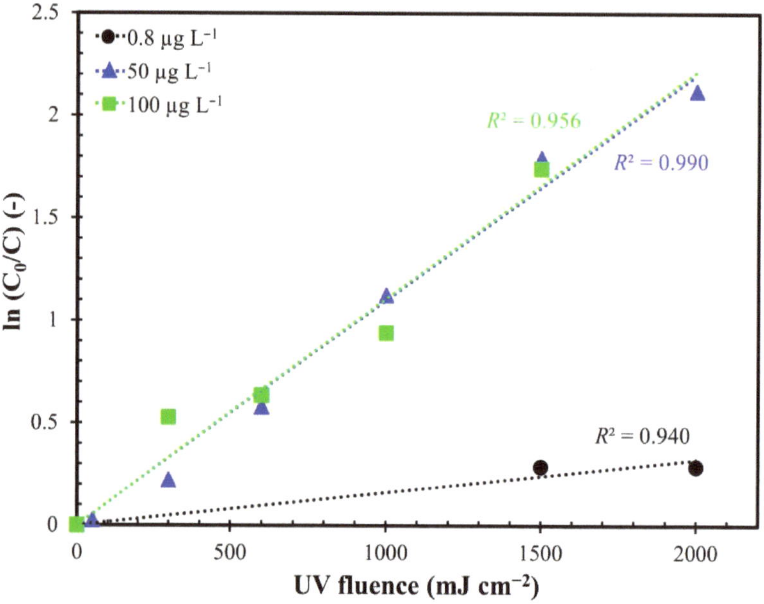

Figure A2. Normalized MC-LR concentration decay following a first-order kinetics model in UV/H_2O_2 processes with different initial MC-LR concentration. The colored lines represented curve fitting. Conditions: H_2O_2 dosage = 0.3 mM; pH = 7.5 and fluence rate = 0.2 mW cm^{-2}.

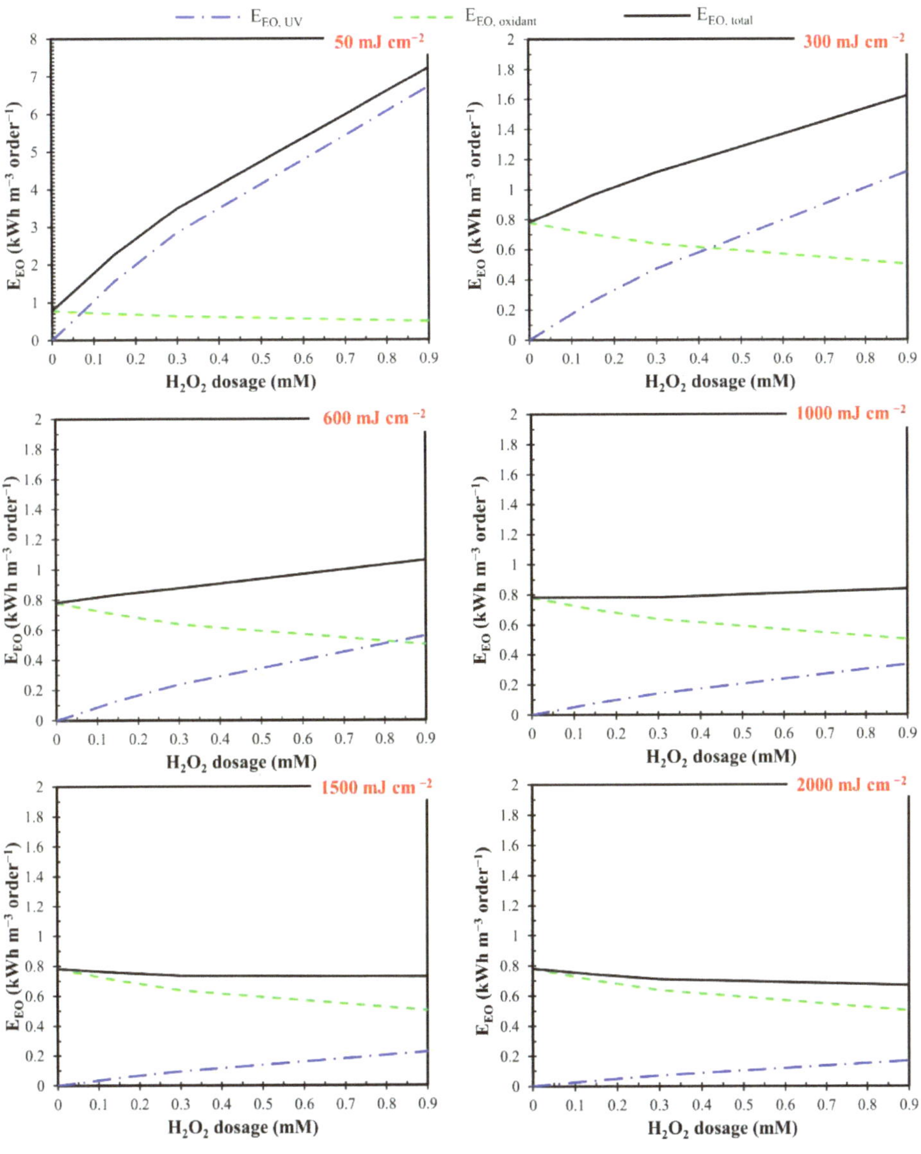

Figure A3. Electrical energy per order (E_{EO}) as function of H_2O_2 dosage with different UV fluence. Conditions: MC-LR$_0$ = 50 μg L^{-1}; pH = 7.5 and fluence rate = 0.2 mW cm^{-2}.

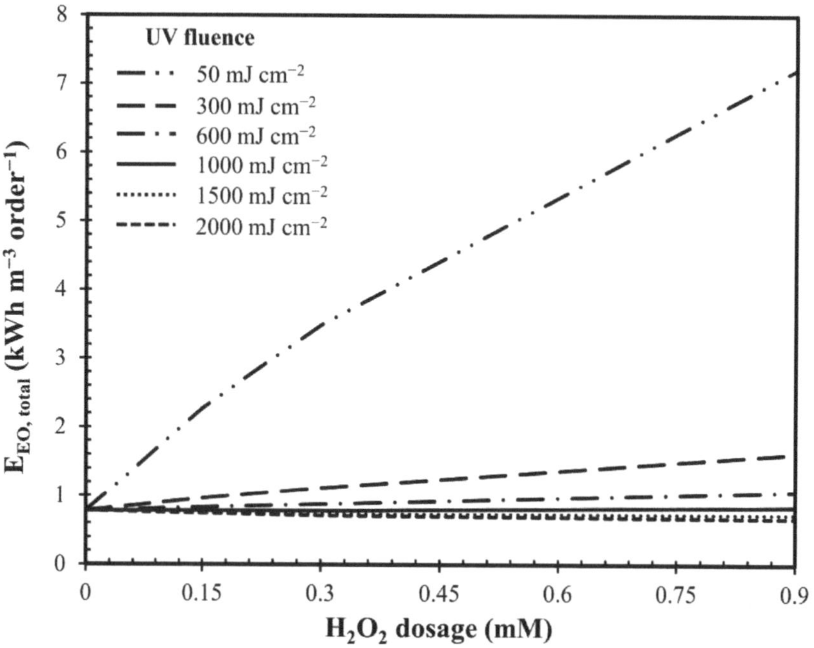

Figure A4. Total electrical energy per order (E_{EO}), by different UV fluences, as a function of H_2O_2 dosage.

Table A1. Kinetic constants and half-life times (HLT) of MC-LR during degradation by UV and UV/H_2O_2 using a first-order kinetics model. Conditions: $MC\text{-}LR_0$ = 50 μg L^{-1}; pH = 7.5; fluence rate = 0.2 mW cm^{-2} and N = number of total data.

	UV	UV/H_2O_2 (0.15 mM)	UV/H_2O_2 (0.30 mM)	UV/H_2O_2 (0.90 mM)
R^2 (-)	0.977	0.974	0.985	0.963
$k_{fluence}$ (cm^2 mJ)	0.0009	0.001	0.0011	0.0014
k_{time} (min^{-1})	0.0108	0.0120	0.0132	0.0168
HLT (min)	64.2	57.8	52.5	41.3
***n* (-)**	12	7	16	7

Table A2. Kinetic constants and half-life times (HLT) of MC-LR during degradation by UV/H_2O_2, with different initial MC-LR concentration, using a first-order kinetics model. Conditions: H_2O_2 dosage = 0.3 mM; pH = 7.5; fluence rate = 0.2 mW cm^{-2} and N = number of total data.

	UV/H_2O_2 (0.8 μg L^{-1})	UV/H_2O_2 (50 μg L^{-1})	UV/H_2O_2 (100 μg L^{-1})
R^2 (-)	0.940	0.990	0.956
$k_{fluence}$ (cm^2 mJ)	0.0002	0.0011	0.0011
k_{time} (min^{-1})	0.0024	0.0132	0.0132
HLT (min)	288.8	52.5	52.5
***n* (-)**	3	16	5

Table A3. Characteristics of the raw water before filtration.

Parameter	Unit of Measure	Average Value
pH	-	7.5
Dissolved oxygen	mg L^{-1}	9.5–9.7
Turbidity	NTU	2–3
Absorbance UV at 254 nm	1 cm^{-1}	0.010–0.020
Suspended solids	mg L^{-1}	0–1
Conductivity at 20 °C	µS cm^{-1}	260–270
Alkalinity	mg HCO_3^{-} L^{-1}	120–130
Bacteria colony count at 22 °C	CFU mL^{-1}	80–90
Total coliforms at 37 °C	MPN 100 mL^{-1}	65–75
Enterococcus	MPN 100 mL^{-1}	2–4
Escherichia coli	MPN 100 mL^{-1}	8–10
Pseudomonas aeruginosa	CFU 250 mL^{-1}	2–4
Clostridium perfringens	CFU 100 mL^{-1}	1–3
Cyanobacterial algae	cells L^{-1}	1,700,000–2,000,000
Total algae	cells L^{-1}	3,500,000–3,600,000

References

1. Sorlini, S.; Collivignarelli, M.C.; Carnevale Miino, M. Technologies for the control of emerging contaminants in drinking water treatment plants. *Environ. Eng. Manag. J.* **2019**, *18*, 2203–2216.
2. WHO. *WHO Guidelines for Drinking-Water Quality: Fourth Edition Incorporating the First Addendum*; World Health Organization: Geneve, Switzerland, 2017; pp. 1–631.
3. Leblanc, P.; Merkley, N.; Thomas, K.; Lewis, N.I.; Békri, K.; Renaud, S.L.; Pick, F.R.; McCarron, P.; Miles, C.O.; Quilliam, M.A. Isolation and Characterization of [D-Leu1]microcystin-LY from Microcystis aeruginosa CPCC-464. *Toxins* **2020**, *12*, 77. [CrossRef] [PubMed]
4. Sorlini, S.; Collivignarelli, M.C.; Abba, A. Control Measures for Cyanobacteria and Cyanotoxins in Drinking Water. *Environ. Eng. Manag. J.* **2018**, *17*, 2455–2463. [CrossRef]
5. Kim, S.; KimiD, S.; Mehrotra, R.; Sharma, A. Predicting cyanobacteria occurrence using climatological and environmental controls. *Water Res.* **2020**, *175*, 115639. [CrossRef] [PubMed]
6. IARC. *IARC Monographs on the Evaluation of Carcinogenic Risks to Humans|Ingested Nitrate and Nitrite, and Cyanobacterial Peptide Toxins*; International Agency for Research on Cancer: Lyon, France, 2010; Volume 94, pp. 1–464.
7. Greer, B.; Meneely, J.P.; Elliott, C.T. Uptake and accumulation of Microcystin-LR based on exposure through drinking water: An animal model assessing the human health risk. *Sci. Rep.* **2018**, *8*, 1–10. [CrossRef]
8. Yang, S.; Chen, L.; Wen, C.; Zhang, X.; Feng, X.; Yang, F. MicroRNA expression profiling involved in MC-LR-induced hepatotoxicity using high-throughput sequencing analysis. *J. Toxicol. Environ. Heal. Part A* **2018**, *81*, 89–97. [CrossRef]
9. Sun, Y.T.; Zheng, Q.; Huang, P.; Guo, Z.; Xu, L.H. Microcystin-lr induces protein phosphatase 2a alteration in a human liver cell line. *Environ. Toxicol.* **2013**, *29*, 1236–1244. [CrossRef]
10. Zhou, L.; Yu, H.; Chen, K. Relationship between microcystin in drinking water and colorectal cancer. *Biomed. Environ. Sci.* **2002**, *15*, 166–171.
11. Wen, C.; Zheng, S.; Yang, Y.; Li, X.; Chen, J.; Wang, X.; Feng, X.; Yang, F. Effects of microcystins-LR on genotoxic responses in human intestinal epithelial cells (NCM460). *J. Toxicol. Environ. Heal. Part A* **2019**, *82*, 1113–1119. [CrossRef]
12. Yang, F.; Cao, L.; Massey, I.Y.; Yang, F. The lethal effects and determinants of microcystin-LR on heart: A mini review. *Toxin Rev.* **2020**, 1–10. [CrossRef]
13. Xiang, L.; Li, Y.W.; Wang, Z.R.; Liu, B.L.; Zhao, H.M.; Li, H.; Cai, Q.Y.; Mo, C.H.; Li, Q. Bioaccumulation and Phytotoxicity and Human Health Risk from Microcystin-LR under Various Treatments: A Pot Study. *Toxins* **2020**, *12*, 523. [CrossRef] [PubMed]
14. Araújo, M.K.C.; Chia, M.A.; Arruda-Neto, J.D.D.T.; Tornisielo, V.L.; Vilca, F.Z.; Bittencourt-Oliveira, M.D.C. Microcystin-LR bioaccumulation and depuration kinetics in lettuce and arugula: Human health risk assessment. *Sci. Total Environ.* **2016**, *566*, 1379–1386. [CrossRef] [PubMed]

15. EC Proposal for a Directive of the European Parliament and of the Council on the Quality of Water Intended for Human Consumption (Recast). Available online: https://eur-lex.europa.eu/resource.html?uri=cellar:8c5065b2-074f-11e8-b8f5-01aa75ed71a1.0016.02/DOC_2&format=PDF (accessed on 1 December 2020).
16. IMH Interministerial Decree. Scheme for the introduction. In *Annex I, Part B, of the Legislative Decree 2 February 2001, n. 31, of the "Microcystin-LR" Parameter and its Parameter Value (in Italian)*; Italian Ministry of Health: Rome, Italy, 2012.
17. Arnich, N. FRANCE: Regulation, Risk Management, Risk Assessment and Research on Cyanobacteria and Cyanotoxins. In *Current Approaches to Cyanotoxin Risk Assessment, Risk Management and Regulations in Different Countries*; Chorus, I., Ed.; Federal Environment Agency (Umweltbundesamt): Dessau-Roßlau, Germany, 2012; pp. 63–70.
18. Government of France. *Order of 11 January 2007 Relating to the Quality Limits and References for Raw Water and Water Intended for Human Consumption Mentioned in Articles R. 1321-2, R. 1321-3, R. 1321-7 and R. 1321-38 of Public Health Code*; Government of France: Paris, France, 2007. (In French)
19. Government of Canada. *Guidelines for Canadian Drinking Water Quality; Water and Air Quality Bureau, Healthy Environments and Consumer Safety Branch, Health Canada*; Government of Canada: Ottawa, ON, Canada, 2020; pp. 1–28.
20. Government of Canada. *Guidelines for Canadian Recreational Water Quality—Cyanobacteria and their Toxins*; Government of Canada: Ottawa, ON, Canada, 2020.
21. Australian Government. *Australian Drinking Water Guidelines*; Australian Government: Canberra, Australia, 2011; pp. 1–1172.
22. Luo, Z.; Li, P.; Cai, D.; Chen, Q.; Qin, P.; Tan, T.; Cao, H. Comparison of performances of corn fiber plastic composites made from different parts of corn stalk. *Ind. Crop. Prod.* **2017**, *95*, 521–527. [CrossRef]
23. Czyżewska, W.; Piontek, M. The Efficiency of Microstrainers Filtration in the Process of Removing Phytoplankton with Special Consideration of Cyanobacteria. *Toxins* **2019**, *11*, 285. [CrossRef]
24. Lürling, M.; Kang, L.; Mucci, M.; Van Oosterhout, F.; Noyma, N.P.; Miranda, M.; Huszar, V.L.; Waajen, G.; Marinho, M.M. Coagulation and precipitation of cyanobacterial blooms. *Ecol. Eng.* **2020**, *158*, 106032. [CrossRef]
25. Lama, S.; Muylaert, K.; Karki, T.B.; Foubert, I.; Henderson, R.K.; Vandamme, D. Flocculation properties of several microalgae and a cyanobacterium species during ferric chloride, chitosan and alkaline flocculation. *Bioresour. Technol.* **2016**, *220*, 464–470. [CrossRef]
26. Jeong, B.; Oh, M.S.; Park, H.M.; Park, C.; Kim, E.J.; Hong, S.W. Elimination of microcystin-LR and residual Mn species using permanganate and powdered activated carbon: Oxidation products and pathways. *Water Res.* **2017**, *114*, 189–199. [CrossRef]
27. Drogui, P.; Daghrir, R.; Simard, M.-C.; Sauvageau, C.; Blais, J.F. Removal of microcystin-LR from spiked water using either activated carbon or anthracite as filter material. *Environ. Technol.* **2011**, *33*, 381–391. [CrossRef]
28. Mashile, P.P.; Mpupa, A.; Nomngongo, P.N. Adsorptive removal of microcystin-LR from surface and wastewater using tyre-based powdered activated carbon: Kinetics and isotherms. *Toxicon* **2018**, *145*, 25–31. [CrossRef]
29. Villars, K.; Huang, Y.; Lenhart, J.J. Removal of the Cyanotoxin Microcystin-LR from Drinking Water Using Granular Activated Carbon. *Environ. Eng. Sci.* **2020**, *37*, 585–595. [CrossRef]
30. Moon, B.R.; Kim, T.K.; Kim, M.K.; Choi, J.; Zoh, K.D. Degradation mechanisms of Microcystin-LR during UV-B photolysis and UV/H_2O_2 processes: Byproducts and pathways. *Chemosphere* **2017**, *185*, 1039–1047. [CrossRef] [PubMed]
31. Park, J.A.; Yang, B.; Jang, M.; Kim, J.H.; Kim, S.B.; Park, H.D.; Park, H.M.; Lee, S.H.; Choi, J.W. Oxidation and molecular properties of microcystin-LR, microcystin-RR and anatoxin-a using UV-light-emitting diodes at 255 nm in combination with H_2O_2. *Chem. Eng. J.* **2019**, *366*, 423–432. [CrossRef]
32. Park, J.A.; Yang, B.; Kim, J.H.; Choi, J.W.; Park, H.D.; Lee, S.H. Removal of microcystin-LR using UV-assisted advanced oxidation processes and optimization of photo-Fenton-like process for treating Nak-Dong River water, South Korea. *Chem. Eng. J.* **2018**, *348*, 125–134. [CrossRef]
33. Zhu, G.; Lu, X.; Yang, Z. Characteristics of UV-MicroO 3 Reactor and Its Application to Microcystins Degradation during Surface Water Treatment. *J. Chem.* **2015**, *2015*, 1–9.

34. Sorlini, S.; Biasibetti, M.; Collivignarelli, M.C.; Crotti, B.M. Reducing the chlorine dioxide demand in final disinfection of drinking water treatment plants using activated carbon. *Environ. Technol.* **2015**, *36*, 1499–1509. [CrossRef]
35. Sorlini, S.; Collivignarelli, M.C.; Canato, M. Effectiveness in chlorite removal by two activated carbons under different working conditions: A laboratory study. *J. Water Supply Res. Technol.* **2015**, *64*, 450–461. [CrossRef]
36. Sorlini, S.; Biasibetti, M.; Gialdini, F.; Collivignarelli, M.C. How can drinking water treatments influence chlorine dioxide consumption and by-product formation in final disinfection? *Water Sci. Technol. Water Supply* **2016**, *16*, 333–346. [CrossRef]
37. Li, W.Y.; Liu, Y.; Sun, X.L.; Wang, F.; Qian, L.; Xu, C.; Zhang, J.P. Photocatalytic degradation of MC-LR in water by the UV/TiO2/H2O2 process. *Water Supply* **2015**, *16*, 34–43. [CrossRef]
38. He, X.; Pelaez, M.; Westrick, J.A.; O'Shea, K.E.; Hiskia, A.; Triantis, T.; Kaloudis, T.; Stefan, M.I.; De La Cruz, A.A.; Dionysiou, D.D. Efficient removal of microcystin-LR by UV-C/H2O2 in synthetic and natural water samples. *Water Res.* **2012**, *46*, 1501–1510. [CrossRef]
39. Wang, X.; Utsumi, M.; Yang, Y.; Li, D.; Zhao, Y.; Zhang, Z.; Feng, C.; Sugiura, N.; Cheng, J.J. Degradation of microcystin-LR by highly efficient AgBr/Ag3PO4/TiO2 heterojunction photocatalyst under simulated solar light irradiation. *Appl. Surf. Sci.* **2015**, *325*, 1–12. [CrossRef]
40. Liu, J.; Ye, J.S.; Ou, H.; Lin, J. Effectiveness and intermediates of microcystin-LR degradation by UV/H2O2 via 265 nm ultraviolet light-emitting diodes. *Environ. Sci. Pollut. Res.* **2017**, *24*, 4676–4684. [CrossRef] [PubMed]
41. Collivignarelli, M.C.; Pedrazzani, R.; Sorlini, S.; Abbà, A.; Bertanza, G. H2O2 Based Oxidation Processes for the Treatment of Real High Strength Aqueous Wastes. *Sustainability* **2017**, *9*, 244. [CrossRef]
42. Li, L.; Gao, N.Y.; Deng, Y.; Yao, J.J.; Zhang, K.J.; Li, H.J.; Yin, D.D.; Ou, H.S.; Guo, J.W. Experimental and model comparisons of H2O2 assisted UV photodegradation of Microcystin-LR in simulated drinking water. *J. Zhejiang Univ. A* **2009**, *10*, 1660–1669. [CrossRef]
43. Vlad, S.; Anderson, W.B.; Peldszus, S.; Huck, P.M. Removal of the cyanotoxin anatoxin-a by drinking water treatment processes: A review. *J. Water Heal.* **2014**, *12*, 601–617. [CrossRef] [PubMed]
44. Loaiza-González, J.M.; Salazar, M.C.L.; Rubio-Clemente, A.; Rodriguez, D.C.; Peñuela, G.; Salazar, C.L.; Rodríguez, D.C.; Peñuela, G.A. Efficiency of the removal of microcystin-LR by UV-radiation and hydrogen peroxide. *Revista Facultad de Ingeniería Universidad de Antioquia* **2019**, 9–19. [CrossRef]
45. Penru, Y.; Guastalli, A.R.; Esplugas, S.; Baig, S. Application of UV and UV/H2O2 to seawater: Disinfection and natural organic matter removal. *J. Photochem. Photobiol. A Chem.* **2012**, *233*, 40–45. [CrossRef]
46. Keen, O.; Bolton, J.; Litter, M.; Bircher, K.; Oppenländer, T. Standard reporting of Electrical Energy per Order (EEO) for UV/H2O2 reactors (IUPAC Technical Report). *Pure Appl. Chem.* **2018**, *90*, 1487–1499. [CrossRef]
47. Schneider, M.; Bláha, L. Advanced oxidation processes for the removal of cyanobacterial toxins from drinking water. *Environ. Sci. Eur.* **2020**, *32*, 1–24. [CrossRef]
48. Grützmacher, G.; Böttcher, G.; Chorus, I.; Bartel, H. Removal of microcystins by slow sand filtration. *Environ. Toxicol.* **2002**, *17*, 386–394. [CrossRef]
49. Jeon, Y.; Li, L.; Calvillo, J.; Ryu, H.; Domingo, J.W.S.; Choi, O.; Brown, J.; Seo, Y. Impact of algal organic matter on the performance, cyanotoxin removal, and biofilms of biologically-active filtration systems. *Water Res.* **2020**, *184*, 116120. [CrossRef]
50. Klassen, N.V.; Marchington, D.; McGowan, H.C. H2O2 Determination by the I3- Method and by KMnO4 Titration. *Anal. Chem.* **1994**, *66*, 2921–2925. [CrossRef]
51. Rahn, R.O.; Bolton, J.; Stefan, M.I. The Iodide/Iodate Actinometer in UV Disinfection: Determination of the Fluence Rate Distribution in UV Reactors. *Photochem. Photobiol.* **2006**, *82*, 611–615. [CrossRef] [PubMed]
52. Liu, W.; Andrews, S.; Stefan, M.I.; Bolton, J.R. Optimal methods for quenching H2O2 residuals prior to UFC testing. *Water Res.* **2003**, *37*, 3697–3703. [CrossRef]
53. Collivignarelli, M.C.; Abbà, A.; Miino, M.C.; Arab, H.; Bestetti, M.; Franz, S. Decolorization and biodegradability of a real pharmaceutical wastewater treated by H_2O_2-assisted photoelectrocatalysis on TiO_2 meshes. *J. Hazard. Mater.* **2020**, *387*, 121668. [CrossRef] [PubMed]
54. Malpass, G.R.P.; Miwa, D.; Mortari, D.; Machado, S.; Motheo, A. Decolorisation of real textile waste using electrochemical techniques: Effect of the chloride concentration. *Water Res.* **2007**, *41*, 2969–2977. [CrossRef] [PubMed]

55. Farkas, J.; Náfrádi, M.; Hlogyik, T.; Pravda, B.C.; Gajda-Schrantz, K.; Hernádi, K.; Alapi, T. Comparison of advanced oxidation processes in the decomposition of diuron and monuron–efficiency, intermediates, electrical energy per order and the effect of various matrices. *Environ. Sci. Water Res. Technol.* **2018**, *4*, 1345–1360. [CrossRef]
56. Zhang, R.; Yang, Y.; Huang, C.-H.; Zhao, L.; Sun, P. Kinetics and modeling of sulfonamide antibiotic degradation in wastewater and human urine by UV/H_2O_2 and UV/PDS. *Water Res.* **2016**, *103*, 283–292. [CrossRef]
57. Yao, H.; Sun, P.; Minakata, D.; Crittenden, J.C.; Huang, C.-H. Kinetics and Modeling of Degradation of Ionophore Antibiotics by UV and UV/H_2O_2. *Environ. Sci. Technol.* **2013**, *47*, 4581–4589. [CrossRef]
58. Sun, P.; Tyree, C.; Huang, C.-H. Inactivation of Escherichia coli, Bacteriophage MS2, and Bacillus Spores under UV/H_2O_2 and UV/Peroxydisulfate Advanced Disinfection Conditions. *Environ. Sci. Technol.* **2016**, *50*, 4448–4458. [CrossRef]

toxins

MDPI

Article

The Efficacy of Hydrogen Peroxide in Mitigating Cyanobacterial Blooms and Altering Microbial Communities across Four Lakes in NY, USA

Mark W. Lusty and Christopher J. Gobler *

School of Marine and Atmospheric Sciences, Stony Brook University, Southampton, NY 11968, USA; mark.lusty@stonybrook.edu
* Correspondence: christopher.gobler@stonybrook.edu

Received: 3 June 2020; Accepted: 22 June 2020; Published: 29 June 2020

check for updates

Abstract: Hydrogen peroxide (H_2O_2) has been proposed as an agent to mitigate toxic cyanobacterial blooms due to the heightened sensitivity of cyanobacteria to reactive oxygen species relative to eukaryotic organisms. Here, experiments were conducted using water from four diverse, eutrophic lake ecosystems to study the effects of H_2O_2 on cyanobacteria and non-target members of the microbial community. H_2O_2 was administered at 4 μg L^{-1} and a combination of fluorometry, microscopy, flow cytometry, and high throughput DNA sequencing were used to quantify the effects on eukaryotic and prokaryotic plankton communities. The addition of H_2O_2 resulted in a significant reduction in cyanobacteria levels in nearly all experiments (10 of 11), reducing their relative abundance from, on average, 85% to 29% of the total phytoplankton community with *Planktothrix* being highly sensitive, *Microcystis* being moderately sensitive, and *Cylindrospermopsis* being most resistant. Concurrently, eukaryotic algal levels increased in 75% of experiments. The bacterial phyla *Actinobacteria*, cyanobacteria, *Planctomycetes*, and *Verrucomicrobia* were most negatively impacted by H_2O_2, with *Actinobacteria* being the most sensitive. The ability of H_2O_2 to reduce, but not fully eliminate, cyanobacteria from the eutrophic water bodies studied here suggests it may not be an ideal mitigation approach in high biomass ecosystems.

Keywords: cyanobacteria; hydrogen peroxide; 16S rRNA; harmful algal blooms

Key Contribution: Use of high throughput DNA sequencing of the 16s RNA gene to quantify the effects of H_2O_2 on cyanobacteria and non-target prokaryotic plankton.

1. Introduction

Cyanobacteria, or blue-green algae, are photosynthetic prokaryotes that are ubiquitous in fresh and marine waterbodies. Blooms of cyanobacteria in eutrophic waters can be associated with light attenuation and hypoxia, and some bloom-forming cyanobacteria are capable of producing a suite of toxins, most commonly the hepatotoxin, microcystin [1]. Consequently, the World Health Organization (WHO) and US EPA have set drinking water and bathing guidance values microcystin [2,3]. In addition to exposure through drinking water and bathing, cyanotoxins can be ingested through the consumption of fish and shellfish [4–6]. These toxins can also affect animals; between 2007 and 2011, there were 67 cases of dog poisonings due to toxic cyanobacteria blooms across the U.S., 38 of which were fatal [7].

The occurrence of harmful cyanobacterial blooms is often linked to excessive anthropogenic eutrophication [5,8,9] but reducing nutrient loads can be a difficult and lengthy process that can involve changing fertilizer and wastewater disposal practices. Hence, there is interest in identifying mitigation approaches that can selectively target and remove toxic cyanobacterial blooms in order to prevent exposure and harm. Hydrogen peroxide (H_2O_2) has been considered for this role [10–12]. As a strong oxidant, it is known for its disinfectant capabilities, is a naturally occurring compound in aquatic systems, and quickly decomposes into water and gaseous oxygen [13]. As H_2O_2 decomposes, it releases hydroxyl radicals, strong reactive oxygen species known to damage cells and inhibit photosynthetic activity by causing damage to photosystem II [14,15]. H_2O_2 has been shown to be specifically detrimental to the growth and function of cyanobacteria and capable of reducing biomass of *Microcystis* and *Planktothrix* by 50% in less than 48 h [10,11]. Cyanobacteria are known to be more sensitive to H_2O_2 than eukaryotic primary producers [13,14,16], and a previous study found *Microcystis aeruginosa* to be ten-times more sensitive than species of green algae and diatoms [13]. This may be due, in part, to the photosystems of cyanobacteria not being protected within an organelle [14]. In addition, unlike cyanobacteria, eukaryotic phytoplankton commonly produce enzymes such as ascorbate peroxidase that break down H_2O_2 and protect them from damage by reactive oxygen species (ROS) such as hydroxyl radicals [17]. A whole lake study examining mesozooplankton abundances, mostly *Daphnia* and *Diaphanosoma*, found that they were unaffected at 2 mg H_2O_2 L^{-1}, a concentration that inhibited the cyanobacterium *Planktothrix* [11].

The effects of hydrogen peroxide on cyanobacteria has been well-documented in laboratory cultures [13,18,19]. However, research assessing the effect of H_2O_2 on other important members of planktonic communities such as picocyanobacteria, eukaryotes, and heterotrophic bacteria has been limited. It is important that effects of H_2O_2 on cyanobacteria and the rest of the prokaryotic and eukaryotic community are understood before H_2O_2 is widely used for mitigation purposes in natural ecosystems.

This project, therefore, sought to understand the effects of H_2O_2 on multiple genera of toxin-producing cyanobacteria (i.e., *Microcystis*, *Dolichospermum*, *Cylindrospermopsis*, and *Planktothrix*) as well as co-occurring plankton including picocyanobacteria, eukaryotic algae, and heterotrophic bacteria. This was done through a series of incubation experiments performed using environmental samples from four contrasting water bodies across Long Island, NY, USA. Microbial communities were assessed using standard (microscopy, fluorometry) and molecular (high throughput amplicon sequencing) approaches to establish a comprehensive assessment of the efficacy of H_2O_2 as a mitigation approach for toxic cyanobacterial blooms.

2. Results

2.1. Fluorometric Response of the Phytoplankton Community

During three experiments utilizing water from Lake Agawam, initial cyanobacterial biomass ranged from 73 to 243 µg Chla L^{-1}, dominated by mixtures of *Microcystis*, *Planktothrix*, and *Dolichospermum*. Cyanobacterial biomass was significantly lower four to six days after exposure to 4 mg H_2O_2 L^{-1} in two of three experiments, with concentrations 52% ($p < 0.001$; Figure 1a) and 43% ($p < 0.005$; Figure 1c) lower than the control. Cyanobacterial biomass was reduced below the New York State Department of Environmental Conservation (NYSDEC) level of concern of 25 µg Chla L^{-1} in the one experiment that had the lowest initial concentration of these experiments (73 µg Chla L^{-1}; Figure 1a). Initial green algal biomass was low (0–0.24 µg Chla L^{-1}) but was significantly higher than the control following exposure to H_2O_2 in one of the experiments ($p < 0.001$; Figure 1a). Unicellular brown algae were fluorometrically undetectable during the Lake Agawam experiments.

Initial cyanobacterial biomass for the three Mill Pond experiments ranged from 45 to 366 µg Chla L^{-1}, dominated by *Microcystis* and *Cylindrospermopsis*. Four or seven days after exposure to H_2O_2, cyanobacterial biomass was significantly lower than the control in all three experiments by 99% ($p < 0.001$; Figure 1d), 93% ($p < 0.001$; Figure 1e), and 95% ($p < 0.001$; Figure 1f), respectively, and below the level of concern of 25 µg Chla L^{-1} in two experiments (Figure 1d,f). Initial biomass of green algae in Mill Pond ranged from below detection to 7.3 µg Chla L^{-1} and rose higher than the control to 76 ± 7 µg Chla L^{-1} ($p < 0.001$; Figure 1d), 30 ± 2 µg Chla L^{-1} ($p < 0.001$; Figure 1e), and 342 ± 18 µg Chla L^{-1} ($p < 0.001$;

Figure 1f), respectively, in all three experiments following H_2O_2 exposure. Initial unicellular brown algae biomass levels were below detection at the start of the three experiments, but values rose significantly to 5 ± 1 µg Chla L^{-1} ($p < 0.005$) and 24 ± 3 µg Chla L^{-1} ($p < 0.001$) in two of the three experiments, but remained undetectable in the third experiment (Figure 1d,f).

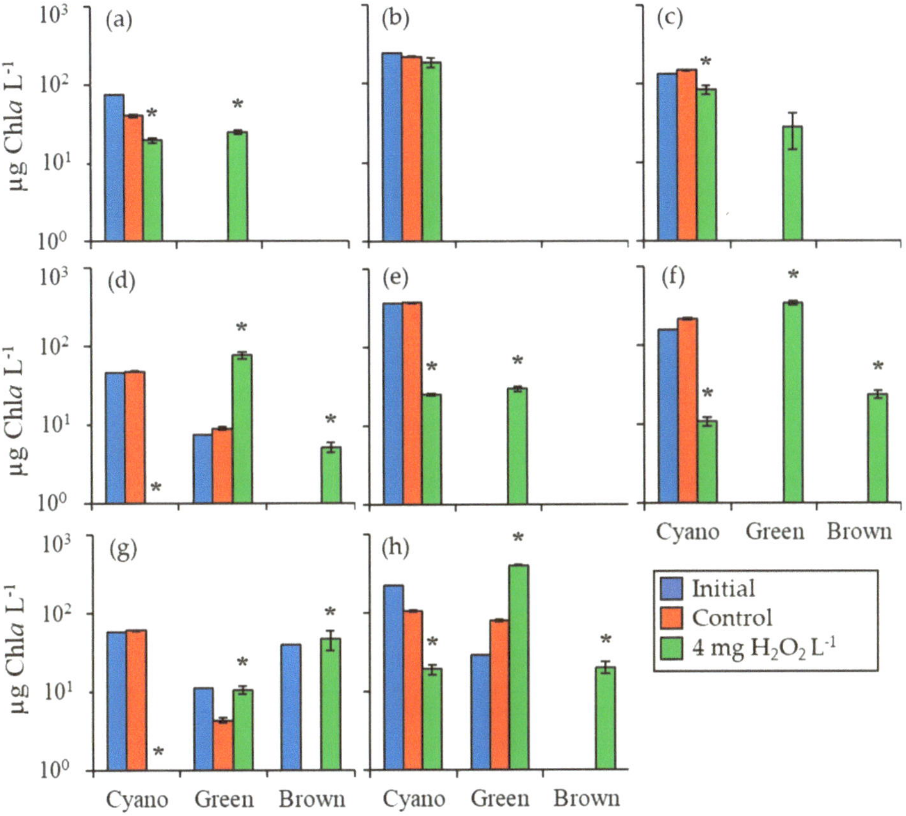

Figure 1. Fluoroprobe biomass measurements for bottle experiments from Lake Agawam (**a**) 7/21/16, (**b**) 10/20/16, (**c**) 6/9/17; Mill Pond (**d**) 7/21/16, (**e**) 10/20/16, (**f**) 6/30/17; (**g**) Georgica Pond 7/21/16; and (**h**) Roth Pond 6/9/17. Asterisks show significant changes ($p < 0.05$) in treatments relative to control. Error bars show standard error.

For the Georgica Pond experiment, initial cyanobacterial biomass was 58 µg Chla L^{-1} being dominated by *Aphanizomenon* (Figure 1g). Four days after treatment with 4 mg H_2O_2 L^{-1}, cyanobacterial biomass was 99.8% lower than the control and nearly 0 µg Chla L^{-1} ($p < 0.001$; Figure 1g). Initial green algal biomass was 11 µg Chla L^{-1} and was 144% higher than the control following exposure to H_2O_2 ($p < 0.01$; Figure 1g). Initial unicellular brown algae biomass was 39 µg Chla L^{-1} and rose to 47 ± 13 µg Chla L^{-1} following H_2O_2 exposure, significantly higher than the control where concentration had fallen ($p < 0.05$; Figure 1g).

For the Roth Pond experiment, initial cyanobacterial biomass was 222 µg Chla L^{-1} and was dominated by *Microcystis* and *Cylindrospermopsis* (Figure 1h). Cyanobacterial biomass was 82% lower than the control ($p < 0.001$) at 19 ± 3 µg Chla L^{-1} in the 4 mg H_2O_2 L^{-1} treatment after six days. Initial green algae biomass was 29 µg Chla L^{-1} but rose in the H_2O_2 treatment to 403 ± 10 µg Chla L^{-1}, 400% higher than the control ($p < 0.001$; Figure 1h). Unicellular brown algae were undetectable at the start of the experiment but were significantly higher than the control at 20 ± 3 μg Chla L^{-1} 6 days after H_2O_2 exposure ($p < 0.005$; Figure 1h).

In summary, during the eight incubation experiments among four locations assessed fluorometrically, cyanobacterial levels were significantly lowered following the addition of H_2O_2 than the control in seven experiments, green algae levels became significantly higher in H_2O_2 treatments relative to the control in six of eight experiments, and unicellular brown algae became significantly higher in H_2O_2 treatments relative to the control in four of eight experiments.

2.2. Detailed Assessment of Planktonic Responses to H_2O_2

Given the strong and significant effects of H_2O_2 on plankton communities during this first set of experiments, three additional experiments were performed utilizing water from three ecosystems with additional analyses performed to more fully assess the response of plankton communities to H_2O_2. In the first of these experiments from Lake Agawam, the initial cyanobacterial biomass was 103 μg Chl*a* L^{-1} and five days following exposure to 4 mg H_2O_2 L^{-1} was significantly lower than the control at 23 ± 2 μg Chl*a* L^{-1} ($p < 0.001$; Figure 2a). Initial green algae biomass concentration was undetectable but rose to 69 ± 4 μg Chl*a* L^{-1} after treatment with H_2O_2, a level significantly higher than the control ($p < 0.001$; Figure 2a). Unicellular brown algae biomass was undetectable at the start of the experiment but rose to levels significantly higher than the control at 0.4 ± 0.2 μg Chl*a* L^{-1} in the H_2O_2 treatment ($p < 0.05$; Figure 2a). Picocyanobacteria (*Cyanobium*) concentrations in Lake Agawam were initially 5340 ± 360 cells mL^{-1} and decreased in the control to 1690 ± 80 cells mL^{-1} (Figure 2b). The H_2O_2 treatment was reduced by less, and was 110% higher relative to the control with a final concentration of 3560 ± 90 cells mL^{-1} ($p < 0.001$; Figure 2b). Eukaryotic algae concentrations were 193% higher in the H_2O_2 treatment than the control at 9090 ± 150 cells mL^{-1} ($p < 0.001$; Figure 2b). The initial concentration of heterotrophic bacteria was 7.0×10^5 cells mL^{-1} and was 44% lower than the control after the addition of H_2O_2 at 3.45×10^5 cells mL^{-1} ($p < 0.001$; Figure 2b). Diatom densities in Lake Agawam were 34 cells mL^{-1} and levels rose to be significantly higher in the treatment ($p < 0.05$) relative to the control to 120 ± 28 cells mL^{-1} (Figure 2c). Green algae concentrations were initially 657 cells mL^{-1} and were nearly six-fold higher in the treatment compared to the control at 3700 ± 50 cells mL^{-1} ($p < 0.001$; Figure 2c). Initial *Microcystis* concentrations were 222 colonies mL^{-1} and were significantly reduced by H_2O_2 to below the control to 34 ± 7 colonies mL^{-1} ($p < 0.001$; Figure 2c). There were 76 *Dolichospermum* chains mL^{-1} at the start of the experiment and concentrations sharply declined to 4 ± 4 chains mL^{-1} following H_2O_2 addition, a level significantly lower than the control ($p < 0.005$; Figure 2c). Finally, following H_2O_2 addition *Planktothrix* concentrations were 50% of the control at 439 ± 43 chains mL^{-1} ($p < 0.01$; Figure 2c).

High throughput sequencing of the 16S rDNA gene indicated that the relative abundance of *Actinobacteria* in Lake Agawam was initially 17 ± 1% and dropped to 5 ± 1% five days after exposure to 4 mg H_2O_2 L^{-1}, significantly lower than the control ($p < 0.001$). *Planctomycetes* was 4 ± 1% initially and was significantly reduced to 2 ± 1%, significantly lower than the control ($p < 0.001$), while *Verrucomicrobia* was 4 ± 1% and was reduced to 0.4 ± 0.1%, significantly lower than the control ($p < 0.001$; Figure 2d). The sequenced relative abundance of *Bacteroidetes* in Lake Agawam was significantly higher than the control after the addition of H_2O_2 at 42 ± 4% compared to an initial of 32 ± 1% ($p < 0.001$; Figure 2d). The sequenced relative abundance of *Proteobacteria* and other less abundant taxa abundances were not significantly altered by H_2O_2 (Figure 2d).

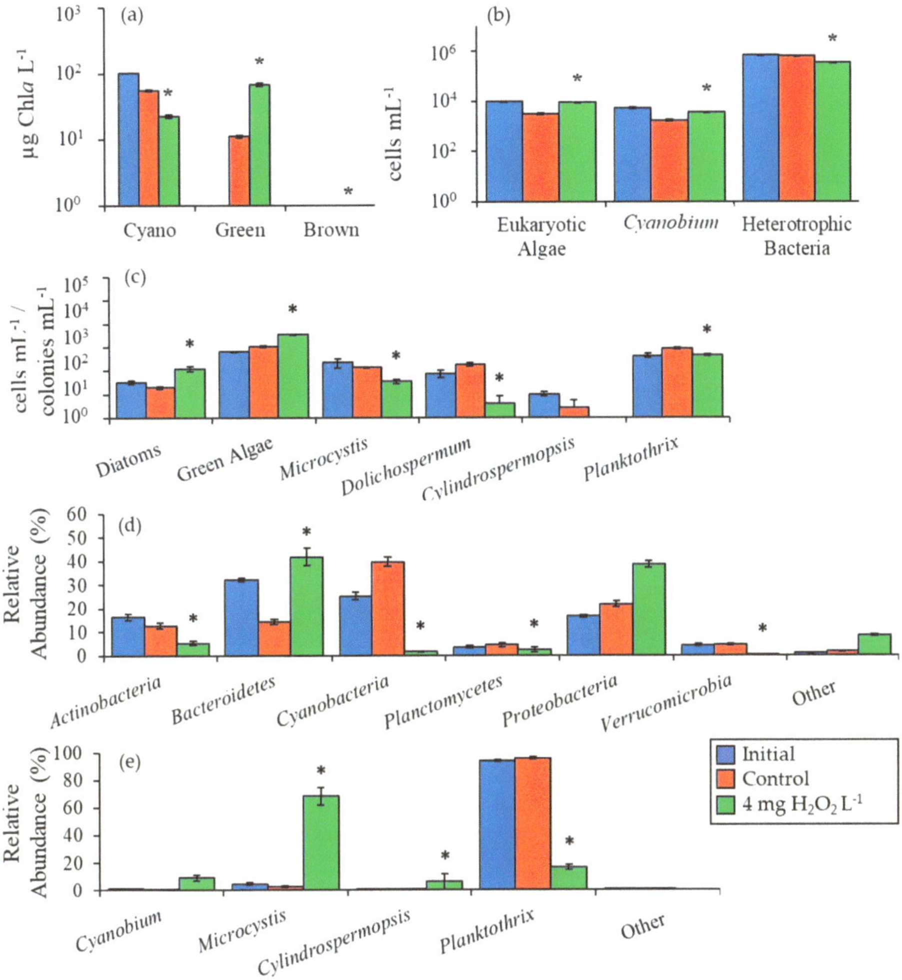

Figure 2. (**a**) Fluoroprobe biomass, (**b**) flow cytometry, (**c**) microscopy, (**d**) phylum level relative abundance, and (**e**) genus level cyanobacteria relative abundance for Lake Agawam experiment, 10/3/16. Asterisks show significant changes ($p < 0.05$) in treatments relative to control. Error bars show standard error.

Among all prokaryotes, the sequenced relative abundance of cyanobacteria was initially 25 ± 2% but was reduced to lower than the control to 2 ± 1% after H_2O_2 exposure ($p < 0.001$). Among the cyanobacteria, *Planktothrix* was the most abundant taxa in the Lake Agawam experiment with an initial relative abundance of 94 ± 1% that was significantly lower than the control at 17 ± 2% following the addition of H_2O_2 ($p < 0.001$; Figure 2e). In contrast, *Microcystis* had an initial abundance of 4 ± 1% that was higher relative to the control at 68 ± 6% in the H_2O_2 treatment ($p < 0.001$; Figure 2e).

Cylindrospermopsis made up only 0.2 ± 0.1% of initial cyanobacterial sequences but was higher than the control at 6 ± 6% following the H_2O_2 addition ($p < 0.001$; Figure 2e). Multiplying cyanobacterial Chl*a* values by sequenced relative abundances provided an estimate of individual biomasses and revealed that *Planktothrix* biomass decreased significantly ($p < 0.001$), while *Microcystis* ($p < 0.001$) and *Cylindrospermopsis* ($p < 0.001$) biomasses increased (Figure 3).

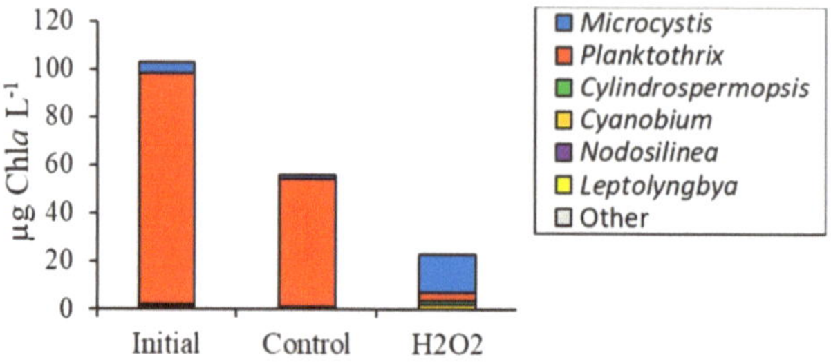

Figure 3. Absolute abundance (relative abundance multiplied by biomass) of cyanobacteria for Lake Agawam experiment, 10/3/16.

For the experiment that provided a detailed assessment from the Mill Pond planktonic community, initial cyanobacterial biomass was 394 μg Chl*a* L^{-1} and was 96% lower than the control after five days following exposure to H_2O_2 at 13 ± 3 μg Chl*a* L^{-1} ($p < 0.001$; Figure 4a). Green algae were below detectable levels in Mill Pond at the start of the experiment but were significantly higher at 60 ± 7 μg Chl*a* L^{-1} ($p < 0.001$) following H_2O_2 treatment; unicellular brown algae were fluorometrically undetectable in this experiment (Figure 4a). Initial eukaryotic algae concentrations were 1.14 ± 0.30 × 10^4 cells mL^{-1} and were unchanged by H_2O_2 but were 52% lower relative to the control ($p < 0.001$; Figure 4b).

Cyanobium and heterotrophic bacteria concentrations in the H_2O_2 treatment were not significantly different from the control (Figure 4b). Levels of diatoms were low in Mill Pond prior to the experiment (< 10 cells mL^{-1}) but increased and were significantly higher than the control after the H_2O_2 addition with a final concentration of 157 ± 35 cells mL^{-1} ($p < 0.01$; Figure 4c). Green algae cell densities were initially 1290 ± 140 cells mL^{-1} and were significantly higher in the H_2O_2 treatment compared to the control at 2.65 ± 0.37 × 10^4 cells mL^{-1} ($p < 0.01$; Figure 4c). Initial *Microcystis* concentrations were 108 ± 4 colonies mL^{-1} and were significantly lower than the control following exposure to H_2O_2 ($p < 0.05$; Figure 4c). *Cylindrospermopsis* initial concentration was 2.45 ± 0.14 × 10^5 chains mL^{-1} and significantly decreased by 100% following exposure to H_2O_2 ($p < 0.001$; Figure 4c) while *Dolichospermum* and *Planktothrix* levels were unaffected.

High throughput sequencing revealed that H_2O_2 caused a significant decline in the sequenced relative abundance of several bacterial groups in Mill Pond including *Actinobacteria*, *Planctomycetes*, and *Verrucomicrobia* ($p < 0.005$; Figure 4d). In contrast, the sequenced relative abundances of *Bacteroidetes* and *Proteobacteria* were higher in the H_2O_2 treatment relative to the control ($p < 0.001$; Figure 4d). Among cyanobacteria identified via sequencing of 16S rDNA, *Cylindrospermopsis* was the dominant operational taxonomic unit (OTU) (88 ± 1% of cyanobacteria sequences) in initial sequences and increased in relative abundance following H_2O_2 addition to 98 ± 1%, significantly higher than the control ($p < 0.001$). In contrast, the sequenced relative abundances of *Microcystis* and *Nodosilinea* were significantly lowered by H_2O_2 relative to the control ($p < 0.001$; Figure 4e). Estimated changes in absolute abundances based on fluorometry and sequencing revealed that, despite the differential sensitivities of differing cyanobacterial groups to H_2O_2, the biomass of *Cylindrospermopsis* ($p < 0.001$), *Microcystis* ($p < 0.01$), and *Nodosilinea* ($p < 0.001$) all significantly declined in the treatment relative to the initial levels and the control (Figure 5).

Finally, during the Roth Pond experiment, initial cyanobacteria biomass was 59 μg Chl*a* L^{-1} and was significantly lower in H_2O_2 treatments at 39 ± 3 μg Chl*a* L^{-1} relative to the control seven days after exposure ($p < 0.01$; Figure 6a). Initial green algae biomass was 141 μg Chl*a* L^{-1} but was not significantly altered by H_2O_2, while unicellular brown algal biomass levels were 100% higher in the H_2O_2 treatment compared to the control at 239 ± 12 μg Chl*a* L^{-1} ($p < 0.01$; Figure 6a). Flow cytometrically quantified *Cyanobium* concentrations in the Roth Pond were 99.8% lower in the H_2O_2 treatment compared to the control ($p < 0.001$) whereas levels of eukaryotic algae were unchanged (Figure 6b). Levels of heterotrophic bacteria were 33% higher than the controls seven days after the addition of H_2O_2 ($p < 0.005$; Figure 6b). Green algae identified microscopically were 111% higher in the H_2O_2 treatment compared to the control at $1.28 \pm 0.06 \times 10^5$ cells mL^{-1} ($p < 0.001$; Figure 6c), while diatom levels were unchanged (Figure 6c). *Microcystis* concentrations were 1320 ± 180 colonies mL^{-1} and were reduced by 100% by H_2O_2 ($p < 0.001$) while *Dolichospermum* concentrations were unchanged (Figure 6c). Final, total concentration of microcystin was 0.56 ± 0.13 µg L^{-1} in the control, and significantly higher at 0.85 ± 0.06 µg L^{-1} in the H_2O_2 treatment ($p < 0.005$).

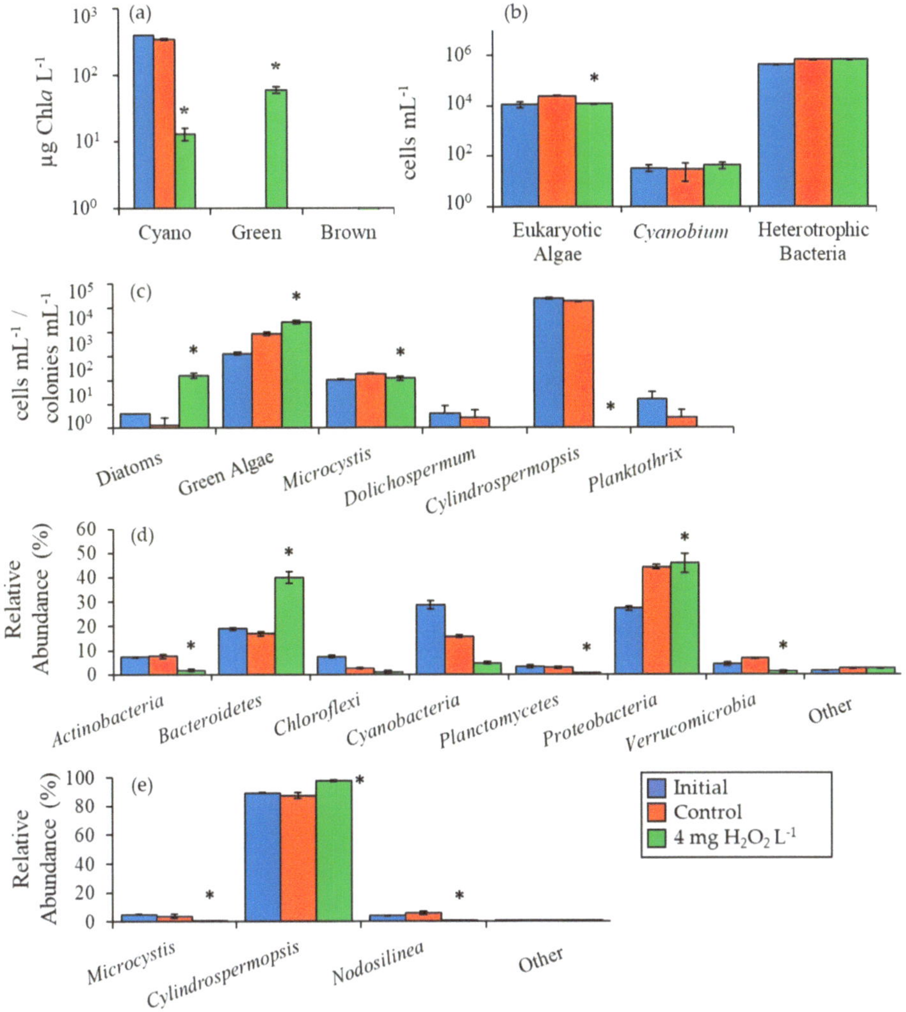

Figure 4. (**a**) Fluoroprobe biomass, (**b**) flow cytometry, (**c**) microscopy, (**d**) phylum level relative abundance, and (**e**) genus level cyanobacteria relative abundance for Mill Pond experiment, 10/3/16. Asterisks show significant changes ($p < 0.05$) in treatments relative to control. Error bars show standard error.

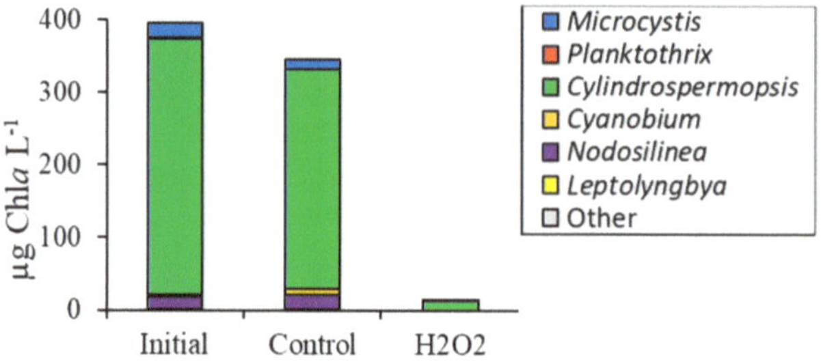

Figure 5. Absolute abundance (relative abundance multiplied by biomass) of cyanobacteria for Mill Pond experiment 10/3/16.

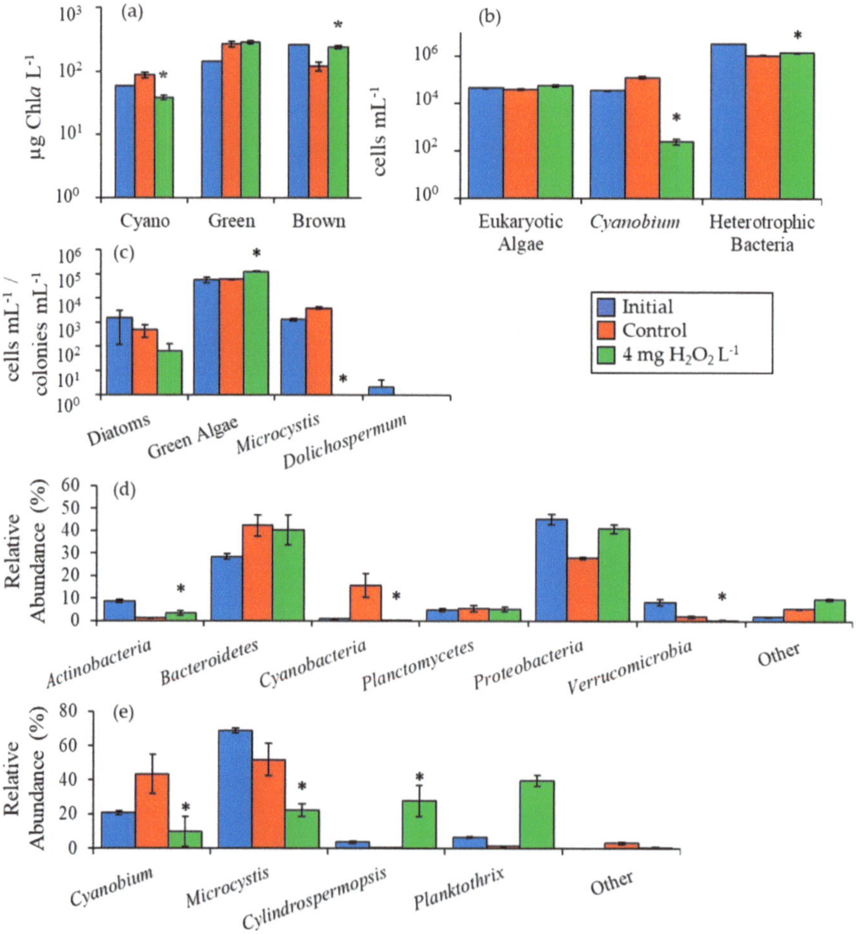

Figure 6. (**a**) Fluoroprobe biomass, (**b**) flow cytometry, (**c**) microscopy, (**d**) phylum level relative abundance, and (**e**) genus level cyanobacteria relative abundance for Roth Pond experiment, 6/30/17. Asterisks show significant changes ($p < 0.05$) in treatments relative to control. Error bars show standard error.

Sequencing of the 16S rDNA indicated *Actinobacteria* were significantly higher in the H_2O_2 treatment compared to the control ($p < 0.001$) whereas cyanobacteria and *Verrucomicrobia* relative abundances were significantly lower than the control ($p < 0.001$; Figure 6e). Among the cyanobacteria, *Microcystis* and *Cyanobium* were the two most abundant genera in Roth Pond and the relative abundance of both was lower in the H_2O_2 treatment compared to the control ($p < 0.001$; Figure 6e). In contrast, *Cylindrospermopsis* relative abundance was initially 4 ± 1% and increased to 28 ± 9% following in the H_2O_2 treatment and was significantly higher than the control ($p < 0.001$). Despite the lower relative and absolute abundances of *Microcystis* during this experiment, the concentration of the cyanotoxin, microcystin, was marginally higher in the H_2O_2 treatment than the control, at 0.9 ± 0.1 µg L^{-1} compared to 0.6 ± 0.1 µg L^{-1} ($p < 0.05$). The individual biomasses (relative abundance multiplied by cyanobacterial Chl*a*) for *Microcystis* ($p < 0.05$) and *Cyanobium* ($p < 0.005$) significantly declined in response to H_2O_2, while total biomass of *Planktothrix* ($p < 0.01$) and *Cylindrospermopsis* ($p < 0.1$) increased (Figure 7).

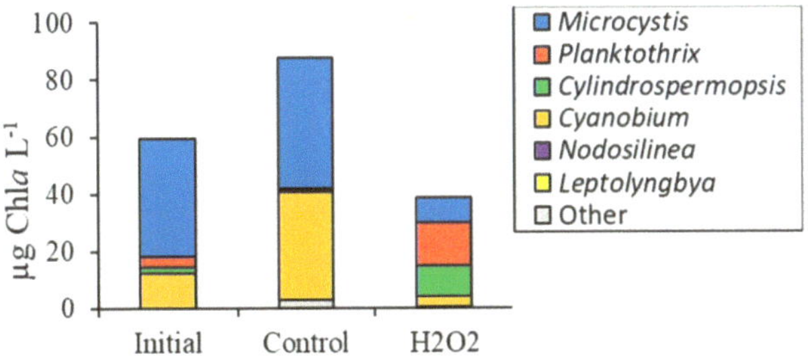

Figure 7. Absolute abundance (relative abundance multiplied by biomass) of cyanobacteria for Roth Pond experiment, 6/30/17.

3. Discussion

During this study, H_2O_2 was used to mitigate cyanobacterial populations in environmental samples from four water bodies. H_2O_2 almost always reduced the levels of cyanobacteria and increased the abundances of eukaryotic algae. Effects on non-photosynthetic prokaryotes were complex, as some bacteria were consistently inhibited by H_2O_2 while others were promoted. Collectively, these findings provide new insight into the complex manner in which H_2O_2 can alter the composition of microbial communities dominated by cyanobacteria.

3.1. Cyanobacteria vs. Eukaryotes

Cyanobacterial biomass and cell abundance were significantly reduced by H_2O_2 in a majority of experiments (91%, 10 of 11), and eukaryotic green and unicellular brown algae were significantly increased in 73% and 55% of experiments, and did not decline significantly in any experiment. The ability of H_2O_2 to significantly reduce concentrations of cyanobacteria relative to eukaryotic algae is consistent with previous observations [11,13,14]. Across all experiments, the fluorometric relative abundance of cyanobacteria was significantly reduced from 85 ± 8% to 29 ± 10% ($p < 0.005$). The trend was also reflected in microscopy, with significant cyanobacterial reductions and significant increases of eukaryotic algae in all experiments quantified. Differences in flow cytometry were more varied, with *Cyanobium* and eukaryotic algae counts each significantly lower than control in one experiment, and significantly higher in another. A lack of ascorbate peroxidases in cyanobacteria, which are common in eukaryotes, has been offered as one possible factor contributing to their greater sensitivity [20,21]. Additionally, photosystem II of cyanobacteria is not protected within a cell organelle, making it physically more susceptible to damage from H_2O_2 [22].

Despite the common reduction in cyanobacterial biomass in experiments, it was reduced to levels below detection limit in only two out of 11 experiments and reduced below 25 µg Chl*a* L^{-1}

(NYSDEC guidance value for cyanobacterial blooms) in one other. The inability to completely eliminate cyanobacteria may result in their resurgence over a short period of time [11]. While microcystin was only quantified once in this study, whole-water concentrations slightly increased (0.6 to 0.9 $\mu g\ L^{-1}$) in that experiment, a finding that slightly contrasts with prior studies [11] but is perhaps not completely surprising given the dynamics of cyanobacteria in that experiment. While total cyanobacterial biomass decreased in that experiment, it was one of the smallest declines in this study (33%). Further, while one microcystin producer, *Microcystis* [1], significantly declined in relative abundance during that experiment, another, *Planktothrix* [1], significantly increased and seemingly contributed to the slightly higher toxicity observed at the end of this experiment.

3.2. Comparing Cyanobacterial Genera

While cyanobacterial biomass and cell densities were reduced in a majority of experiments, the reductions varied between genera. Although cyanobacteria do not usually produce ascorbate peroxidases, they are capable of producing a suite of other anti-oxidant enzymes, including catalases, peroxidases, and peroxiredoxins [21]. While peroxiredoxins are ubiquitously present across cyanobacteria, other enzymes can vary between genera and strains, which may account for some of the variability in relative reductions observed [21].

Microcystis was the most ubiquitous of the cyanobacteria during this study and was significantly reduced in microscopic counts to levels less than control in all experiments, but was reduced below detection in only one. *Microcystis* was reduced to a significantly lower relative and absolute abundance compared to the unamended controls in two of the three experiments where sequencing was performed and was higher with H_2O_2 in the experiment where *Planktothrix* was dominant. In contrast to *Microcystis*, *Cyanobium* and *Cylindrospermopsis* were more resistant to H_2O_2. *Cyanobium* was the second most abundant cyanobacteria in Roth Pond, behind *Microcystis*, and cell concentrations were significantly lower relative to the control in one experiment, and significantly higher in another. The sequenced relative abundance of *Cyanobium* was significantly lower in one of the three experiments. A prior laboratory study found *Cyanobium* four-fold less sensitive to H_2O_2 than *Microcystis* [13]. *Cylindrospermopsis* concentrations in microscopy were detected in two experiments, and were significantly reduced in only one (Figure 4c). The sequenced relative abundance of *Cylindrospermopsis* in H_2O_2 treatments was higher than the control in all three experiments, including one where it was dominant over *Microcystis*, though it was still reduced in absolute abundance (Figures 4c and 5).

The heightened sensitivity of *Microcystis* to H_2O_2 relative to other genera may be related to its deficient antioxidant systems as some strains of *Microcystis* lack typical cyanobacterial catalases [21]. Importantly, however, while *Microcystis* densities were significantly reduced by initial treatment with H_2O_2, it often remained one of the most abundant cyanobacterial genera in experiments where it dominated and was never fully eliminated. *Microcystis* commonly forms large globular colonies with cells embedded in and surrounded by polysaccharide mucous [23]. The extracellular polymeric substances of this mucous have strong H_2O_2 scavenging abilities and provide an antioxidant buffer for the cells within [24]. This additional protection may partly explain the perseverance of *Microcystis* during experiments. In addition, it has been shown that some strains of *Microcystis* incapable of microcystin synthesis have high levels of thioredoxin and peroxiredoxin, enzymes involved in H_2O_2 degradation [19] making these strains more likely to survive repeated H_2O_2 doses than toxic strains. Hence, the persistence of *Microcystis* may be partly facilitated by shifts among differing strains.

The relative H_2O_2 resistance of *Cylindrospermopsis* among cyanobacteria may be due, in part, to their ability to produce superoxide dismutase, catalase, and ascorbate peroxidase [25]. The ability to produce ascorbate peroxidase in *Cylindrospermopsis* is somewhat unique, as it was believed to be lacking in other cyanobacteria, and therefore contributes to the reduced effectiveness of peroxide against this genus [17,20]. In addition, *Cylindrospermopsis* can produce single cell akinetes which it uses to survive unfavorable conditions [26]. Akinete production may also account for differences between gene detection and microscopic counts where *Cylindrospermopsis* reduced was below microscopic detection, as the visible trichomes may have fragmented into akinetes or small, morphologically unidentifiable fragments [27]. It is unlikely the 16S rDNA of *Cylindrospermopsis* persisted after the cells were destroyed by H_2O_2, as DNA often degrades within 24 h in freshwater [28,29].

Planktothrix dominated the Lake Agawam experiment that was examined in detail and displayed the largest relative decline in sequenced abundance of any microbe, dropping from 94% to 17% in the H_2O_2 treatment. In the Roth Pond experiment, *Planktothrix* increased slightly in relative and absolute abundance of sequences but was below detection levels in microscopic counts. H_2O_2 has been shown to be effective in controlling *Planktothrix* in lakes and ponds at similar concentrations to those used here [11,30,31]. This experiment where *Planktothrix* declined by nearly 80% in relative abundance was also the only instance when the sequenced relative abundance of *Microcystis* among the cyanobacteria increased following treatment with H_2O_2, despite its absolute decline, supporting the hypothesis that *Planktothrix* is more sensitive to H_2O_2 than *Microcystis* [30,32], and in assessing all available literature [11,31,32], may be the most sensitive of the cyanobacterial genera to H_2O_2.

The collective response of these experiments leads evidence to support an 'open niche' hypothesis [33]. with regard to the effects of H_2O_2 on cyanobacteria. While different cyanobacteria are likely differentially sensitive to H_2O_2, it also seems that the effects of H_2O_2 are somewhat conditional upon the original community composition. That is, in many cases the dominant cyanobacterial genus is most reduced by H_2O_2, perhaps by providing the most organic surface area for the H_2O_2 to react with, allowing genera at lower relative abundances to fill the niche left open by the formerly dominant genera.

3.3. Heterotrophic Bacteria

The net effect of H_2O_2 on bacteria during this study was inconclusive, as concentrations of heterotrophic bacteria were significantly lower than the control in one of three experiments, and significantly higher in another. In contrast, high throughput sequencing of 16S rDNA revealed that, beyond with changes in total bacterial densities, there were pronounced shifts within prokaryotic communities following treatment with H_2O_2. The prokaryote taxa identified here were categorized at the phylum level; some traits discussed below may not be indicative of all taxa within a phylum, especially for the very diverse *Proteobacteria,* but offer some insight to potential strategies employed.

Consistent with fluorometric and microscopic evaluations, the sequenced relative abundance of cyanobacteria made up an average 18 ± 9% of initial abundances and was significantly reduced in two of three experiments. The relative abundance of *Actinobacteria,* which made up an average 11 ± 3% of initial abundances, was likewise reduced in two of three experiments of experiments, demonstrating that strains of bacteria within this phylum are susceptible to H_2O_2. In lakes, *Actinobacteria* are small, thin walled, free-living ultramicrobia (< 0.1 μm^3) that are abundant in the epilimnion [34]. *Actinobacteria* are obligate aerobes [35], and densities decrease with decreasing oxygen levels and depth [34,36]. They are defense specialists that are relatively resistant to grazing, but have slow growth rates compared to other bacteria [34,37]. These slow growth rates may make them less likely to recover from initial populations declines induced by H_2O_2.

Planctomycetes and Verrucomicrobia, which have a close phylogenetic relationship [38], made up a smaller portion of the total bacterial community (< 5%) but also appeared susceptible to H_2O_2, significantly decreasing in sequenced relative abundance in two of three experiments, and in all experiments, respectively. Verrucomicrobia are usually found throughout the water column and are associated with high-nutrient environments and algal blooms [34]. Prior mesocosm experiments have shown Verrucomicrobia to strongly increase in response to Microcystis degradation [39]. Planctomycetes similarly are capable of breaking down high-molecular weight organic compounds [40,41]. Both phyla are also capable of producing bifunctional catalase-peroxidases [42]. Peroxide resistance, benefitting from organic carbon, and declining Microcystis abundances should have all theoretically promoted these groups. The absence of such a response in their sequenced relative abundance may have been a function of a larger increase or lesser decline in other bacterial groups or a stronger negative effect of H_2O_2 relative to any benefit derived from algal organic matter.

Proteobacteria increased significantly in sequenced relative abundances in one of three experiments. *Proteobacteria* were one of the most abundant prokaryotes, making up 30 ± 8% of initial abundances. *Proteobacteria* have a relatively short generation time, allowing them to respond rapidly to changing conditions [34] and are copiotrophic, able to assimilate small organic acids as well as degrade complex organic compounds [34,40,41]. *Proteobacteria* also contain catalase-peroxidases and manganese catalases that protect them from oxidative stress [42].

The average initial relative abundance of *Bacteroidetes* was 27 ± 4%, and values were significantly higher following H_2O_2 exposure in two of three experiments. *Bacteroidetes* are usually particle associated and chemoorganotrophic [34], and are capable of degrading complex, high-molecular weight organic compounds [34,40,41]. *Bacteroidetes* become abundant when DOC or algae-derived DOC are high [34], and *Bacteroidetes* abundances often increase during the degradation of *Microcystis* blooms [39]. Their ability to benefit from H_2O_2 in two experiments and to be unaffected in a third may relate to their exploitation of DOC released by lysing cyanobacteria and/or their association with particles that might scavenge some H_2O_2 and thereby protect attached cells.

The findings presented here contrast slightly with Lin et al. (2018) who similarly performed mesocosm experiments and analyzed the 16S rDNA for bacterial community shifts in response to 8mg H_2O_2 L^{-1}. While *Firmicutes* increased in sequenced relative abundance in that study, *Proteobacteria* and *Bacteroidetes* were reduced [32], whereas these two groups increased or were generally unchanged in the present study. Those results, however, emanated from a single experiment that was performed in winter in Dianchi Lake, China, where the temperature was 10 °C [32]. Moreover, changes in relative abundance can be complex as changes in relative abundance of any one group is also dependent on the response of other groups.

3.4. Comparison Among Ecosystems

The reactivity of H_2O_2 is influenced by its rate of decay, which can be affected by oxidation-reduction processes and the organic matter content of water bodies [43,44]. The nearest measure of organic matter concentration this study performed was algal biomass. Georgica Pond samples had the lowest total biomass with a concentration of 97 μg Chl*a* L^{-1}, and a cyanobacterial biomass of 73 μg Chl*a* L^{-1}, which was reduced 99.8% by H_2O_2. Lake Agawam was the next densest, with an average total biomass of 164 μg Chl*a* L^{-1}, 138 μg Chl*a* L^{-1} of which was cyanobacteria, which was reduced by 53%. Mill Pond experiments had an average total biomass of 267 μg Chl*a* L^{-1}, with an average cyanobacterial concentration of 241 μg Chl*a* L^{-1}, which was reduced by 96%. Roth Pond was the densest and most mixed, with an average total biomass of 486 μg Chl*a* L^{-1}, and an average cyanobacterial concentration of 140 μg Chl*a* L^{-1}, which was reduced by 67%. Across these four systems, there was no significant relationship between effectiveness of H_2O_2 in reduction of cyanobacterial biomass and total biomass (Figure 8a), and, therefore, other factors may have had a greater influence on efficacy. For some individual systems, specifically Lake Agawam and Roth Pond, increasing levels of algal biomass were significantly correlated with decreasing reductions on cyanobacterial biomass during H_2O_2 treatments (Lake Agawam, $p < 0.005$, Figure 8b; Roth Pond, $p < 0.001$, Figure 8c), meaning H_2O_2 became less effective at controlling cyanobacteria as total algal biomass increased. This trend was not detected in water from Mill Pond or Georgica Pond. Regardless, the findings for Lake Agawam and Roth Pond samples suggest that the efficacy of H_2O_2 in controlling cyanobacteria can depending on the levels of total algal biomass, but that the relationship is partly conditional upon other factors that may differ independently across ecosystems including levels of total organic carbon in the water column and the inventory of organic matter within sediments.

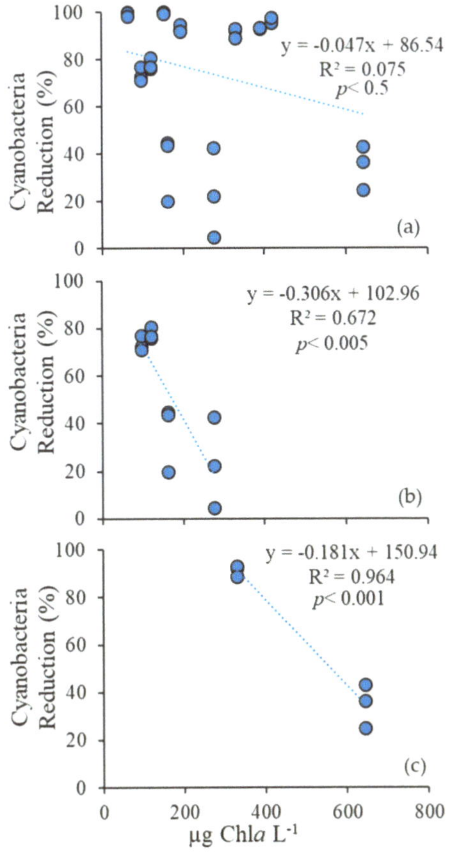

Figure 8. Regression plots for percent cyanobacterial reduction per initial unit biomass (μg Chl*a* L^{-1}) following the addition of 4 mg L^{-1} H_2O_2 for all experiments (**a**), Lake Agawam experiments (**b**), and Roth Pond experiments (**c**).

Across all experiments, initial reductions in cyanobacterial biomass persisted during the four-to-seven-day incubations. Other long-term studies from lakes and wastewater stabilization ponds dominated by *Planktothrix* found the effects of H_2O_2 persisted for five to seven weeks [11,31], and in a mixed assemblage of *Microcystis* and *Planktothrix* persisted for three weeks [17]. Longer observations would be required to determine the ability of the cyanobacteria to recover and the time frame within which such a recovery would occur.

4. Conclusions

This study demonstrated that H_2O_2 administered at a moderate dose (4 mg L^{-1}) consistently inhibits cyanobacteria and promotes the growth of eukaryotic algae, primarily green algae. H_2O_2 significantly reduced heterotrophic bacterial densities, with the phylum *Actinobacteria* most consistently reduced. Other bacterial phyla also declined, but were relatively less impacted, potentially due to their antioxidant enzymes and recovery fueled by use of cyanobacterial-derived organic matter, with some phyla increasing in relative abundance following the addition of H_2O_2. H_2O_2 did not successfully reduce cyanobacteria levels below detection or guidance levels in a majority of experiments at the concentrations used here (4 mg L^{-1}). While the elimination of cyanobacteria might have been achieved with higher doses of H_2O_2, prior research has suggested that such levels may cause collateral damage on non-target organisms [11,12]. The use of bottle experiments here, however, likely maximized contact between H_2O_2 and the plankton community, while minimizing scavenging by sediments. Future, larger scale and whole ecosystem experiments may provide deeper insight with regard the true effect and effectiveness of H_2O_2 in altering plankton community structure and mitigating cyanobacterial blooms.

5. Materials and Methods

Bottle incubation experiments were performed within four study systems: Georgica Pond (latitude, longitude = 40.938671, −72.230216), Mill Pond (40.914848, −72.358037), Lake Agawam (40.875074, −72.391942), and Roth Pond (40.911734, −73.123772). Lake Agawam, Mill Pond, and Georgica Pond are shallow (2–3 m) natural freshwater (Lake Agawam, Mill Pond) or brackish (Georgica Pond; salinity 0–20 PSU) water bodies located on Long Island's south shore and are 0.24 km^2, 0.49 km^2, and 1.17 km^2, respectively, and known to experience repeated cyanobacterial blooms [45,46]. Roth Pond is a man-made, 3×10^{-3} km^2, 1 m deep water body on the campus of Stony Brook University, in Stony Brook, NY, USA.

This study presents 11 experiments performed in two rounds (Table 1). The first round of experiments consisted of three experiments from Lake Agawam, three experiments from Mill Pond, one experiment from Roth Pond, and one experiment from Georgica Pond that were evaluated for changes in phytoplankton communities in response to H_2O_2 using a bbe Moldaenke (Kiel, Germany) Fluoroprobe. The second round of experiments were performed at Lake Agawam, Mill Pond, and Roth Pond and involved in-depth and detailed investigations of the prokaryotic and eukaryotic plankton communities' responses to H_2O_2 using fluorometry, microscopy, flow cytometry, and high-throughput amplicon sequencing.

Table 1. Overview of experiments and analysis performed for each waterbody.

Waterbody	Dates	Analysis
Lake Agawam	7/21/16 10/20/16 6/9/17	Fluorescence only
	10/3/16	Full [1]
Mill Pond	7/21/16 10/20/16 6/30/17	Fluorescence only
	10/3/16	Full [1]
Georgica Pond	7/21/16	Fluorescence only
Roth Pond	6/9/17	Fluorescence only
	6/30/17	Full [1]

[1] Fluorescence, microscopy, flow cytometry, DNA sequencing.

Surface water was collected using 20 L carboys from each site experiencing a cyanobacterial bloom defined as > 25 µg Chl*a* L^{-1} from cyanobacteria as quantified on a bbe Fluoroprobe according to the (NYSDEC; details below). Bloom water was transferred from the carboys into 4-L polycarbonate bottles using acid-washed Tygon tubing placed at the bottom of the bottle to reduce bubbling and disturbance of the plankton. Each bottle experiment consisted of two treatments, with three replicate bottles each: an unamended control, and 4 mg H_2O_2 L^{-1} treatment achieved via the addition of a 3% w/v H_2O_2 solution. These H_2O_2 levels have been previously shown to reduce levels of *Planktothrix* but not eukaryotes in a matter of days in European lakes [11,12]. Bottles were incubated for four-to-seven days under ambient light and temperature conditions in an outdoor flow-through table, constantly flushed with water from Old Fort Pond, Southampton, NY, which maintains water temperatures comparable to the shallow lakes and ponds studied here (Figure 9). Initial and final timepoint samples were obtained and preserved for microscopy (5% Lugol's iodine), flow cytometry (10% buffered formalin stored at −80 °C), and the extraction of DNA (50 mL onto a 0.2 μm, 47 mm Isopore polycarbonate filter frozen at −80 °C). Lugol's iodine preserved samples were analyzed using a Sedgewick Rafter slide to quantify cyanobacteria at the genus level as well as eukaryotic algae that were broadly categorized as unicellular green algae (chlorophytes) or diatoms. Formalin-preserved samples were used to quantify the abundances of phycocyanin-containing pico-cyanobacteria, pico- and nano-eukaryotic phytoplankton, and SYBR Green I-stained heterotrophic bacteria on a CytoFLEX flow cytometer (Beckman Coulter, Indianapolis, IN, USA) based on fluorescence patterns and particle size [47]. Pico-cyanobacteria are identified here as *Cyanobium* based on high-throughput sequencing identification results. Fluorescence measurements were made on initial and final live samples on a BBE Fluoroprobe which categorizes algal biomass (Chl*a*) based on the fluorescence signatures of green algae, cyanobacteria, and unicellular brown algae (diatoms, dinoflagellates, raphidophytes) [48]. Whole-water samples (1 mL) were collected and frozen for one experiment for analysis of the cyanobacterial toxin, microcystin, via an Abraxis Microcystins ELISA assay which included lysing cells for a total concentration [49]. Differences between treatments and controls for each parameter measured in the experiments were assessed via a one-way ANOVA. Values below detection were entered as zero.

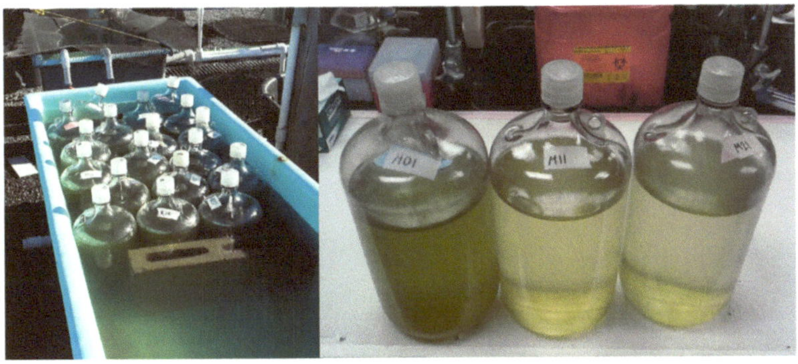

Figure 9. Outdoor flow-through table, and 4 L bottles used for incubation experiments. Example of pigment change between control and H_2O_2 treatments.

DNA Extraction, Sequencing, and Analysis

DNA barcoding analysis was performed to assess changes in community composition of cyanobacteria and other prokaryotes. Samples were initially heated in a water bath to 65° for 10 min to aid in lysing, and extractions were performed using a Qiagen DNeasy® PowerWater® Kit. Double-stranded DNA was quantified on a Qubit® fluorometer using a dsDNA BR Assay kit. Aliquots were normalized to an equal quantity of DNA, and sent to Molecular Research Laboratories (Shallowater, TX, USA) for amplicon sequencing. Paired-end sequencing was performed on an Illumina MiSeq (2 × 300bp) following the manufacturer's guidelines. The 16S rRNA gene V4 variable region (~252bp) was amplified using universal primers 515F: 5′-GTG YCA GCM GCC GCG GTAA-3′ [50] and 806RB: 5′-GGA CTA CNV GGG TWT CTA AT-3′ [51]. For each sample, an identifying barcode was placed on the forward primer and a 30 cycle PCR using the HotStarTaq Plus Master Mix Kit (Qiagen, Valencia, CA, USA) was performed. The following PCR conditions were used: 94 °C for 3 min, followed by 28 cycles of 94 °C for 30 s, 53 °C for 40 s and 72 °C for 1 min, and a final elongation step at 72 °C for 5 min. Samples were purified using calibrated Ampure XP beads and subsequently used to prepare an Illumina DNA library.

Sequence data was processed using the Quantitative Insights into Microbial Ecology QIIME 1 (v1.9.1) and QIIME 2 (v2019.1.0) following the "Moving Pictures" pipeline (QIIME, http://qiime.org; [52]). Raw sequences were depleted of sample barcodes in QIIME 1. Paired-end reads were demultiplexed in QIIME 2 using the DEMUX plugin, and were depleted of primers using the Cutadapt Plugin. The library was filtered for chimeric sequences, denoised and dereplicated using the DADA2 plugin. A naïve Bayes classifier was trained using the SILVA rRNA (16S SSU) v132 reference database at 99% similarity, and was used with the q2-feature-classifier and classify-sklearn plugins to assign taxonomies. The dataset was filtered to remove mitochondria and chloroplast features. Prokaryotic OTUs were examined at the phylum level and were expressed as relative abundance of each. Cyanobacteria OTUs were then examined at the genus level, also expressed as relative abundance. For each set, prokaryote OTUs and cyanobacteria OTUs, relative abundances exceeding 5% were compared, and the remaining were grouped as "other". Statistical analysis of diversity, as well as shifts in relative abundance were performed in QIIME 2. Absolute abundances of cyanobacterial genera were estimated by multiplying the Fluoroprobe quantified levels of cyanobacterial fluorescence by the relative abundance of each genera exceeding 5% of the total OTU reads. Raw sequencing data was deposited to the NCBI Sequence Read Archive (SRA) under accession number PRJNA642309.

Author Contributions: Conceptualization, M.W.L. and C.J.G.; Methodology, M.W.L. and C.J.G.; Validation, C.J.G.; Formal analysis, M.W.L. and C.J.G.; Investigation, M.W.L.; Resources, C.J.G.; Data curation, M.W.L.; Writing—Original draft preparation, M.W.L.; Writing—Review and editing, C.J.G.; Visualization, M.W.L.; Supervision, C.J.G; Project administration, C.J.G.; Funding acquisition, C.J.G. All authors have read and agreed to the published version of the manuscript.

Funding: This research was funded by the New York State Department of Environmental Conservation, the Tamarind Foundation, and NOAA's MERHAB program.

Acknowledgments: We acknowledge the support of the New York State Department of Environmental Conservation for support of this project, the New York State Center for Clean Water Technology in making measurements of hydrogen peroxide, and Solitude Lake Management for performing whole lake applications of hydrogen peroxide.

Conflicts of Interest: The authors declare no conflicts of interest.

References

1. Backer, L.C.; McGillicuddy, D.J., Jr. Harmful algal blooms: At the interface between coastal oceanography and human health. *Oceanography* **2006**, *19*, 94. [CrossRef] [PubMed]
2. Bartram, J.; Chorus, I. *Toxic Cyanobacteria in Water: A Guide to their Public Health Consequences, Monitoring and Management*; CRC Press: London, UK, 1999.
3. United States Environmental Protection Agency. *Drinking Water Health Advisory for the Cyanobacterial Microcystin Toxins*; United States Environmental Protection Agency: Washington DC, USA, 2015.
4. Bukaveckas, P.A.; Lesutienė, J.; Gasiūnaitė, Z.R.; Ložys, L.; Olenina, I.; Pilkaitytė, R.; Pūtys, Ž.; Tassone, S.; Wood, J. Microcystin in aquatic food webs of the Baltic and Chesapeake Bay regions. *Estuar. Coast. Shelf Sci.* **2017**, *191*, 50–59. [CrossRef]
5. Ibelings, B.W.; Chorus, I. Accumulation of cyanobacterial toxins in freshwater "seafood" and its consequences for public health: A review. *Environ. Pollut.* **2007**, *150*, 177–192. [CrossRef] [PubMed]
6. Poste, A.E.; Hecky, R.E.; Guildford, S.J. Evaluating microcystin exposure risk through fish consumption. *Environ. Sci. Technol.* **2011**, *45*, 5806–5811. [CrossRef]
7. Backer, L.C.; Landsberg, J.H.; Miller, M.; Keel, K.; Taylor, T.K. Canine cyanotoxin poisonings in the United States (1920s–2012): Review of suspected and confirmed cases from three data sources. *Toxins* **2013**, *5*, 1597–1628. [CrossRef]
8. Harke, M.J.; Steffen, M.M.; Gobler, C.J.; Otten, T.G.; Wilhelm, S.W.; Wood, S.A.; Paerl, H.W. A review of the global ecology, genomics, and biogeography of the toxic cyanobacterium, Microcystis spp. *Harmful Algae* **2016**, *54*, 4–20. [CrossRef]
9. Paerl, H.W. Nuisance phytoplankton blooms in coastal, estuarine, and inland waters 1. *Limnol. Oceanogr.* **1988**, *33*, 823–843. [CrossRef]
10. Barrington, D.J.; Ghadouani, A.; Ivey, G.N. Environmental factors and the application of hydrogen peroxide for the removal of toxic cyanobacteria from waste stabilization ponds. *J. Environ. Eng.* **2011**, *137*, 952–960. [CrossRef]

11. Matthijs, H.C.; Visser, P.M.; Reeze, B.; Meeuse, J.; Slot, P.C.; Wijn, G.; Talens, R.; Huisman, J. Selective suppression of harmful cyanobacteria in an entire lake with hydrogen peroxide. *Water Res.* **2012**, *46*, 1460–1472. [CrossRef]
12. Matthijs, H.C.; Jančula, D.; Visser, P.M.; Maršálek, B. Existing and emerging cyanocidal compounds: New perspectives for cyanobacterial bloom mitigation. *Aquat. Ecol.* **2016**, *50*, 443–460. [CrossRef]
13. Drábková, M.; Admiraal, W.; Maršálek, B. Combined exposure to hydrogen peroxide and light selective effects on cyanobacteria, green algae, and diatoms. *Environ. Sci. Technol.* **2007**, *41*, 309–314. [CrossRef] [PubMed]
14. Barrington, D.J.; Ghadouani, A. Application of hydrogen peroxide for the removal of toxic cyanobacteria and other phytoplankton from wastewater. *Environ. Sci. Technol.* **2008**, *42*, 8916–8921. [CrossRef] [PubMed]
15. Samuilov, V.; Timofeev, K.; Sinitsyn, S.; Bezryadnov, D. H_2O_2-induced inhibition of photosynthetic O_2 evolution by *Anabaena* variabilis cells. *Biochemistry* **2004**, *69*, 926–933.
16. Drabkova, M.; Maršálek, B.; Admiraal, W. Photodynamic therapy against cyanobacteria. *Environ. Toxicol. Int. J.* **2007**, *22*, 112–115. [CrossRef] [PubMed]
17. Barrington, D.J.; Reichwaldt, E.S.; Ghadouani, A. The use of hydrogen peroxide to remove cyanobacteria and microcystins from waste stabilization ponds and hypereutrophic systems. *Ecol. Eng.* **2013**, *50*, 86–94. [CrossRef]
18. Zilliges, Y.; Kehr, J.-C.; Meissner, S.; Ishida, K.; Mikkat, S.; Hagemann, M.; Kaplan, A.; Borner, T.; Dittmann, E. The Cyanobacterial Hepatotoxin Microcystin Binds to Proteins and Increases the Fitness of *Microcystis* under Oxidative Stress Conditions. *PLoS ONE* **2011**, *6*, e17615. [CrossRef]
19. Schuurmans, J.M.; Brinkmann, B.W.; Makower, A.K.; Dittmann, E.; Huisman, J.; Matthijs, H.C. Microcystin interferes with defense against high oxidative stress in harmful cyanobacteria. *Harmful Algae* **2018**, *78*, 47–55. [CrossRef]
20. Passardi, F.; Zamocky, M.; Favet, J.; Jakopitsch, C.; Penel, C.; Obinger, C.; Dunand, C. Phylogenetic distribution of catalase-peroxidases: Are there patches of order in chaos? *Gene* **2007**, *397*, 101–113. [CrossRef]
21. Bernroitner, M.; Zamocky, M.; Furtmüller, P.G.; Peschek, G.A.; Obinger, C. Occurrence, phylogeny, structure, and function of catalases and peroxidases in cyanobacteria. *J. Exp. Bot.* **2009**, *60*, 423–440. [CrossRef]
22. Barroin, G.; Feuillade, M. Hydrogen peroxide as a potential algicide for Oscillatoria rubescens DC. *Water Res.* **1986**, *20*, 619–623. [CrossRef]
23. Lürling, M.; Meng, D.; Faassen, E.J. Effects of hydrogen peroxide and ultrasound on biomass reduction and toxin release in the cyanobacterium, Microcystis aeruginosa. *Toxins* **2014**, *6*, 3260–3280. [CrossRef] [PubMed]
24. Gao, L.; Pan, X.; Zhang, D.; Mu, S.; Lee, D.-J.; Halik, U. Extracellular polymeric substances buffer against the biocidal effect of H2O2 on the bloom-forming cyanobacterium Microcystis aeruginosa. *Water Res.* **2015**, *69*, 51–58. [CrossRef] [PubMed]
25. Schrader, K.K.; Dayan, F.E. Antioxidant enzyme activities in the cyanobacteria *Planktothrix* agardhii, *Planktothrix* perornata, Raphidiopsis brookii, and the green alga Selenastrum capricornutum. In *Handbook on Cyanobacteria: Biochemistry, Biotechnology, and Applications*; Nova Science Publishers: Hauppauge, NY, USA, 2010; pp. 473–483.
26. Moore, D.; O'Donohue, M.; Garnett, C.; Critchley, C.; Shaw, G. Factors affecting akinete differentiation in *Cylindrospermopsis* raciborskii (Nostocales, Cyanobacteria). *Freshw. Biol.* **2005**, *50*, 345–352. [CrossRef]
27. Moustaka-Gouni, M.; Kormas, K.A.; Vardaka, E.; Katsiapi, M.; Gkelis, S. Raphidiopsis mediterranea Skuja represents non-heterocytous life-cycle stages of *Cylindrospermopsis* raciborskii (Woloszynska) Seenayya et Subba Raju in Lake Kastoria (Greece), its type locality: Evidence by morphological and phylogenetic analysis. *Harmful Algae* **2009**, *8*, 864–872. [CrossRef]
28. Nielsen, K.M.; Johnsen, P.J.; Bensasson, D.; Daffonchio, D. Release and persistence of extracellular DNA in the environment. *Environ. Biosaf. Res.* **2007**, *6*, 37–53. [CrossRef]
29. Zulkefli, N.; Kim, K.-H.; Hwang, S.-J. Effects of Microbial Activity and Environmental Parameters on the Degradation of Extracellular Environmental DNA from a Eutrophic Lake. *Int. J. Environ. Res. Public Health* **2019**, *16*, 3339. [CrossRef]
30. Yang, Z.; Buley, R.P.; Fernandez-Figueroa, E.G.; Barros, M.U.G.; Rajendran, S.; Wilson, A.E. Hydrogen peroxide treatment promotes chlorophytes over toxic cyanobacteria in a hyper-eutrophic aquaculture pond. *Environ. Pollut.* **2018**, *240*, 590–598. [CrossRef]

31. Sinha, A.K.; Eggleton, M.A.; Lochmann, R.T. An environmentally friendly approach for mitigating cyanobacterial bloom and their toxins in hypereutrophic ponds: Potentiality of a newly developed granular hydrogen peroxide-based compound. *Sci. Total Environ.* **2018**, *637*, 524–537. [CrossRef]
32. Lin, L.; Shan, K.; Xiong, Q.; Zhou, Q.; Li, L.; Gan, N.; Song, L. The ecological risks of hydrogen peroxide as a cyanocide: Its effect on the community structure of bacterioplankton. *J. Oceanol. Limnol.* **2018**, *36*, 2231–2242. [CrossRef]
33. Smayda, T.J.; Reynolds, C.S. Strategies of marine dinoflagellate survival and some rules of assembly. *J. Sea Res.* **2003**, *49*, 95–106. [CrossRef]
34. Newton, R.J.; Jones, S.E.; Eiler, A.; McMahon, K.D.; Bertilsson, S. A guide to the natural history of freshwater lake bacteria. *Microbiol. Mol. Biol. Rev.* **2011**, *75*, 14–49. [CrossRef] [PubMed]
35. Den Hengst, C.D.; Buttner, M.J. Redox control in actinobacteria. *Biochim. Biophys. Acta (BBA)-Gen. Subj.* **2008**, *1780*, 1201–1216. [CrossRef] [PubMed]
36. Taipale, S.; Jones, R.I.; Tiirola, M. Vertical diversity of bacteria in an oxygen-stratified humic lake, evaluated using DNA and phospholipid analyses. *Aquat. Microb. Ecol.* **2009**, *55*, 1–16. [CrossRef]
37. Šimek, K.; Horňák, K.; Jezbera, J.; Nedoma, J.; Vrba, J.; Straškrábová, V.; Macek, M.; Dolan, J.R.; Hahn, M.W. Maximum growth rates and possible life strategies of different bacterioplankton groups in relation to phosphorus availability in a freshwater reservoir. *Environ. Microbiol.* **2006**, *8*, 1613–1624. [CrossRef]
38. Wagner, M.; Horn, M. The Planctomycetes, Verrucomicrobia, Chlamydiae and sister phyla comprise a superphylum with biotechnological and medical relevance. *Curr. Opin. Biotechnol.* **2006**, *17*, 241–249. [CrossRef]
39. Shao, K.; Gao, G.; Chi, K.; Qin, B.; Tang, X.; Yao, X.; Dai, J. Decomposition of *Microcystis* blooms: Implications for the structure of the sediment bacterial community, as assessed by a mesocosm experiment in Lake Taihu, China. *J. Basic Microbiol.* **2013**, *53*, 549–554. [CrossRef]
40. Nold, S.C.; Zwart, G. Patterns and governing forces in aquatic microbial communities. *Aquat. Ecol.* **1998**, *32*, 17–35. [CrossRef]
41. DeLong, E.F.; Franks, D.G.; Alldredge, A.L. Phylogenetic diversity of aggregate-attached vs. free-living marine bacterial assemblages. *Limnol. Oceanogr.* **1993**, *38*, 924–934. [CrossRef]
42. Zámocký, M.; Gasselhuber, B.; Furtmüller, P.G.; Obinger, C. Molecular evolution of hydrogen peroxide degrading enzymes. *Arch. Biochem. Biophys.* **2012**, *525*, 131–144. [CrossRef]
43. Cooper, W.J.; Zepp, R.G. Hydrogen peroxide decay in waters with suspended soils: Evidence for biologically mediated processes. *Can. J. Fish. Aquat. Sci.* **1990**, *47*, 888–893. [CrossRef]
44. Häkkinen, P.J.; Anesio, A.M.; Granéli, W. Hydrogen peroxide distribution, production, and decay in boreal lakes. *Can. J. Fish. Aquat. Sci.* **2004**, *61*, 1520–1527. [CrossRef]
45. Gobler, C.J.; Davis, T.W.; Coyne, K.J.; Boyer, G.L. Interactive influences of nutrient loading, zooplankton grazing, and microcystin synthetase gene expression on cyanobacterial bloom dynamics in a eutrophic New York lake. *Harmful Algae* **2007**, *6*, 119–133. [CrossRef]
46. Davis, T.W.; Berry, D.L.; Boyer, G.L.; Gobler, C.J. The effects of temperature and nutrients on the growth and dynamics of toxic and non-toxic strains of *Microcystis* during cyanobacteria blooms. *Harmful Algae* **2009**, *8*, 715–725. [CrossRef]
47. Kang, Y.; Koch, F.; Gobler, C.J. The interactive roles of nutrient loading and zooplankton grazing in facilitating the expansion of harmful algal blooms caused by the pelagophyte, Aureoumbra lagunensis, to the Indian River Lagoon, FL, USA. *Harmful Algae* **2015**, *49*, 162–173. [CrossRef]
48. Jankowiak, J.; Hattenrath-Lehmann, T.; Kramer, B.J.; Ladds, M.; Gobler, C.J. Deciphering the effects of nitrogen, phosphorus, and temperature on cyanobacterial bloom intensification, diversity, and toxicity in western Lake Erie. *Limnol. Oceanogr.* **2019**, *64*, 1347–1370. [CrossRef]
49. Harke, M.J.; Gobler, C.J. Daily transcriptome changes reveal the role of nitrogen in controlling microcystin synthesis and nutrient transport in the toxic cyanobacterium, *Microcystis* aeruginosa. *BMC Genom.* **2015**, *16*, 1068. [CrossRef] [PubMed]
50. Parada, A.E.; Needham, D.M.; Fuhrman, J.A. Every base matters: Assessing small subunit rRNA primers for marine microbiomes with mock communities, time series and global field samples. *Environ. Microbiol.* **2016**, *18*, 1403–1414. [CrossRef] [PubMed]

51. Klindworth, A.; Pruesse, E.; Schweer, T.; Peplies, J.; Quast, C.; Horn, M.; Glöckner, F.O. Evaluation of general 16S ribosomal RNA gene PCR primers for classical and next-generation sequencing-based diversity studies. *Nucleic Acids Res.* **2012**, *41*, e1. [CrossRef]
52. Caporaso, J.G.; Kuczynski, J.; Stombaugh, J.; Bittinger, K.; Bushman, F.D.; Costello, E.K.; Fierer, N.; Peña, A.G.; Goodrich, J.K.; Gordon, J.I.; et al. QIIME allows analysis of high-throughput community sequencing data. *Nat. Methods* **2010**, *7*, 335–336. [CrossRef] [PubMed]

toxins

MDPI

Article

Diversity Assessment of Toxic Cyanobacterial Blooms during Oxidation

Saber Moradinejad [1,*], Hana Trigui [1], Juan Francisco Guerra Maldonado [1], Jesse Shapiro [2], Yves Terrat [2], Arash Zamyadi [3,4], Sarah Dorner [1] and Michèle Prévost [1]

1 Department of Civil, Geological, and Mining Engineering, Polytechnique Montréal, Montréal, QC H3T 1J4, Canada; Hana.trigui@polymtl.ca (H.T.); Juan-francisco.guerra-maldonado@polymtl.ca (J.F.G.M.); Sarah.dorner@polymtl.ca (S.D.); Michele.prevost@polymtl.ca (M.P.)
2 Department of biological science, Université de Montréal, Montréal, QC H2V 0B3, Canada; Jesse.shapiro@umontreal.ca (J.S.); Yves.terrat@umontreal.ca (Y.T.)
3 Water Research Australia (WaterRA), Adelaide, SA 5001, Australia; Arash.zamyadi@waterra.com.au
4 BGA Innovation Hub and Water Research Centre, School of Civil and Environmental Engineering, University of New South Wales (UNSW), Sydney, NSW 2052, Australia
* Correspondence: saber.moradinejad@polymtl.ca

Received: 18 September 2020; Accepted: 18 November 2020; Published: 20 November 2020

Abstract: Fresh-water sources of drinking water are experiencing toxic cyanobacterial blooms more frequently. Chemical oxidation is a common approach to treat cyanobacteria and their toxins. This study systematically investigates the bacterial/cyanobacterial community following chemical oxidation (Cl_2, $KMnO_4$, O_3, H_2O_2) using high throughput sequencing. Raw water results from high throughput sequencing show that *Proteobacteria, Actinobacteria, Cyanobacteria* and *Bacteroidetes* were the most abundant phyla. *Dolichospermum, Synechococcus, Microcystis* and *Nostoc* were the most dominant genera. In terms of species, *Dolichospermum sp.90* and *Microcystis aeruginosa* were the most abundant species at the beginning and end of the sampling, respectively. A comparison between the results of high throughput sequencing and taxonomic cell counts highlighted the robustness of high throughput sequencing to thoroughly reveal a wide diversity of bacterial and cyanobacterial communities. Principal component analysis of the oxidation samples results showed a progressive shift in the composition of bacterial/cyanobacterial communities following soft-chlorination with increasing common exposure units (CTs) (0–3.8 mg·min/L). Close cyanobacterial community composition (*Dolichospermum* dominant genus) was observed following low chlorine and mid-$KMnO_4$ (287.7 mg·min/L) exposure. Our results showed that some toxin producing species may persist after oxidation whether they were dominant species or not. Relative persistence of *Dolichospermum sp.90* was observed following soft-chlorination (0.2–0.6 mg/L) and permanganate (5 mg/L) oxidation with increasing oxidant exposure. Pre-oxidation using H_2O_2 (10 mg/L and one day contact time) caused a clear decrease in the relative abundance of all the taxa and some species including the toxin producing taxa. These observations suggest selectivity of H_2O_2 to provide an efficient barrier against toxin producing cyanobacteria entering a water treatment plant.

Keywords: cyanobacteria; diversity; oxidation; high throughput sequencing; *Dolichospermum*; *Microcystis*

Key Contribution: This study provides a diversity assessment of cyanobacterial bloom following oxidation. Changes, shifts and oxidant persistence potential of cyanobacterial community during oxidation were quantified and reported.

1. Introduction

The occurrence of cyanobacterial blooms in fresh-water bodies has been enhanced due to eutrophication and temperature increases [1,2]. Cyanobacterial blooms may produce and release taste and odour compounds as well as cyanotoxins into water bodies. More than 40 species of cyanobacteria are known as potentially toxic species [3,4]. Microcystin (MC), anatoxin (ATX-a), saxitoxin (STX), cylindrospermopsin (CYN) and β-Methylamino-L-alanine (BMAA) are the five major groups of cyanotoxins due to their toxicity and frequent occurrence around the world [5].

Conventional drinking water treatment plants are often challenged by the removal of cyanobacteria and cyanotoxins [6–9] including both low risk (low cell numbers in the source) and high-risk water treatment plants (high cell numbers in the source). Several studies have evaluated the removal of the cyanobacteria and their harmful metabolites using different oxidants such as chlorine, ozone and, potassium permanganate [10–15]. Pre-oxidation dampens the cyanobacterial shock before entering the drinking water treatment plant and can limit the accumulation of cyanobacteria within the plant. Chlorine and ozone are also used as primary disinfectants providing an additional oxidation barrier after the filtration to remove cyanobacterial harmful metabolites (cyanotoxins) [16,17].

Dynamic and complex behaviour of cyanobacterial blooms, including cyanotoxin production and release under various environmental conditions represents a treatment challenge. The oxidation efficiency of cyanobacteria blooms may vary according to water quality parameters (such as pH, Dissolved Organic Carbon (DOC) and the presence of other bacterial communities), cyanobacterial community shape, potential agglomeration, and the growth phase [14,18–20]. Cyanobacteria treatment efficiency improvement requires an understanding of the cyanobacterial composition structure in response to treatment processes [21]. Molecular methods have been used to study the fate of the microbial community, including cyanobacteria, within different conditions [5,22]. Molecular methods can overcome the challenges of microscopic cell counts such as time and qualified person requirements, as well as changes in biovolumes during analyses [23–25].

Molecular methods such as high throughput sequencing have been deployed to study cyanobacterial communities and identify cyanotoxin biosynthesis genes [5,26–34]. Diversity of the cyanobacterial community and toxigenic cyanobacteria was assessed based on the Operational Taxonomy Units (OTUs) derived from 16S rRNA gene amplification (metabarcoding) [5]. Limited studies have applied high throughput sequencing to monitor the fate of cyanobacteria during the treatment processes. Xu, Pei [35] studied the microbial community of the sludge in six different drinking water treatment plants using high throughput sequencing. Results showed that cyanobacteria were the most dominant phylum in two treatment plants with a higher level of nutrients in raw water. *Planktothrix*, *Microcystis* and *Cyindrospermopsis* were the most abundant genera and were positively correlated with the nutrient levels in raw water. Pei, Xu [36] used 16S rRNA sequencing to study the shifts in the microbial community in clarifier sludge following coagulation by $FeCl_3$, $AlCl_3$ and PAFC (Polyaluminium Ferric Chloride). Results revealed selective removal of the different bacterial species, as the relative abundance of the *Microcystis*, *Rhodobacter*, *Phenylobacterium* and *Hydrogenophaga* decreased in $AlCl_3$ sludge compare to the $FeCl_3$ and PAFC. Lower *Microcystis* abundance could be related to high Al toxicity or large and high-density floc in $FeCl_3$ and PAFC, which plays a protective role for microorganisms [36]. Lusty and Gobler (2020) used 16S rRNA to evaluate the mitigation of cyanobacterial blooms using H_2O_2. Results showed relative persistence of *Cyanobium* and *Cylindrospermopsis* to a moderate H_2O_2 dose (4 mg/L); *Plankthotrix* and *Microcystis* (abundant genus) were the most sensitive genera, respectively [37].

High throughput sequencing has been widely applied to study bacterial/cyanobacterial communities. Fewer studies focused on the diversity of bacterial/cyanobacterial communities during water treatment processes. However, no study has focused on the cyanobacterial community following chemical oxidation using high throughput sequencing. Understanding the shifts and the potential selective persistence in cyanobacterial communities following oxidation is important for choosing an efficient oxidant. Thus, the objective of this study was to assess the structural composition of the cyanobacteria community following oxidation (with Cl_2, O_3, $KMnO_4$, H_2O_2) using high throughput metagenomic shotgun sequencing over the seasonal bloom period.

2. Results and Discussion

2.1. Cyanobacterial Bloom Characteristics Throughout Sampling

The cyanobacteria bloom samples were collected on 5 days of the bloom period (1, 13, 15, 21 and 29 August 2018) from Missisquoi Bay (Lake Champlain) close to the water intake of the drinking water treatment plant. Cyanobacterial bloom characteristics are presented in the Table 1. DOC (Dissolved Organic Carbon) and pH did not demonstrate considerable variation throughout the sampling period. Total cell counts and biovolumes follow the same decreasing and increasing trend during the sampling period. For both parameters, the highest values were found at the beginning of the sampling period (1 August), followed by a significant drop on 13 August, where cyanobacterial cell counts decreased from 3.3×10^5 to 7.8×10^4 cells/mL and the biovolumes from 30.6 mm^3/L to 4.6 mm^3/L. The second drop was observed between 15 August and 21 August, from 1.4×10^5 to 6.8×10^4 cells/mL for cell count and from 9.4 to 0.3 mm^3/L for the biovolume. According to algal cell abundance descriptors (biovolume and cell counts) during the bloom period, the main peaks of cyanobacterial bloom occurred on 1 August and to a lower extent on 15 August. The observed cell count exceeds the alert level of the 6.5×10^4 cells/mL for drinking water treatment plants [38], except for the 29 August with 5.4×10^4 cells/mL.

Table 1. Cyanobacterial bloom characteristics.

Sampling Date	DOC (mg/L)	pH	Cell Count (cells/mL)	Biovolume (mm^3/L)
1 August 2018	5.9	7.6	3.3×10^5	30.6
13 August 2018	5.8	7.3	7.8×10^4	4.6
15 August 2018	5.5	7.4	1.4×10^5	9.4
21 August 2018	4.9	7.4	6.8×10^4	0.3
29 August 2018	5.6	7.5	5.4×10^4	2.1

2.2. Variation of the Cyanobacterial Bloom Composition

The diversity and community variation of bacterial and cyanobacterial communities during the bloom sampling period were studied using comparative metagenomics reads levels of phylum, order and genus. The number of reads for taxonomic data was normalized by relative abundance (Figure 1).

Analysis of relative abundance of the bacterial community at the phylum level, based on high throughput sequencing data, showed that Proteobacteria, Actinobacteria, Cyanobacteria, Bacteroidetes, Firmicutes and Verrucomicrobia were the six most abundant phyla throughout the sampling period (Figure 1a). As expected, Proteobacteria was by far the most abundant phylum at the beginning of the sampling period (1 August) as has been observed by Pei et al. (2017) [36]. The high relative abundance of Proteobacteria, especially in the beginning of sampling and Bacteroidetes at the end of the sampling period, may indicate contamination of the sampling point with human/animal-associated fecal markers [39–41]. For the rest of the sampling dates (13, 15 and 21 August), Proteobacteria remained the predominant phylum, but at a lower extent than what was observed in the first and last days of sampling. The cyanobacteria phylum accounts for 5 to 10% of total relative abundance assigned to the phylum level in all samples. The cyanobacteria relative abundance started at 5% of the phyla, followed by an increase in the middle of the sampling (13 August and 15 August) to 10%. By the end of August, the cyanobacterial contribution in the whole bacterial community decreased to 5%.

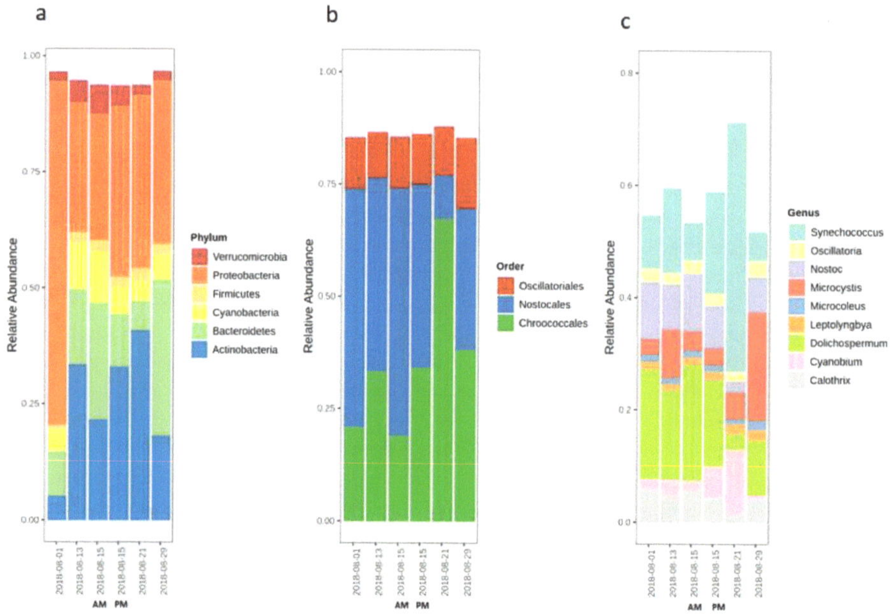

Figure 1. Identity of major detected bloom-associated cyanobacterial community members during the sampling period: (**a**) relative abundance of the different phylum, (**b**) the relative abundance of orders belonging to cyanobacterial phylum, (**c**) relative abundance of genera belonging to the *Nostocales*, *Chrooccocales* and *Oscillatoriales* orders.

At the order level, the cyanobacterial community was dominated by members of the *Chroococcales*, *Nostocales* and *Oscillatoriales* during the cyanobacterial bloom period (from 1 August and 29 August) (Figure 1b). The relative abundance of the *Nostocales* and *Chroococcales* varied between the different dates and even within the same day (15 August). On the other hand, *Oscillatoriales* relative abundance remained steady (Figure 1b).

Analysis at the genus level showed that within the *Chroococcales* order, the predominant genera were *Microcystis* and *Synechococcus*. *Synechococcus* was the dominant genus within *Chroococcales* until 21 August of the sampling period, followed by *Microcystis*, which became the dominant genus on 29 August (Figure 1c). The predominant genera in the *Oscillatoriales and Nostocales* orders were *Oscillatoria* and *Dolichospermum* (formerly known as *Anabaena*), respectively (Figure 1c). Depending on the sampling date, *Synechococcus*, *Microcystis* and *Dolichospermum* were the predominant genera during the bloom period. Our observations are consistent with short/long term investigations of cyanobacterial bloom dynamics using taxonomic cell count and metagenomics [14,21,42–44]. *Microcystis* and *Dolichospermum* share a spatio-temporal niche during blooms and often are dominant within the cyanobacterial community, but have distinct environmental preferences; for example, they have different responses to nutrients [45–47]. During this study, the abundant genus shifted from *Nostocales* members, dominated by *Dolichospermum* for the first three weeks, to the *Microcystis* genus by the end of the sampling period (29 August). The shift to *Microcystis* dominance can probably be attributed to species-specific associations between *Microcystis spp.* and associated bacteria referred to as the epibiont phenomenon. Some species within *Proteobacteria*, *Bacteroidetes* and other phyla can shelter in the *Microcystis* mucilage to avoid being grazed [48]. These species play essential roles in enhancing the environmental adaptation of *Microcystis* within the cyanobacterial bloom, like maintaining redox balance and coping with oxidative stress [48]. Our results are in accordance with the previously reported results that *Proteobacteria*, *Bacteriodetes* tend to dominate in the *Microcystis* mucilage (blooms and culture-dependent studies) [49,50]. Moreover, allelopathy may influence the successional dominance of *Microcystis* and *Dolichospermum* within aquatic systems, whereby organisms produce bioactive compounds (allelochemicals) in the environment to positively or negatively influence the growth of neighboring species [51]. The effects of these allelochemicals on cyanobacterial community distribution within the environment are connected with nutrient availability and environmental conditions [52–54]. Competition experiments between toxic *Microcystis* and *Dolichospermum* strains, based on the lab coculture-dependent method, showed that *Microcystis* significantly inhibited the growth of *Dolichospermum,* whereas the effects of *Dolichospermum* on *Microcystis* were minimal [53,55]. *Dolichospermum* biovolume and biomass were sharply reduced after exposure to *Microcystis* strains [56]. Further investigation is required to explain the interaction and succession of *Microcystis* and *Dolichospermum* in cyanobacterial blooms.

An interesting insight was gained considering within-day changes in bacterial community composition (morning—AM and afternoon—PM) for 15 August (Figure 1). Based on the relative abundance analysis, a shift in bacterial composition at the phylum level was observed between the two samples on 15 August. In the afternoon, *Proteobacteria* and *Actinobacteria* increased, while *Bacteroidetes* and *Cyanobacteria* decreased as compared to their initial composition at the start of the sampling in the morning. At the order level, the most abundant order shifted from *Nostocales* in the morning to *Chroococcolas* in the afternoon. At the genus level, the changes were more evident for *Synechococcus* (increase) and *Dolichospermum* (decrease). These within-day changes in bacterial composition reflect the variation in stratification and mixing patterns of the cyanobacterial community in Missisquoi bay, as shown by Ndong et al. (2014) Ndong, Bird [57]. These findings highlight the importance of timing the sample collection by considering the stratification variation in the diel cycle (morning, noon, afternoon).

2.3. Impact of Oxidation on Cyanobacterial Diversity

The impact of oxidation on cyanobacterial diversity was assessed by using the impact of increased CT (the terminology used in water treatment is CT, which represents the product of the oxidant concentration (C) and contact time (T) to inactivate microorganisms) on two different samples, first on 1 August, dominated by *Dolichospermum* and the second on 29 August dominated by the *Microcystis* genus. The oxidants considered Cl_2, $KMnO_4$, O_3, H_2O_2 differ widely in their mode of action and their persistence [58]. As a result, CT values may vary over three orders of magnitude from 0.1 mg·min/L Cl_2 and a maximum of 7035 mg·min/L for H_2O_2.

Using non-normalized data is a source of bias in statistical analyses of the effects of oxidants on the bacterial/cyanobacterial communities. Thus, the data were normalized prior to analysis. Using the common exposure unit (CT) is not representative to compare the effects of different oxidants simultaneously. A normalized oxidant exposure (relative CT) is used to compare different oxidants among each other (Relative CT for each oxidant; max CT = 1 and min CT = 0, the exposure points in between were calculated accordingly); for example, max CT obtained for H_2O_2, on 1 August, is 7035 mg·min/L and is considered as relative CT = 1, the control condition is considered as relative CT = 0.

The doses (concentration) and contact time (T) were selected using the product CT, which is the foundation of disinfection and oxidation in drinking water. The choice of CT also considered prior evidence showing differences in species' resistance to the oxidant exposure (CT) [6,10,12,14–16,18,42,59,60]. However, all previously reported observations were based on taxonomic cell counts and not on high throughput sequencing. The applied dosages reflect drinking water industry practices using regulated treatment processes.

2.3.1. Cyanobacterial Composition

To assess the cyanobacterial bloom composition following oxidation using $KMnO_4$, Cl_2 and H_2O_2, samples were taken on 1 August (Figure 2) and 29 August (Figure S1). *Dolichospermum* and *Microcystis* were the dominant genera on 1 August and 29 August, respectively.

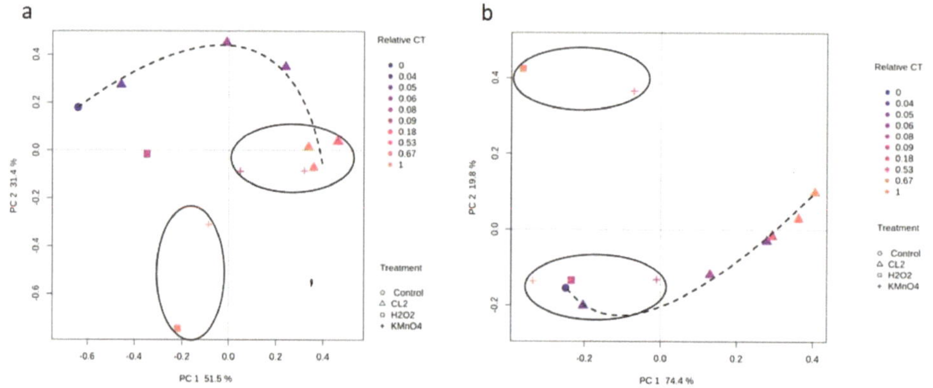

Figure 2. Principal components analysis (PCA) of the normalized relative abundance of comparative metagenomics reads in 1 August sample. Data are plotted following the genus-level classification (**a**) PCA analysis of bacterial community following oxidation using different common exposure units (CT), (**b**) PCA of the cyanobacterial community following oxidation using different CT.

Figure 2 shows the dissimilarity among groups of bacterial communities following oxidation at different exposures (relative CT) using Principal Component Analysis (PCA) on 1 August. Samples that appear more closely together within a PCA are assumed to be more similar in bacterial and cyanobacterial composition (Figure 2). For the bacterial community, principal axis one and principal axis two for PCA represent 51.5 and 31.4% of the variation among the samples, respectively. High-relative CT exerted clustering of two groups, the first one includes samples after Cl_2 and $KMnO_4$ oxidation and the second one encompasses $KMnO_4$ and H_2O_2 oxidation. Chlorinated samples showed a clear progressive shift (as the relative CT increased) in the bacterial community.

On the other hand, the $KMnO_4$ and H_2O_2 induce large shifts as relative CT increases. Bacterial and cyanobacterial composition similarity following the different oxidation on 29 August (*Microcystis*) is presented in Figure S1. Like the 1 August result, a large shift in the bacterial community composition following H_2O_2 oxidation was observed for 29 August. On the other hand, $KMnO_4$ results showed similar bacterial composition on 29 August.

For the cyanobacterial community (1 August), principal axis one, and principal axis two for PCA represent 74.4 and 19.8% of the variation among the samples, respectively (Figure 2b). A clear trend in the cyanobacterial composition variation is observed following chlorination. Cyanobacterial composition after exposure to Cl_2 (low relative CT < 0.05), $KMnO_4$ (low and high relative CT), H_2O_2 (low relative CT $= 0.08$) oxidation clustered in the same group with the control condition. High-relative (CT $= 1$) using H_2O_2 and (mid-relative $= 0.53$) $KMnO_4$ are grouped in a distinct second cluster, revealing similar cyanobacterial assemblages. This is in contrast with cyanobacterial/bacterial community trends following chlorination that display a progressive shift. Differences between the observed trends for the three oxidants were expected because of different mechanisms of actions and kinetics, persistence and selectivity. Cell count based studies show progressive shifts following oxidation [12,14]. Moreover, selective cyanobacteria oxidation has been demonstrated in the lab [11,60], in the field [37,61], and in drinking water treatment plants [8,62].

Our results enlighten the different oxidation impacts on the bacterial community between the Cl_2 and H_2O_2 oxidation. For 29 August, which was dominated by the *Microcystis* genus, $KMnO_4$ results revealed similar cyanobacterial composition. However, a large variation in cyanobacterial composition following H_2O_2 was observed (e.g., the samples from the 1 August, Figure 3).

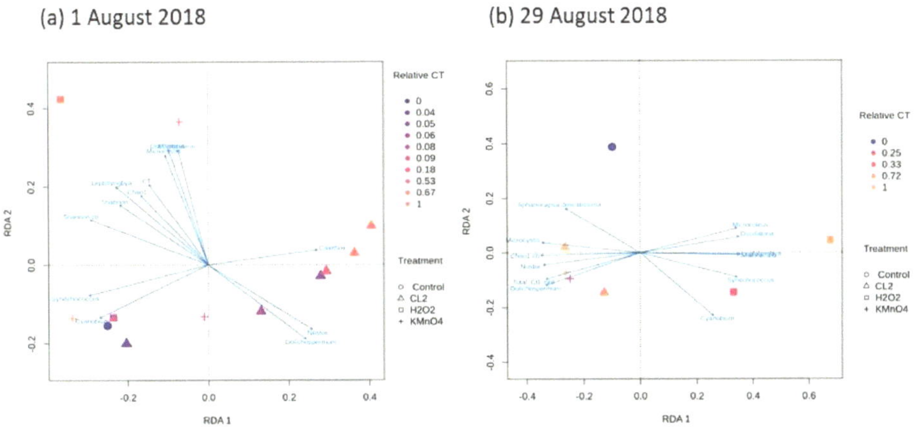

Figure 3. Redundancy analysis (RDA) of oxidant effect on cyanobacterial diversity and the cyanobacterial community at genus level Cl_2 (0.6 mg/L), $KMnO_4$ (5 mg/L), H_2O_2 (10 mg/L) (**a**) 1 August 2018 (**b**) 29 August 2018.

Redundancy analysis (RDA) was used to explore the correlation between the relative abundance of nine dominant cyanobacterial genera (60% of the cyanobacterial community) observed in 1 August or 29 August samples and the different oxidant exposures (Figure 3). The relationship between the CT and diversity indexes (Shannon and Chao1) varies between the two dates. On 1 August, both diversity indices increase with CT, while on 29 August only the Richness index (Chao1) increased with CT. In general, the diversity of the cyanobacterial community increased due to oxidation of the dominated genus, as discussed further in detail for each oxidant.

For 1 August (*Dolichospermum* dominant), RDA analysis establishes an inverse correlation between the *Dolichospermum* genus and a wide range of chlorine exposures with relative CT > 0.04. Chlorination appears to have a lesser impact on the *Dolichospermum* genus than the *Microcystis*, possibly because of its abundance. On the other hand, no such relationship is observed between *Dolichospermum* and $KMnO_4$ exposure (relative CT = 1) and high H_2O_2 exposure (relative CT = 1), suggesting that these oxidants at these exposure levels had a negative impact on the *Dolichospermum* genus persistence within the community. Furthermore, for $KMnO_4$ at relative CT > 0.53, *Microcystis*, *Synechococcus*, and *Leptolyngbya* are less impacted by oxidation. Figure 3b shows the effect of the different oxidants on the cyanobacterial community of 29 August, when *Microcystis* genus is dominant. An immediate shift from the control is observed for all oxidants and for all relative CT. Unlike 1 August, the Shannon diversity is no longer correlated with *Dolichospermum*. These differences could reflect the morphological differences between the dominant genera (*Microcystis* vs. *Dolichospermum)* as a unicellular aggregate that are less resistant to oxidation as compared to filaments [63]. Chlorine and $KMnO_4$ exposures had a low impact on *Microcystis* and *Dolichospermum*. On the other hand, any H_2O_2 exposure causes a reduction in *Dolichospermum* and *Microcystis*, more so on 29 August. Removal of *Microcystis* using H_2O_2 was shown by Lusty and Gobler. (2020) and is in accordance with our observations of diverging correlation between Chao1 and *Microcystis* [37].

Figure S2 shows the cyanobacterial composition following oxidation at genus level. Results show no significant variation in the relative abundance of *Dolichospermum* and *Microcystis* following Cl_2, $KMnO_4$ oxidation. Thus, no relative persistence to oxidation was observed at the genus level. On the other hand, high H_2O_2 exposure caused a decline in both *Dolichospermum* and *Microcystis* genus, which demonstrates the effect of high H_2O_2 exposure on cyanobacterial removal.

2.3.2. Effect of Oxidation on Cyanobacterial Community Richness and Diversity

The diversity of the cyanobacterial composition following oxidation (1 August) at the species level is presented in Figure 4 (top 25 most abundant species). *Dolichospermum sp.90* was the dominant species in control conditions, and its relative abundance increases following chlorination and decreases after $KMnO_4$ and H_2O_2. Although no trends in relative abundance were seen at the genus level (Figure S2), a similar trend can be seen for the three species present *Dolichospermum sp.90, Dolichospermum cylindrica* and, *Dolichospermum sp. PCC7108*. Regardless of the oxidant *Dolichospermum sp.90* remains the dominant species. Following the $KMnO_4$ and H_2O_2 oxidation, *Dolichospermum sp.90* was still the dominant species; its relative abundance declined compared to the control. The relative abundance of *Microcystis aeruginosa* is not impacted by chlorine but increases with $KMnO_4$ and H_2O_2 exposure, confirming selective removal of *Microcystis* showed by Lusty and Gobler. (2020) [37].

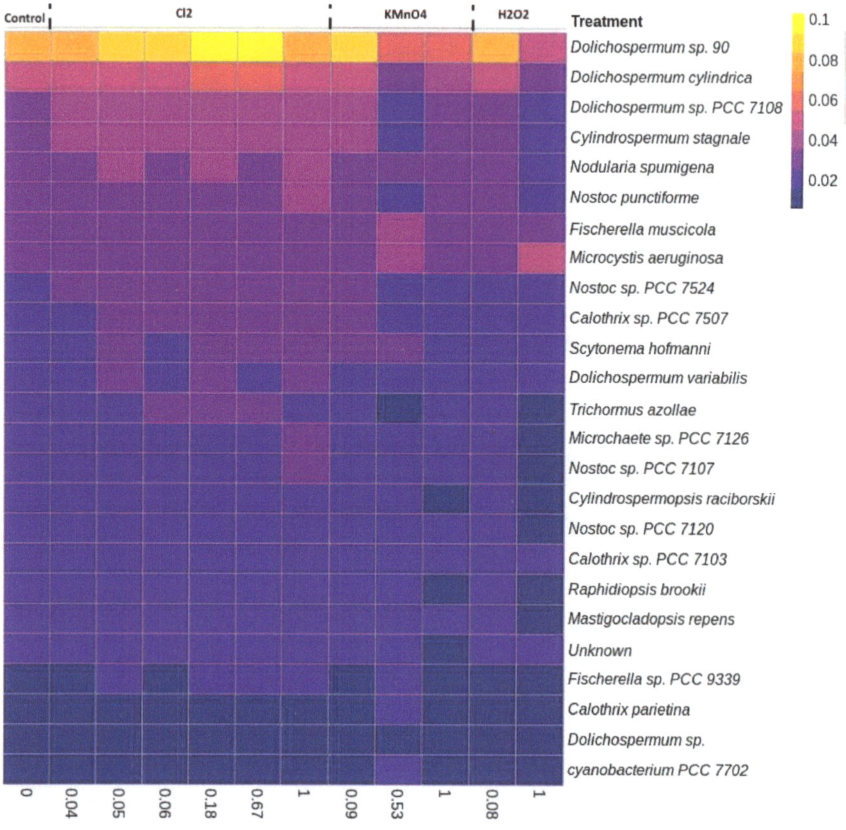

Figure 4. Cyanobacterial species heat map following the oxidation using Cl_2 (0.6 mg/L), $KMnO_4$ (5 mg/L), H_2O_2 (10 mg/L) (1 August 2018).

In the 29 August samples, *Microcystis aeruginosa* was the most abundant species (Figure S3); its relative abundance increased after chlorination and $KMnO_4$ oxidation (as the chlorine exposure increased) and decreased after H_2O_2 oxidation as compared to the control condition. Despite the very low relative abundance of *Dolichospermum sp.90, Dolichospermum cylindrica* and *Dolichospermum sp. PCC7108*, similar trends were observed on 1 August (when *Dolichospermum* was abundant).

The community richness and diversity indices for each treatment for 1 August and 29 August samples are illustrated in Figure 5 and Figure S4. Shannon and Chao1 show a small decline following chlorination in comparison with the control, while they increase slightly following $KMnO_4$ exposure. The total cell numbers following $KMnO_4$ decreased by up to 63% for high $KMnO_4$ exposure (Figure S7). A remarkable decrease in richness is observed at high relative CT of H_2O_2, while the diversity (Shannon index) increases. The decline in the richness index could be the result of some less abundant species no longer identified. Indeed, total cell counts following H_2O_2 relative CT = 1, decreased by more than 50% (Figure S6). The same trends in the alpha diversity measurements are observed in the last week of the sampling (29 August), where the *Microcysits* were the most abundant genus (Figure S4).

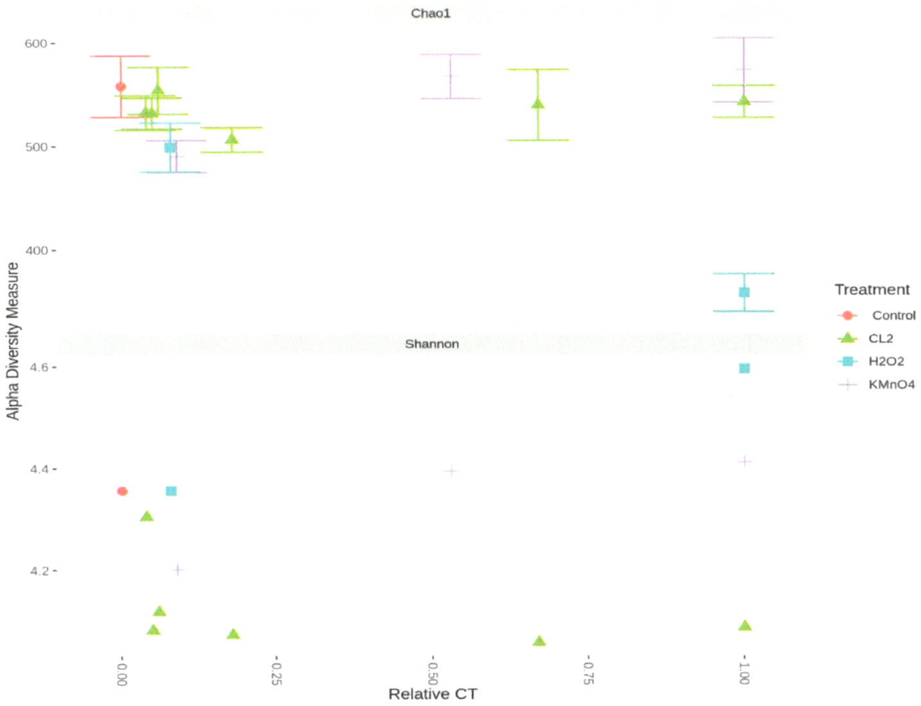

Figure 5. Alpha diversity measures of the cyanobacterial community following oxidation Cl_2 (0.6 mg/L), $KMnO_4$ (5 mg/L), H_2O_2 (10 mg/L) (1 August 2018).

2.4. *Cyanobacterial Community Assessment Following Oxidation; Longitudinal Study*

The induced changes of cyanobacterial composition structure following the oxidation (Cl_2, $KMnO_4$, O_3 and, H_2O_2) are assessed separately. The analysis was performed at the genus and species level.

2.4.1. Chlorination (Cl_2)

The chlorination experiments were conducted on 1 August (Figure 6a), and 29 August (Figure 6b). In the first chlorination trial (1 August), the abundant genera were *Dolichospermum* and *Nostoc*, representing approximatively 20% and 10% of the cyanobacterial community, respectively (Figure 6a). Figure 6 shows a limited effect of chlorination on the relative abundance of all cyanobacteria genera except *Synechococcus*. In addition, taxonomic cell counts show decrease (up to 30%) in total cyanobacteria cell counts for the trial on 1 August and limited variation for the 29 August trial (15% variation) (Figure S5). In terms of the species, *Dolichospermum sp.90* and *Dolichospermum cylindrica* were dominant for the 1 August trial. For the 29 August trial, *Microcystis aeruginosa* was the dominant species, followed by *Dolichospermum sp.90* (Figure S6). In all trials, as chlorination exposure increased, the relative abundance of the abundant species, either *Dolichospermum sp.90 or Microcystis aeruginosa*, increased slightly. Moreover, the relative abundance of *Dolichospermum sp.90* increased as *Microcystis aeruginosa* did on 29 August. Chlorination results show that *Dolichospermum* species and *Microcystis aeruginosa* are relatively more persistent than the other species.

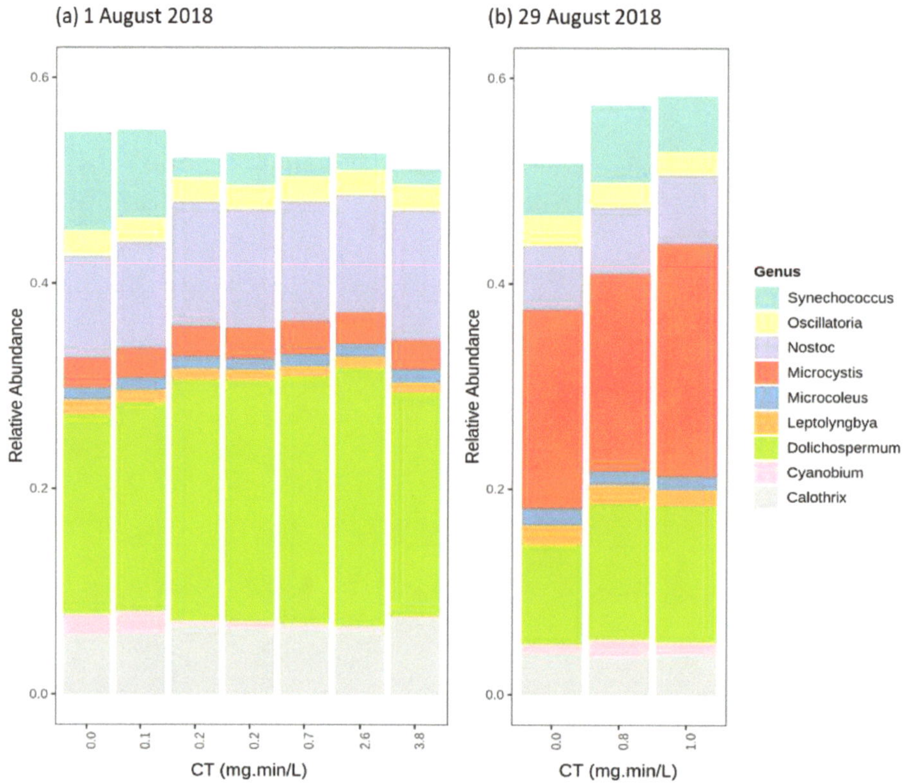

Figure 6. The relative abundance of the most abundant genus following chlorination (0.6 mg/L and 0.2 mg/L) (**a**) 1 August 2018 (*Dolichospermum* genus abundant) (**b**) 29 August 2018 (*Microcystis* genus abundant).

2.4.2. Potassium Permanganate ($KMnO_4$)

The relative abundance of the different genera following oxidation using permanganate shows limited variation for both $KMnO_4$ tests (Figure 7). Total cell counts decreased (up to 57%) in the first trial (1 August) and remained stable in the second $KMnO_4$ trial (less than 1% variation) (29 August). Total cell counts following the first $KMnO_4$ trial showed a decrease at 278 mg·min/L (Figure S7). The relative abundance of the *Dolichospermum sp.90* increased slightly, whether it was the abundant species or not, suggesting the relative persistence of the *Dolichospermum sp.90* during $KMnO_4$ oxidation (Figure S8).

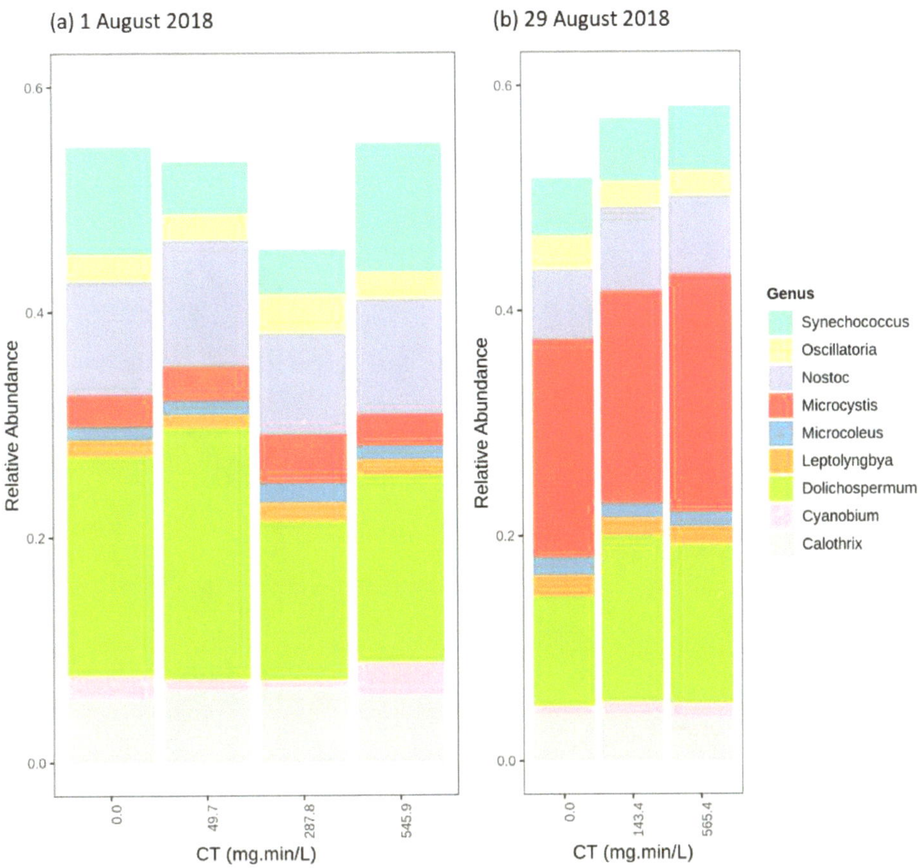

Figure 7. Relative abundance of the most abundant genus following $KMnO_4$ (5 mg/L) oxidation (**a**) 1 August 2018 (*Dolichospermum* genus abundant) (**b**) 29 August 2018 (*Microcystis* genus abundant).

2.4.3. Ozonation (O_3)

The first and second ozonation trials were performed on 15 August (*Dolichospermum* most abundant genus) and 21 August (*Synechococcus* most abundant genus). Cyanobacterial community results following ozonation in the second trial (at genus level—Figure 8) showed a decline of the relative abundance of *Synechococcus*, followed by an increase for *Microcystis*, as compared to the control condition. However, no significant variation was observed for the relative abundance of the different genera in the first ozonation trials. Furthermore, total cyanobacteria cell counts revealed no significant change for both ozonation tests (up to 15%) (Figure S9). In the control condition of the 15 August ozonation trial, *Microcystis aeruginosa* was the dominant species, followed by *Dolichospermum sp.90* (Figure S10). *Dolichospermum sp.90* was not the dominant species, but it remained intact following ozonation. Although *Cyanobioum gracil* and *Synechococcus sp.* were the dominant species in the 21 August trial control, *Microcystis aeruginosa* became the dominant species following ozonation. At the low dosage applied, only secondary oxidation radical by-products are likely to react with cyanobacterial cells. Under these soft ozonation conditions, unicellular *Cyanobium gracil* and *Synechococcus sp.* cells were more susceptible than *Microcystis aeruginosa*.

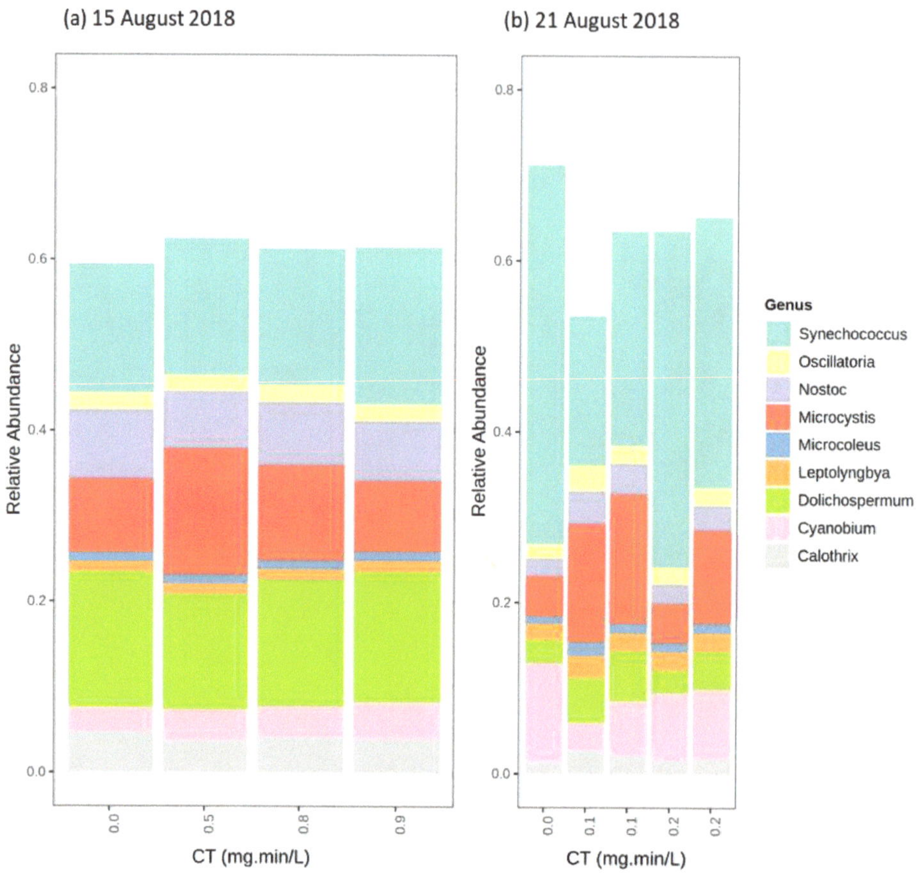

Figure 8. Relative abundance of the most abundant genus following O_3 (0.3 mg/L and 0.1 mg/L) oxidation (**a**) 15 August 2018 (*Dolichospermum* genus Abundant), (**b**) 21 August 2018 (*Synechococcus* genus abundant).

2.4.4. Hydrogen Peroxide (H_2O_2)

The *Dolichospermum/Dolichospermum sp.90* and *Microcystis/Microcystis aeruginosa* were the most abundant genus/species on 1 August and 29 August H_2O_2 oxidation, respectively. The relative abundance of the *Dolichospermum* declined following H_2O_2 exposure of CT = 7035 mg·min/L. *Microcystis* relative abundance decreases by more than 10% at 8442 mg·min/L exposure of H_2O_2 on 29 August experiment (Figure 9). At the same H_2O_2 exposure, *Dolichospermum* decreased by 5%. Total cyanobacteria cell counts declined following H_2O_2 oxidation for both dates: 52% decrease on 1 August and 49% decrease on 29 August (Figure S11). The relative abundance of the dominant species (*Dolichospermum sp.90* 1 August and *Microcystis aeruginosa* 29 August) decreases after H_2O_2 oxidation. The relative abundance of *Microcystis aeruginosa* in the H_2O_2 trial on 1 August increased after high H_2O_2 exposure. *Microcystis aeruginosa* was susceptible to H_2O_2 oxidation when abundant (as the relative abundance decreased), but it persists as an abundant species. Our results are in accordance with Lusty and Gobler. (2020) [37]. Figure S12 unveils higher susceptibility of *Dolichospermum* species to high H_2O_2 exposure as compared to *Microcystis aeruginosa*. Our results show the effect of high H_2O_2 exposure on the cyanobacteria species, which is in accordance with previous studies [61,64].

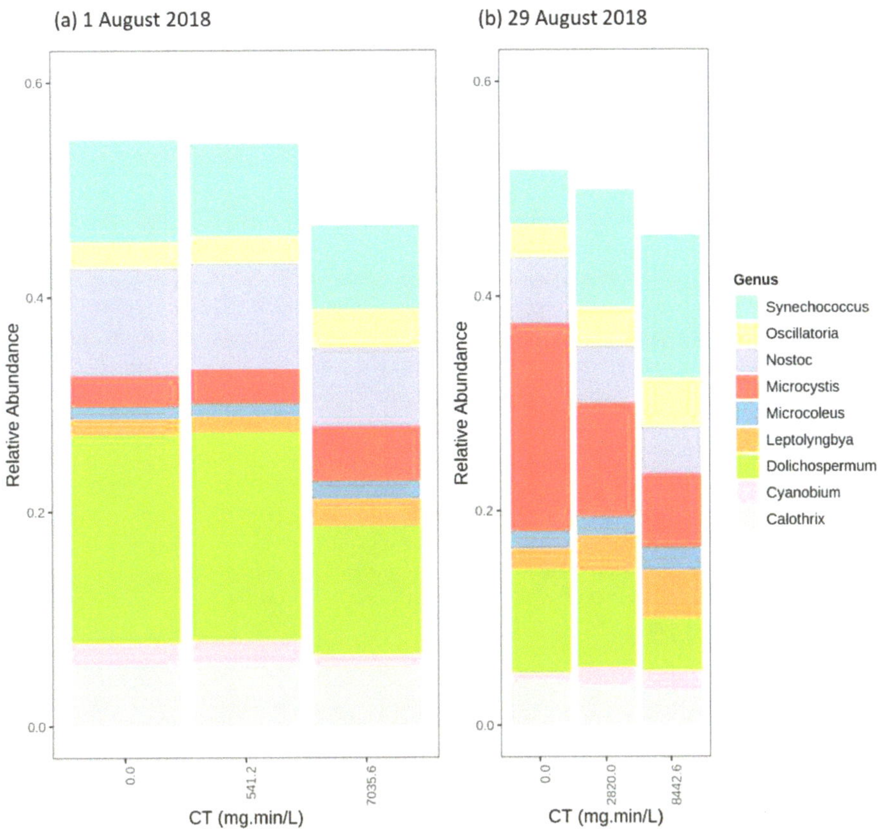

Figure 9. Relative abundance of the most abundant genus following H_2O_2 (10 mg/L) oxidation on (**a**) 1 August 2018 (*Dolichospermum* genus abundant) (**b**) 29 August 2018 (*Microcystis* genus abundant).

Cyanobacterial composition analysis at species level exhibits relative persistence of different *Dolichospermum* species and *Microcystis aeruginosa* when it is abundant following soft-chlorination, soft ozonation and permanganate oxidation. This result is in accordance with the previous studies, which were mainly based on the cell count and the lab-cultured species [10,12,14,58,59]. Our results highlight the ability of hydrogen peroxide to decrease the relative abundance of different taxa, including the toxin-producing taxa of interest. H_2O_2 selectivity provides an efficient barrier against toxic cyanobacteria entering a drinking water treatment plant.

2.5. Comparison of the Microscopic Cell Count vs. High Throughput Sequencing Results

The taxonomic cell count results (genus) from the chlorination on 1 August and ozonation on 15 August are presented in Figure S13. The observed genera from microscopic cell counts do not completely match with high throughput sequencing results as *Aphanocapsa* and *Aphanothece* were only reported by taxonomic cell count. Misclassification at the genus level is less common than for species level; microscopic cell counts showed that *Dolichospermum spiroids*, *Aphanocapsa delicatissma*, *Aphanotheche clathrate brevis* and *Aphanocapsa holistca* were the most abundant species following chlorination and ozonation. However, high throughput sequencing did not identify them as an abundant species. The abundant species from high throughput sequencing were *Microcystis aeruginosa*, *Dolichospermum Sp.90* and, *Dolichospermum cylindrica* (Figures S6 and S10). Despite the potential of the microscopic cell count to provide absolute quantitative data such as cells/mL and biovolume, it has some drawbacks.

The differences in community composition structure retrieved from the microscopic taxonomic cell counts and high throughput sequencing could be the result of the limitations inherent to these methods. For high throughput sequencing, these limitations include incomplete DNA sequencing libraries or using different libraries to identify genus and species. In the case of microscopic taxonomic cell counts, several sources of uncertainty and error have been identified [65,66]. The morphological similarity among cyanobacteria taxa may lead to overlooking or misidentifying cyanobacteria under the microscope, especially low abundant species [5,28]. Few genomes might be available for some taxa identified by taxonomic cell count (e.g., *Aphanocapsa* and *Aphanothece*), which may result in low relative abundance of these taxa using high throughput sequencing. Additionally, oxidation may further hinder the ability to identify cells because of its impact on the cell structure. Oxidation at higher dosages can cause significant morphological deformation of the cyanobacteria species, especially for H_2O_2 and $KMnO_4$ in our study [63]. Flow cytometry conducted using the method described by Moradinejad et al. (2019) showed partial membrane damage (up to 80%) under the soft-oxidation conditions performed in this study (data not shown) [63].

This study is the first to provide insights into the impact of different pre-oxidants (doses and contact time) on the diversity and cyanobacterial community composition. The experimental design focused on CTs at which the shifts occurred, in order to provide actionable results for water utilities. Additional investigation would be beneficial to extend our observations to other water bodies. In addition, using assembly/binning method is suggested for future studies to provide a more accurate view of cyanobacterial genus/species.

3. Conclusions

A comparison of the microscopic vs. high throughput sequencing results demonstrates the ability and robustness of high throughput sequencing to fully reveal a wide diversity of cyanobacterial communities in response to oxidant stress.

Results from longitudinal sampling over a bloom period of 4 weeks by high throughput sequencing highlight quick composition and abundance shifts in the cyanobacterial communities that occur within a day. High throughput sequencing revealed clearer shifts during a bloom from an initial dominance of *Dolichospermum/Dolichospermum Sp.90* toward a late summer dominance by *Microcystis/ Microcystis aeruginosa*.

Overall, pre-oxidation caused deeper changes in the diversity of whole bacterial communities, especially proteobacteria, than was observed for the cyanobacterial community. Such changes should be considered when assessing the impact of using oxidants for onsite source control.

Depending on the oxidants used, alpha diversity indexes (Shannon and Chao1) showed that oxidation resulted in different structural composition shifts within bacterial and cyanobacterial communities.

Soft-chlorination using dosages 0.2 and 0.6 mg/L that were typically used for pre-oxidation caused a progressive shift in the bacterial and cyanobacterial communities with increasing CTs (0–3.8 mg·min/L). The results using $KMnO_4$ (5 mg/L) and H_2O_2 (10 mg/L) induced larger and distinct shifts in the structural composition of bacterial and cyanobacterial communities.

Regardless of the significant differences in community distribution shifts caused by different oxidants, some toxin producing species could persist after oxidation whether they were dominant species or not.

Soft-chlorination results revealed that *Dolichospermum sp.90* was relatively persistent with increasing CTs whether it was the dominant species or not. Relative persistence of *Dolichospermum sp.90* was also observed within $KMnO_4$ oxidation with increasing $KMnO_4$ exposure, regardless of the dominant species.

Soft-ozonation using dosage 0.1–0.3 mg/L results showed the relative persistence of *Microcystis aeruginosa* (the dominant species) with increasing ozone exposures (0–0.9 mg·min/L).

Only pre-oxidation with H_2O_2 (10 mg/L) caused a clear decrease in the relative abundance of all taxa and some species including the toxin-producing taxa of interest. As such, H_2O_2 would provide an effective first barrier against toxin producing cyanobacteria entering the drinking water treatment plant.

Selection of the most effective pre-oxidant for drinking water purposes should be made considering its impact on the cyanobacterial community structure/diversity, prevention of disinfection by-product formation and other drivers such as color and the presence of taste odor compounds.

4. Material and Methods

4.1. Sampling Site Description

The oxidation tests were conducted using natural bloom samples. Cyanobacterial bloom samples were collected and transferred to the laboratory from the Bedford water treatment plant intake (Missisquoi Bay—Lake Champlain) in southern Quebec, Canada (45°02′22.0″ N 73°04′40.5″ W). Sampling was performed during the bloom season from 1 August to 29 August 2018. Several cyanobacterial blooms have been documented for the Lake Champlain in previous years [14,21,42,43].

4.2. Chemicals and Reagents

A free chlorine stock of 2000 mg/L was freshly prepared from sodium hypochlorite (5.25%) on the day of the experiment. Free chlorine residual was measured using an N,N-diethyl-p-phenylenediamine (DPD) colorimetric method based on Standard Methods (SM) 4500-Cl G [67]. Samples were dosed at room temperature (22 °C). A stock of sodium thiosulfate 3000 mg/L was used to quench the chlorine samples at a dose of 1.1 mg/L per 1 mg/L chlorine.

A potassium permanganate ($KMnO_4$) stock solution of 5000 mg/L was prepared by dissolving $KMnO_4$ crystals into the ultrapure water. A DPD colorimetric method ratio of 0.891 $KMnO_4/Cl_2$ SM 4500-Cl G was used to determine the $KMnO_4$ residual [67]. An amount of 1.2 mg/L sodium thiosulfate per 1 mg/L $KMnO_4$ was used to quench further oxidation with $KMnO_4$.

A bench-scale ozone generator (details in [14]) was used to prepare an ozone stock solution 50–60 mg/L. SM 4500-O_3 was used to determine the stock and ozone residual concentration [14]. An ozone stock solution (50–60 mg/L) was prepared with gaseous ozone using a bench-scale ozone generator (additional details in [14]). Ozone stock concentration and residual ozone in water samples were measured using SM 4500-O_3 [14]. An amount of 1.6 mg/L sodium thiosulfate per 1 mg/L ozone was used to quench the ozonation samples.

Stabilized hydrogen peroxide (30%, Sigma Aldrich, MO, USA) was used to prepare the Hydrogen peroxide (H_2O_2) stock (10 g/L). A colorimetric test kit (Chemetrics K-5510, Midlands, VA, USA) was performed to measure the hydrogen peroxide residual. Moreover, 1.2mg/L sodium thiosulfate per 1 mg/L H_2O_2 was used to quench further oxidation with H_2O_2.

The selected oxidant doses and contact times are presented in Table 2. The doses (concentration) and contact time have been selected based on the common pre-oxidation doses used in drinking water treatment intake.

Oxidant exposures, concentration vs. contact time, (CT) were calculated using Equation (1):

$$CT = \int_0^t [Oxidant]dt = \frac{C_0}{k_{decay}}(e^{k_{decay}t} - 1) \quad (1)$$

where k_{decay} (min^{-1}) is the first-order decay rate, t (min) is the exposure time, and C_0 (mg/L) is the initial concentration of oxidant at time zero. The selection of oxidant dose was based on the commonly used pre-oxidation doses in the operation of the drinking water treatment plant.

Table 2. Experimental plan.

Oxidant	Water Type	Oxidant Dose (mg/L)	Contact Time
Cl_2	Real Bloom, Missisquoi Bay, Quebec, Canada	0.2 0.6	1 min 2 min 5 min 10 min 60 min 120 min
O_3		0.1 0.3	1 min 2 min 5 min 10 min
$KMnO_4$		5	30 min 60 min 120 min
H_2O_2		10	6 h 24 h

4.3. DNA Extraction, Metagenomic Preparation, Sequencing and Bioinformatic Analysis

Total nucleic acid was extracted from the frozen filters (after filtration, samples were quickly transferred to −80 °C before DNA extraction) using an RNeasy PowerWater Kit (Qiagen Group, Germantwon, MD, USA) with modification. Before the extraction, 200 μL of nuclease-free water and 5 μL of TATAA Universal DNA spike II (TATAA Biocenter AB) were added to the filters to evaluate DNA extraction yields using RT-qPCR. RNeasy PowerWater Isolation kit solution PM1 was used to lyse the cells along with Dithiothreitol (DTT), which prevents disulfide bonds forming residues of proteins. A total volume of 60 μL nuclease-free water provided with the kit was used to elute the total nucleic acid, of which 30 μL of DNA (with minimum 1 ng of DNA) extracts were stored at −20 °C. DNA was subsequently purified with the Zymo Kit (Zymo Research, Irvine, CA, USA) according to the manufacturer's instructions. Each DNA sample was resuspended in 60 μL of nuclease-free water and quantified with a Qubit v.2.0 fluorometer (Life Technologies, Burlington, ON, Canada). A volume of 30 μL DNA was sent for pyrosequencing (Roche 454 FLX instrumentation with Titanium chemistry) to the Genome Quebec.

An Illumina NovaSeq 6000 platform using S4 flow cells was applied to sequence DNA libraries. A home-made bioinformatic pipeline was used for further analysis of Paired-end raw reads of 150 base pairs (bp) as follows. First, raw reads trimming quality was performed using the SolexaQA v3.1.7.1, default parameters [68]. Further analyses were carried out on the trimmed reads shorter than 75 nt. An in-house script, based on the screening of identical leading 20 bp, was used to remove artificial duplicates. Gene fragments were predicted using FragGeneScan-Plus v3.0 based on the trimmed high-quality reads [69]. Then, predicted fragments of protein were clustered at 90% similarity level using cd-hit v4.8.1 [70]. A diamond engine was used for similarity search on the M5nr database based on a representative of each cluster (https://github.com/MG-RAST/myM5NR). Best hits (minimal e-value of 1×10^{-5}) combined with the last common ancestor approach were used to assess the taxonomic affiliation of protein fragments. It should be mentioned that the annotation process uses a read-mapping process of small gene fragments of encoding proteins on a large database of proteins. One gene fragments encoding protein in the database could match with multiple species or strains.

4.4. DOC and Cyanobacteria Cell Count

Pre-rinsed 0.45 μm membrane filters (Supor 45 μm, 47 m, PES PALL, Port Washington, NY, USA) and carbon-free glass vials were used for Dissolved Organic Carbon (DOC) samples. DOC measurements were performed via a 5310 total organic carbon analyzer (Sievers Analytical Instruments,

Boulder, CO, USA). An inverted microscope with 20× magnification was used to perform cell count samples preserved with Lugol's Iodine [71,72].

4.5. Statistical Analysis

All analyses were performed using a custom bioinformatics pipeline implemented in R (v.3.6,2, RStudio, Inc., Boston, MA, USA), phyloseq (V.1.28.0) to visualize the community composition at phylum (all bacteria reads), order, and genus (cyanobacteria reads) [73]. The twenty-five most abundant cyanobacteria species were visualized using pheatmap (v.1.0.12) [74]. Then, the alpha diversity metrics were estimated using phyloseq's estimate richness function (Shannon and Chao1). Taxonomic data were normalized by the centred log-ratio transformation using easy CODA (v.0.31.1) [75]. The beta-diversity was analyzed using the vegan package (v.2.5-6), where the similarity matrices were calculated based on the Euclidean distance [76]. The homogeneity of variances of normalized data related to each oxidant was analyzed before building the model. A Redundancy Analysis (RDA) constrained ordination to each oxidant applied to the cyanobacteria genus and tested by the permutation test (>95% significance).

Supplementary Materials: The following are available online at http://www.mdpi.com/2072-6651/12/11/728/s1. Figure S1: Principal components analysis (PCA) of the normalized relative abundance of comparative metagenomics reads in 29 August 2018 sample. Data are plotted following the genus-level classification (a) PCA analysis of bacterial community following oxidation using different CT (b) PCA of the cyanobacterial community following oxidation using different CT. Figure S2: Relative abundance of the most abundant genus following the oxidation using Cl_2, $KMnO_4$, H_2O_2 (1 August 2018 abundant: *Dolichospermum*). Figure S3: Cyanobacterial Species heat map following the oxidation using Cl_2, $KMnO_4$, H_2O_2 (29 August 2018 abundant: *Microcystis*). Figure S4: Alpha diversity measures of cyanobacterial community following oxidation Cl_2, $KMnO_4$, H_2O_2 (29 August 2018, abundant genus: *Microcystis*). Figure S5: Total cyanobacteria cell counts following chlorination for 1 August 2018 trial and 29 August 2018 trial. Figure S6: Cyanobacterial species heat map following the chlorination (a) 1 August 2018 trial, (b) 29 August 2018 trial. Figure S7: Total cyanobacteria cell counts following the permanganate oxidation for 1 August 2018 trial and 29 August 2018 trial. Figure S8: Cyanobacterial Species heat map following the oxidation using $KMnO_4$ (a) 1 August 2018 trial, (b) 29 August 2018 trial. Figure S9: Total cyanobacteria cell counts following the O_3 oxidation for 15 August 2018 trial and 21 August 2018 trial. Figure S10: Cyanobacterial Species heat map following the oxidation using O_3 (a) 15 August 2018 trial, (b) 21 August 2018 trial. Figure S11: Total cyanobacteria cell counts following the H_2O_2 oxidation for 1 August 2018 trial and 29 August 2018 trial. Figure S12: Cyanobacterial Species heat map following the oxidation using H_2O_2 (a) 1 August 2018 trial (b) 29 August 2018 trial. Figure S13: Relative abundance of cyanobacteria species (via light Microscopy) following oxidation (a) O_3 second trial (15 August 2018) (b) Cl_2 first trial (1 August 2018).

Author Contributions: Conceptualization, S.M., H.T., A.Z., S.D. and M.P.; Formal analysis, S.M., H.T., J.F.G.M. and Y.T.; Funding acquisition, J.S., S.D. and M.P.; Methodology, S.M., H.T., J.F.G.M., J.S., Y.T., A.Z., S.D. and M.P.; Software, J.F.G.M.; Supervision, A.Z., S.D. and M.P.; Visualization, S.M., H.T., J.F.G.M., S.D. and M.P.; Writing—original draft, S.M., H.T. and J.F.G.M.; Writing—review and editing, S.M., H.T., J.F.G.M., J.S., Y.T., A.Z., S.D. and M.P. All authors have read and agreed to the published version of the manuscript.

Funding: This research was funded by Genome Canada and Genome Quebec: Algal Blooms, Treatment, Risk Assessment, Prediction and Prevention through Genomics (ATRAPP) Project, Grant number Genome Canada/UM RQ000607 and the APC was funded by Genome Canada and Genome Quebec (ATRAPP project).

Acknowledgments: The authors acknowledge support from Algal Blooms, Treatment, Risk Assessment, Prediction and Prevention through Genomics (ATRAPP), the authors thank the staff at NSERC Industrial Chair on Drinking Water at Polytechnique Montreal. The authors thank the staff at Microbial Evolutionary Genomics (Shapiro lab) at Universite de Montreal.

Conflicts of Interest: The authors declare no conflict of interest.

References

1. Wells, M.L.; Trainer, V.L.; Smayda, T.J.; Karlson, B.S.; Trick, C.G.; Kudela, R.M.; Ishikawa, A.; Bernard, S.; Wulff, A.; Anderson, D.M.; et al. Harmful algal blooms and climate change: Learning from the past and present to forecast the future. *Harmful Algae* **2015**, *49*, 68–93. [CrossRef] [PubMed]
2. Paerl, H.W.; Paul, V.J. Climate change: Links to global expansion of harmful cyanobacteria. *Water Res.* **2012**, *46*, 1349–1363. [CrossRef]
3. Al-Sammak, M.A.; Hoagland, K.D.; Cassada, D.; Snow, D.D. Co-occurrence of the cyanotoxins BMAA, DABA and anatoxin-a in Nebraska reservoirs, fish, and aquatic plants. *Toxins* **2014**, *6*, 488–508. [CrossRef] [PubMed]

4. Westrick, J.A.; Szlag, D.C.; Southwell, B.J.; Sinclair, J. A review of cyanobacteria and cyanotoxins removal/inactivation in drinking water treatment. *Anal. Bioanal. Chem.* **2010**, *397*, 1705–1714. [CrossRef]
5. Casero, M.C.; Velázquez, D.; Medina-Cobo, M.; Quesada, A.; Cirés, S. Unmasking the identity of toxigenic cyanobacteria driving a multi-toxin bloom by high-throughput sequencing of cyanotoxins genes and 16S rRNA metabarcoding. *Sci. Total Environ.* **2019**, *665*, 367–378. [CrossRef] [PubMed]
6. Zamyadi, A.; Ho, L.; Newcombe, G.; Bustamante, H.; Prévost, M. Fate of toxic cyanobacterial cells and disinfection by-products formation after chlorination. *Water Res.* **2012**, *46*, 1524–1535. [CrossRef] [PubMed]
7. Pazouki, P.; Prévost, M.; McQuaid, N.; Barbeau, B.; De Boutray, M.-L.; Zamyadi, A.; Dorner, S. Breakthrough of cyanobacteria in bank filtration. *Water Res.* **2016**, *102*, 170–179. [CrossRef] [PubMed]
8. Zamyadi, A.; Dorner, S.; Ndong, M.; Ellis, D.; Bolduc, A.; Bastien, C.; Prévost, M. Low-risk cyanobacterial bloom sources: Cell accumulation within full-scale treatment plants. *J. Am. Water Work. Assoc.* **2013**, *105*, E651–E663. [CrossRef]
9. Almuhtaram, H.; Cui, Y.; Zamyadi, A.; Hofmann, R. Cyanotoxins and Cyanobacteria Cell Accumulations in Drinking Water Treatment Plants with a Low Risk of Bloom Formation at the Source. *Toxins* **2018**, *10*, 430. [CrossRef] [PubMed]
10. Fan, J.; Daly, R.; Hobson, P.; Ho, L.; Brookes, J. Impact of potassium permanganate on cyanobacterial cell integrity and toxin release and degradation. *Chemosphere* **2013**, *92*, 529–534. [CrossRef] [PubMed]
11. Fan, J.; Ho, L.; Hobson, P.; Daly, R.; Brookes, J.D. Application of Various Oxidants for Cyanobacteria Control and Cyanotoxin Removal in Wastewater Treatment. *J. Environ. Eng.* **2014**, *140*, 04014022. [CrossRef]
12. Coral, L.A.; Zamyadi, A.; Barbeau, B.; Bassetti, F.J.; Lapolli, F.R.; Prevost, M. Oxidation of *M. aeruginosa* and *A. flos-aquae* by ozone: Impacts on cell integrity and chlorination by-product formation. *Water Res.* **2013**, *47*, 2983–2994. [CrossRef] [PubMed]
13. Ding, J.; Shi, H.; Timmons, T.; Adams, C. Release and Removal of Microcystins from Microcystis during Oxidative-, Physical-, and UV-Based Disinfection. *J. Environ. Eng.* **2010**, *136*, 2–11. [CrossRef]
14. Zamyadi, A.; Coral, L.A.; Barbeau, B.; Dorner, S.; Lapolli, F.R.; Prévost, M. Fate of toxic cyanobacterial genera from natural bloom events during ozonation. *Water Res.* **2015**, *73*, 204–215. [CrossRef]
15. Zamyadi, A.; Greenstein, K.E.; Glover, C.M.; Adams, C.; Rosenfeldt, E.; Wert, E.C. Impact of Hydrogen Peroxide and Copper Sulfate on the Delayed Release of Microcystin. *Water* **2020**, *12*, 1105. [CrossRef]
16. Zamyadi, A.; Ho, L.; Newcombe, G.; Daly, R.I.; Burch, M.; Baker, P.; Prévost, M. Release and Oxidation of Cell-Bound Saxitoxins during Chlorination of *Anabaena circinalis* Cells. *Environ. Sci. Technol.* **2010**, *44*, 9055–9061.e9. [CrossRef]
17. Vlad, S.; Anderson, W.B.; Peldszus, S.; Huck, P.M. Removal of the cyanotoxin anatoxin-a by drinking water treatment processes: A review. *J. Water Health* **2014**, *12*, 601–617. [CrossRef]
18. He, X.; Wert, E.C. Colonial cell disaggregation and intracellular microcystin release following chlorination of naturally occurring Microcystis. *Water Res.* **2016**, *101*, 10–16. [CrossRef]
19. Wert, E.C.; Dong, M.M.; Rosario-Ortiz, F.L. Using digital flow cytometry to assess the degradation of three cyanobacteria species after oxidation processes. *Water Res.* **2013**, *47*, 3752–3761. [CrossRef]
20. Merel, S.; Walker, D.; Chicana, R.; Snyder, S.A.; Baurès, E.; Thomas, O. State of knowledge and concerns on cyanobacterial blooms and cyanotoxins. *Environ. Int.* **2013**, *59*, 303–327. [CrossRef]
21. Zamyadi, A.; Dorner, S.; Sauvé, S.; Ellis, D.; Bolduc, A.; Bastien, C.; Prévost, M. Species-dependence of cyanobacteria removal efficiency by different drinking water treatment processes. *Water Res.* **2013**, *47*, 2689–2700. [CrossRef]
22. Zhu, L.; Zuo, J.; Song, L.; Gan, N. Microcystin-degrading bacteria affect mcyD expression and microcystin synthesis in *Microcystis* spp. *J. Environ. Sci.* **2016**, *41*, 195–201. [CrossRef] [PubMed]
23. Zamyadi, A.; Romanis, C.; Mills, T.; Neilan, B.; Choo, F.; Coral, L.A.; Gale, D.; Newcombe, G.; Crosbie, N.D.; Stuetz, R.M.; et al. Diagnosing water treatment critical control points for cyanobacterial removal: Exploring benefits of combined microscopy, next-generation sequencing, and cell integrity methods. *Water Res.* **2019**, *152*, 96–105. [CrossRef] [PubMed]
24. Hawkins, P.R.; Holliday, J.; Kathuria, A.; Bowling, L. Change in cyanobacterial biovolume due to preservation by Lugol's Iodine. *Harmful Algae* **2005**, *4*, 1033–1043. [CrossRef]
25. American Water Works Association (AWWA). *M57-Algae Source to Treatment*; American Water Works Association: Denver, CO, USA, 2010.

26. Berry, M.A.; Davis, T.W.; Cory, R.M.; Duhaime, M.B.; Johengen, T.H.; Kling, G.W.; Marino, J.A.; Uyl, P.A.D.; Gossiaux, D.; Dick, G.J.; et al. Cyanobacterial harmful algal blooms are a biological disturbance to Western Lake Erie bacterial communities. *Environ. Microbiol.* **2017**, *19*, 1149–1162. [CrossRef] [PubMed]
27. Lezcano, M.Á.; Velázquez, D.; Quesada, A.; El-Shehawy, R. Diversity and temporal shifts of the bacterial community associated with a toxic cyanobacterial bloom: An interplay between microcystin producers and degraders. *Water Res.* **2017**, *125*, 52–61. [CrossRef] [PubMed]
28. Kim, K.H.; Yoon, Y.; Hong, W.-Y.; Kim, J.; Cho, Y.-C.; Hwang, S.-J. Application of metagenome analysis to characterize the molecular diversity and saxitoxin-producing potentials of a cyanobacterial community: A case study in the North Han River, Korea. *Appl. Biol. Chem.* **2018**, *61*, 153–161. [CrossRef]
29. Eldridge, S.L.C.; Wood, T.M. Annual variations in microcystin occurrence in Upper Klamath Lake, Oregon, based on high-throughput DNA sequencing, qPCR, and environmental parameters. *Lake Reserv. Manag.* **2019**, *36*, 31–44. [CrossRef]
30. Casero, M.C.; Ballot, A.; Agha, R.; Quesada, A.; Cirés, S. Characterization of saxitoxin production and release and phylogeny of sxt genes in paralytic shellfish poisoning toxin-producing Aphanizomenon gracile. *Harmful Algae* **2014**, *37*, 28–37. [CrossRef]
31. Woodhouse, J.N.; Kinsela, A.S.; Collins, R.N.; Bowling, L.C.; Honeyman, G.L.; Holliday, J.K.; Neilan, B.A. Microbial communities reflect temporal changes in cyanobacterial composition in a shallow ephemeral freshwater lake. *ISME J.* **2016**, *10*, 1337–1351. [CrossRef]
32. Pessi, I.S.; Maalouf, P.D.C.; Laughinghouse, H.D.; Baurain, D.; Wilmotte, A. On the use of high-throughput sequencing for the study of cyanobacterial diversity in Antarctic aquatic mats. *J. Phycol.* **2016**, *52*, 356–368. [CrossRef] [PubMed]
33. Scherer, P.I.; Millard, A.D.; Miller, A.; Schoen, R.; Raeder, U.; Geist, J.; Zwirglmaier, K. Temporal Dynamics of the Microbial Community Composition with a Focus on Toxic Cyanobacteria and Toxin Presence during Harmful Algal Blooms in Two South German Lakes. *Front. Microbiol.* **2017**, *8*, 2387. [CrossRef] [PubMed]
34. Willis, A.; Woodhouse, J.N. Defining Cyanobacterial Species: Diversity and Description Through Genomics. *Crit. Rev. Plant Sci.* **2020**, *39*, 101–124. [CrossRef]
35. Xu, H.; Pei, H.; Jin, Y.; Ma, C.; Wang, Y.; Sun, J.; Li, H. High-throughput sequencing reveals microbial communities in drinking water treatment sludge from six geographically distributed plants, including potentially toxic cyanobacteria and pathogens. *Sci. Total Environ.* **2018**, *634*, 769–779. [CrossRef] [PubMed]
36. Pei, H.; Xu, H.; Wang, J.; Jin, Y.; Xiao, H.; Ma, C.; Sun, J.; Li, H. 16S rRNA Gene Amplicon Sequencing Reveals Significant Changes in Microbial Compositions during Cyanobacteria-Laden Drinking Water Sludge Storage. *Environ. Sci. Technol.* **2017**, *51*, 12774–12783. [CrossRef]
37. Lusty, M.W.; Gobler, C.J. The Efficacy of Hydrogen Peroxide in Mitigating Cyanobacterial Blooms and Altering Microbial Communities across Four Lakes in NY, USA. *Toxins* **2020**, *12*, 428. [CrossRef]
38. Newcombe, G.; House, J.; Ho, L.; Baker, P.; Burch, M. *Management Strategies for Cyanobacteria (Blue-Green Algae): A Guide for Water Utilities*; The Cooperative Research Centre for Water Quality and Treatment: Adelaïde, Australia, 2010; p. 112.
39. Lee, C.; Marion, J.; Cheung, M.; Lee, C.S.; Lee, J. Associations among Human-Associated Fecal Contamination, Microcystis aeruginosa, and Microcystin at Lake Erie Beaches. *Int. J. Environ. Res. Public Health* **2015**, *12*, 11466–11485. [CrossRef]
40. Vadde, K.K.; Feng, Q.; Wang, J.; McCarthy, A.J.; Sekar, R. Next-generation sequencing reveals fecal contamination and potentially pathogenic bacteria in a major inflow river of Taihu Lake. *Environ. Pollut.* **2019**, *254*, 113108. [CrossRef]
41. Ballesté, E.; Blanch, A.R. Persistence of Bacteroides Species Populations in a River as Measured by Molecular and Culture Techniques. *Appl. Environ. Microbiol.* **2010**, *76*, 7608–7616. [CrossRef]
42. Zamyadi, A.; MacLeod, S.L.; Fan, Y.; McQuaid, N.; Dorner, S.; Sauvé, S.; Prévost, M. Toxic cyanobacterial breakthrough and accumulation in a drinking water plant: A monitoring and treatment challenge. *Water Res.* **2012**, *46*, 1511–1523. [CrossRef]
43. McQuaid, N.; Zamyadi, A.; Prévost, M.; Bird, D.F.; Dorner, S. Use of in vivophycocyanin fluorescence to monitor potential microcystin-producing cyanobacterial biovolume in a drinkingwater source. *J. Environ. Monit.* **2011**, *13*, 455–463. [CrossRef] [PubMed]

44. Tromas, N.; Fortin, N.; Bedrani, L.; Terrat, Y.; Cardoso, P.; Bird, D.; Greer, C.W.; Shapiro, B.J. Characterising and predicting cyanobacterial blooms in an 8-year amplicon sequencing time course. *ISME J.* **2017**, *11*, 1746–1763. [CrossRef] [PubMed]
45. Andersson, A.; Höglander, H.; Karlsson, C.; Huseby, S. Key role of phosphorus and nitrogen in regulating cyanobacterial community composition in the northern Baltic Sea. *Estuar. Coast. Shelf Sci.* **2015**, *164*, 161–171. [CrossRef]
46. Fortin, N.; Munoz-Ramos, V.; Bird, D.; Lévesque, B.; Whyte, L.G.; Greer, C.W. Toxic Cyanobacterial Bloom Triggers in Missisquoi Bay, Lake Champlain, as Determined by Next-Generation Sequencing and Quantitative PCR. *Life* **2015**, *5*, 1346–1380. [CrossRef]
47. Harke, M.J.; Davis, T.W.; Watson, S.B.; Gobler, C.J. Nutrient-Controlled Niche Differentiation of Western Lake Erie Cyanobacterial Populations Revealed via Metatranscriptomic Surveys. *Environ. Sci. Technol.* **2016**, *50*, 604–615. [CrossRef]
48. Li, Q.; Lin, F.; Yang, C.; Wang, J.; Lin, Y.; Shen, M.; Park, M.S.; Li, T.; Zhao, J. A Large-Scale Comparative Metagenomic Study Reveals the Functional Interactions in Six Bloom-Forming Microcystis-Epibiont Communities. *Front. Microbiol.* **2018**, *9*, 746. [CrossRef]
49. Shi, L.; Cai, Y.; Li, P.; Yang, H.; Liu, Z.; Kong, L.; Yü, Y.; Kong, F. Molecular Identification of the Colony-Associated Cultivable Bacteria of the CyanobacteriumMicrocystis aeruginosaand Their Effects on Algal Growth. *J. Freshw. Ecol.* **2009**, *24*, 211–218. [CrossRef]
50. Maruyama, T.; Kato, K.; Yokoyama, A.; Tanaka, T.; Hiraishi, A.; Park, H.-D. Dynamics of microcystin-degrading bacteria in mucilage of Microcystis. *Microb. Ecol.* **2003**, *46*, 279–288. [CrossRef]
51. Wacklin, P.; Hoffmann, L.; Komarek, J. Nomenclatural validation of the genetically revised cyanobacterial genus Dolichospermum (RALFS ex BORNET et FLAHAULT) comb. nova. *Fottea* **2009**, *9*, 59–64. [CrossRef]
52. Van Wichelen, J.; Vanormelingen, P.; Codd, G.A.; Vyverman, W. The common bloom-forming cyanobacterium Microcystis is prone to a wide array of microbial antagonists. *Harmful Algae* **2016**, *55*, 97–111. [CrossRef]
53. Li, Y.; Li, D. Competition between toxic Microcystis aeruginosa and nontoxic Microcystis wesenbergii with Anabaena PCC. *Environ. Boil. Fishes* **2012**, *24*, 69–78.
54. Kardinaal, W.; Janse, I.; Agterveld, M.K.-V.; Meima, M.; Snoek, J.; Mur, L.; Huisman, J.; Zwart, G.; Visser, P. Microcystis genotype succession in relation to microcystin concentrations in freshwater lakes. *Aquat. Microb. Ecol.* **2007**, *48*, 1–12. [CrossRef]
55. Zhang, X.-W.; Fu, J.; Song, S.; Zhang, P.; Yang, X.-H.; Zhang, L.-R.; Luo, Y.; Liu, C.-H.; Zhu, H.-L. Interspecific competition between Microcystis aeruginosa and Anabaena flos-aquae from Taihu Lake, China. *Z. für Nat. C* **2014**, *69*, 53–60. [CrossRef]
56. Chia, M.A.; Jankowiak, J.G.; Kramer, B.J.; Goleski, J.A.; Huang, I.-S.; Zimba, P.V.; Bittencourt-Oliveira, M.D.C.; Gobler, C.J. Succession and toxicity of Microcystis and Anabaena (Dolichospermum) blooms are controlled by nutrient-dependent allelopathic interactions. *Harmful Algae* **2018**, *74*, 67–77. [CrossRef] [PubMed]
57. Ndong, M.; Bird, D.; Nguyen-Quang, T.; De Boutray, M.-L.; Zamyadi, A.; Vinçon-Leite, B.; Lemaire, B.J.; Prévost, M.; Dorner, S. Estimating the risk of cyanobacterial occurrence using an index integrating meteorological factors: Application to drinking water production. *Water Res.* **2014**, *56*, 98–108. [CrossRef] [PubMed]
58. Fan, J.; Hobson, P.; Ho, L.; Daly, R.; Brookes, J.D. The effects of various control and water treatment processes on the membrane integrity and toxin fate of cyanobacteria. *J. Hazard. Mater.* **2014**, *264*, 313–322. [CrossRef]
59. Zamyadi, A.; Fan, Y.; Daly, R.I.; Prévost, M. Chlorination of Microcystis aeruginosa: Toxin release and oxidation, cellular chlorine demand and disinfection by-products formation. *Water Res.* **2013**, *47*, 1080–1090. [CrossRef]
60. Fan, J.; Ho, L.; Hobson, P.; Brookes, J.D. Evaluating the effectiveness of copper sulphate, chlorine, potassium permanganate, hydrogen peroxide and ozone on cyanobacterial cell integrity. *Water Res.* **2013**, *47*, 5153–5164. [CrossRef]
61. Matthijs, H.C.; Visser, P.M.; Reeze, B.; Meeuse, J.; Slot, P.C.; Wijn, G.; Talens, R.; Huisman, J. Selective suppression of harmful cyanobacteria in an entire lake with hydrogen peroxide. *Water Res.* **2012**, *46*, 1460–1472. [CrossRef]
62. Zamyadi, A.; Henderson, R.K.; Stuetz, R.; Newcombe, G.; Newtown, K.; Gladman, B. Cyanobacterial management in full-scale water treatment and recycling processes: Reactive dosing following intensive monitoring. *Environ. Sci. Water Res. Technol.* **2016**, *2*, 362–375. [CrossRef]

63. Moradinejad, S.; Glover, C.M.; Mailly, J.; Seighalani, T.Z.; Peldszus, S.; Barbeau, B.; Dorner, S.; Prévost, M.; Zamyadi, A. Using Advanced Spectroscopy and Organic Matter Characterization to Evaluate the Impact of Oxidation on Cyanobacteria. *Toxins* **2019**, *11*, 278. [CrossRef] [PubMed]
64. Zhou, S.; Shao, Y.; Gao, N.; Deng, Y.; Qiao, J.; Ou, H.; Deng, J. Effects of different algaecides on the photosynthetic capacity, cell integrity and microcystin-LR release of Microcystis aeruginosa. *Sci. Total Environ.* **2013**, *463*, 111–119. [CrossRef] [PubMed]
65. Park, J.; Kim, Y.; Kim, M.; Lee, W.H. A novel method for cell counting of Microcystis colonies in water resources using a digital imaging flow cytometer and microscope. *Environ. Eng. Res.* **2018**, *24*, 397–403. [CrossRef]
66. Xiao, X.; Sogge, H.; Lagesen, K.; Tooming-Klunderud, A.; Jakobsen, K.S.; Rohrlack, T. Use of High Throughput Sequencing and Light Microscopy Show Contrasting Results in a Study of Phytoplankton Occurrence in a Freshwater Environment. *PLoS ONE* **2014**, *9*, e106510.
67. American Public Health Association (APHA); American Water Works Association (AWWA); Water Environment Federation (WEF). *Standard Methods for the Examination of Water and Wastewater*; American Public Health Association: Washington, DC, USA, 2012; Volume 5, p. 1360.
68. Cox, M.P.; Peterson, D.A.; Biggs, P.J. SolexaQA: At-a-glance quality assessment of Illumina second-generation sequencing data. *BMC Bioinform.* **2010**, *11*, 1–6. [CrossRef]
69. Kim, D.; Hahn, A.S.; Wu, S.-J.; Hanson, N.W.; Konwar, K.M.; Hallam, S.J. *FragGeneScan-Plus for Scalable High-Throughput Short-Read Open Reading Frame Prediction*; IEEE CIBCB Conference: Niagara Falls, ON, Canada, 2015; pp. 1–8.
70. Fu, L.; Niu, B.; Zhu, Z.; Wu, S.; Li, W. CD-HIT: Accelerated for clustering the next-generation sequencing data. *Bioinformatics* **2012**, *28*, 3150–3152. [CrossRef]
71. Lund, J.W.G. A Simple Counting Chamber for Nannoplankton. *Limnol. Oceanogr.* **1959**, *4*, 57–65. [CrossRef]
72. Planas, D.; Desrosiers, M.; Groulx, S.; Paquet, S.; Carignan, R. Pelagic and benthic algal responses in eastern Canadian Boreal Shield lakes following harvesting and wildfires. *Can. J. Fish. Aquat. Sci.* **2000**, *57*, 136–145. [CrossRef]
73. McMurdie, P.J.; Holmes, S.P. phyloseq: An R Package for Reproducible Interactive Analysis and Graphics of Microbiome Census Data. *PLoS ONE* **2013**, *8*, e61217.
74. Raivo, K. *Pheatmap: Pretty Heatmaps*, R package version 1.0.12; GitHub, Inc.: San Francisco, CA, USA, 2019.
75. Greenacre, M. *Compositional Data Analysis in Practice*, 1st ed.; CRC Press: Cleveland, OH, USA, 2018.
76. Oksanen, J.; Blanchet, F.G.; Kindt, R.; Legendre, P.; Minchin, P.R.; O'hara, R.B.; Simpson, G.L.; Solymos, P.; Stevens, M.H.H.; Wagner, H. *Community Ecology Package*, R Package 'Vegan', version 2.5-0; GitHub, Inc.: San Francisco, CA, USA, 2016; p. 285.

toxins

MDPI

Article

Delayed Release of Intracellular Microcystin Following Partial Oxidation of Cultured and Naturally Occurring Cyanobacteria

Katherine E. Greenstein [1], Arash Zamyadi [2,3], Caitlin M. Glover [4], Craig Adams [5], Erik Rosenfeldt [6] and Eric C. Wert [1,*]

1 Southern Nevada Water Authority (SNWA), P.O. Box 99954, Las Vegas, NV 89193-9954, USA; katie.greenstein@gmail.com
2 Water Research Australia (WaterRA), Adelaide, SA 5001, Australia; arash.zamyadi@waterra.com.au
3 BGA Innovation Hub and Water Research Centre, School of Civil and Environmental Engineering, University of New South Wales (UNSW), Sydney, NSW 2052, Australia
4 Department of Civil Engineering, McGill University, Montreal, QC H3A 0G4, Canada; caitlinmeara@gmail.com
5 Department of Civil Engineering, Saint Louis University, St. Louis, MO 63103, USA; craig.adams@slu.edu
6 Hazen and Sawyer, Raleigh, NC 27607, USA; erosenfeldt@hazenandsawyer.com
* Correspondence: eric.wert@snwa.com

Received: 8 April 2020; Accepted: 13 May 2020; Published: 20 May 2020

check for updates

Abstract: Oxidation processes can provide an effective barrier to eliminate cyanotoxins by damaging cyanobacteria cell membranes, releasing intracellular cyanotoxins, and subsequently oxidizing these toxins (now in extracellular form) based on published reaction kinetics. In this work, cyanobacteria cells from two natural blooms (from the United States and Canada) and a laboratory-cultured *Microcystis aeruginosa* strain were treated with chlorine, monochloramine, chlorine dioxide, ozone, and potassium permanganate. The release of microcystin was measured immediately after oxidation ($t \leq 20$ min), and following oxidant residual quenching (stagnation times = 96 or 168 h). Oxidant exposures (CT) were determined resulting in complete release of intracellular microcystin following chlorine (21 mg-min/L), chloramine (72 mg-min/L), chlorine dioxide (58 mg-min/L), ozone (4.1 mg-min/L), and permanganate (391 mg-min/L). Required oxidant exposures using indigenous cells were greater than lab-cultured *Microcystis*. Following partial oxidation of cells (oxidant exposures ≤ CT values cited above), additional intracellular microcystin and dissolved organic carbon (DOC) were released while the samples remained stagnant in the absence of an oxidant (>96 h after quenching). The delayed release of microcystin from partially oxidized cells has implications for drinking water treatment as these cells may be retained on a filter surface or in solids and continue to slowly release cyanotoxins and other metabolites into the finished water.

Keywords: cyanobacteria; oxidation; stagnation; microcystin; water treatment; quenching

Key Contribution: This work provides oxidant exposure (CT) guidance regarding complete, partial, and delayed release of intracellular microcystins for five oxidants. Differences between laboratory cultured and naturally occurring cyanobacteria cells were quantified and reported.

1. Introduction

Toxic cyanobacteria blooms are a public health risk when present in drinking water supplies. Many countries have issued health advisories for cyanotoxins, including microcystin, a hepatotoxic class of cyanotoxins with many different variants, or congeners [1–3]. In response to these cyanotoxin health advisories, drinking water utilities often develop cyanobacteria bloom (frequently termed "harmful algal bloom" (HAB)) management plans to prevent the persistence of cyanotoxins beyond treatment [4,5]. Furthermore, several guidance documents emerged with best practices for managing cyanobacteria blooms in water treatment [5–9]. Recommendations include switching water sources and removing cells through conventional treatment processes (i.e., dissolved air flotation, sedimentation).

Many utilities use pre-oxidation in drinking water treatment to achieve a multitude of objectives including invasive species control (e.g., quagga mussel), prevention of biofilm growth, taste and odor control, oxidation of iron or manganese, improved solids removal, and disinfection [10]. Current guidance also urges caution or recommends eliminating pre-oxidation processes to minimize the risk of cell lysis and the corresponding release of intracellular cyanotoxin before the removal of intact cells (e.g., through coagulation/flocculation/sedimentation and/or filtration). Despite this guidance, some utilities cannot eliminate pre-oxidation in order to comply with disinfection requirements (e.g., *Giardia* inactivation) [11]. Therefore, guidance for the application of pre-oxidants during a cyanobacteria bloom is warranted. While previous studies have generally assessed the complete oxidation of cyanobacteria cells and release of cyanotoxins, particularly with respect to microcystin [12–19], low oxidant exposures (concentration × time (CT)) resulting in incomplete cell oxidation continue to require further examination. Furthermore, delayed release has been observed up to 120 min after the chlorination of cells [17]. Additional research is needed to understand the risk of additional intracellular cyanotoxin release following partial cyanobacteria cell oxidation and oxidant quenching processes using different oxidants (ozone, chloramine, chlorine dioxide, potassium permanganate) in water treatment plants.

Here, microcystin release from both laboratory-cultured and naturally occurring cyanobacteria was assessed with exposure to drinking water pre-oxidation. By normalizing the oxidant dose to the dissolved organic carbon (DOC) concentration of each water, chlorine, monochloramine, chlorine dioxide, ozone, and potassium permanganate were compared in three cyanobacteria suspensions. The objectives of the research were to: (1) establish a normalized framework (i.e., oxidant dose: background DOC ratio, oxidant exposure over time) where the complete release of intracellular microcystin and DOC may be expected following a 20 min reaction time, and (2) examine the delayed release of intracellular microcystin following partial cell damage at low oxidant exposures with stagnation times ranging from 5 min to 7 days. This work improves the guidance regarding continued intracellular microcystin release from partially damaged cells that can impact downstream water treatment processes (i.e., filter surface, solids retention basins).

2. Results and Discussion

2.1. Oxidation (Time ≤ 20 min) Induced Degradation of Pigment and Release of DOC

Across the three waters tested, all five oxidants were applied at either five or six oxidant: DOC dose ratios for contact times ≤ 20 min. The oxidant decay rates during this time period (Table S1) were calculated for each dose ratio, with decay curves shown in the SI (Figures S1–S10). Along with the decay rates, oxidant CT values were calculated for each oxidant using the trapezoid rule with two residual oxidant values. CT values are shown on the secondary axis of Figure 1A–E and in Table S2. Ozone CT values were nominally 0 mg-min/L after the USA bloom was dosed with 0.05–0.25 O_3: DOC dose ratios and after the CA bloom was dosed with of 0.07–0.1 O_3: DOC due to the rapid decay of ozone.

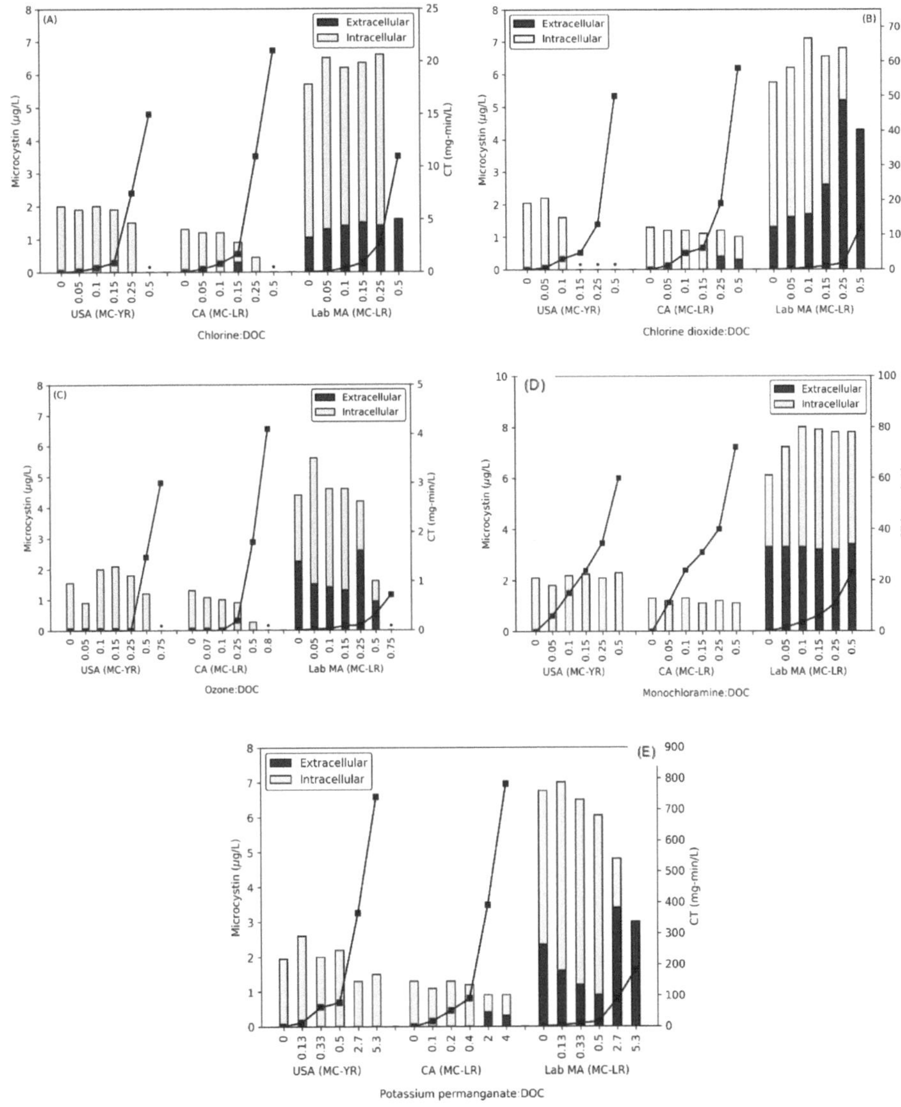

Figure 1. Impact of (**A**) chlorine, (**B**) chlorine dioxide, (**C**) ozone, (**D**) monochloramine, and (**E**) potassium permanganate on the release of intracellular MCs (MC-LR or MC-YR) from the USA bloom, the CA bloom, and the *M. aeruginosa* (lab MA) water (t ≤ 20 min). CT is shown on the secondary axis as black squares. * Concentrations were below the MRL for both extracellular and intracellular MC.

While cell viability was not directly monitored, chlorophyll-a (chl-*a*) (for the USA bloom and lab-cultured *M. aeruginosa*) and phycocyanin (PC) fluorescence (for the CA bloom) were used as proxies for cell damage. The chl-*a* or PC fluorescence were used to calculate cell damage rates with results shown in Table 1 and Table S3. In general, the cell damage rates followed the trend of ozone >> chlorine ≈ chlorine dioxide > potassium permanganate >> monochloramine. The USA and CA bloom waters had significantly lower decay rates for all oxidants as compared to the lab-cultured *M. aeruginosa*. This phenomenon is due to the resilience of natural (indigenous) cells with presence of multiple species and the interference from background organic matter, compared to lab-cultured species, which has been observed previously [16,20–22]. However, at the low pre-oxidation doses, ozonation (1700 $M^{-1}s^{-1}$) and chlorine dioxide (200 $M^{-1}s^{-1}$) rates for the lab-cultured *M. aeruginosa* were significantly lower than those rates observed in past work (1.1×10^5 $M^{-1}s^{-1}$ and 4900 $M^{-1}s^{-1}$) [23]. The potassium permanganate rate (9.1 $M^{-1}s^{-1}$) was comparable, though lower, than the value generated (36 $M^{-1}s^{-1}$) for another unicellular species, *Pseudanabaena sp.*, likely due to differences in cell-specific reactivity as similar experimental conditions were applied [24]. These data illustrate that during low pre-oxidation doses or the first part of the cell-damage curve, the rate is significantly lower. A caveat in comparing the pigment results from the three waters is that the PC fluorescence (used for the CA bloom) likely captured both intra- and extra-cellular pigment, whereas the extracted chl-*a* (used for the USA bloom and lab-cultured water) included only intracellular pigment [25].

Table 1. Cell damage (k_{damage}) and total MC decay (k_{total}) after oxidation ($t \leq 20$ min) for the USA and CA blooms. Rates with R^2 below 0.75 were excluded (e.g., monochloramine). * PC fluorescence was measured instead of extracted chl-*a*.

Oxidant	Water	k_{damage} ($M^{-1}s^{-1}$) (R^2)	k_{total} ($M^{-1}s^{-1}$) (R^2)
Cl_2	USA	25 (0.96)	74.6 (0.91)
	CA	133 * (0.98)	82.9 (0.96)
ClO_2	USA	64 (0.76)	-
	CA	20.9 * (0.84)	11.5 (0.82)
$KMnO_4$	USA	-	-
	CA	2.69 * (0.84)	1.22 (0.73)
Ozone	USA	273 (0.91)	143 (0.99)
	CA	245 * (0.75)	56 (0.75)

While pigment levels are a proxy for cell-viability, the release of DOC is an indicator of cell lysis. At the highest oxidant:DOC dose ratios for chlorine (0.5), monochloramine (0.5), chlorine dioxide (0.5), potassium permanganate (4/5.3), and ozone (0.75/0.80) DOC increased by less than 1.5 mg/L during the ≤20 min exposure period. In each of the three waters, ozone produced the greatest releases of 0.91 mg/L in the USA bloom, 1.20 mg/L in the CA bloom, and 0.30 mg/L in the lab-cultured *M. aeruginosa* water. After exposure to chlorine, monochloramine, chlorine dioxide, and potassium permanganate, DOC releases were consistent across the three waters at 0.30–0.65 mg/L for USA bloom, 0.1–0.2 mg/L for the CA bloom, and 0.2–0.26 mg/L for the lab-cultured *M. aeruginosa*. A potential consequence of the release of DOC during pre-oxidation is that these processes may result in the formation of disinfection byproducts during the initial oxidation period or later during secondary disinfection as has been demonstrated in previous work [17,21,26]. The presence of released DOC may also interfere with the degradation of extracellular MC as it consumes oxidant residual during the exposure period and is a factor contributing to the difficulty in modeling the degradation of MC as it moves from intracellular to extracellular [27].

2.2. *Impact of Oxidants on Release of Microcystin ($t \leq 20$ min)*

To monitor the release of MC congeners from the USA bloom (MC-YR), CA bloom (MC-LR) and the lab-cultured *M. aeruginosa* (MC-LR), the total and extracellular MC were measured (Figure 1A–E). Prior to oxidation, the highest concentration of total MC was observed in the lab-cultured *M. aeruginosa* sample, 5.7 ± 0.86 µg/L MC-LR, whereas the bloom waters contained significantly less at 1.9 ± 0.22 µg/L MC-YR in the USA bloom, and 1.3 µg/L MC-LR in the CA bloom. The extracellular MC started at concentrations below the method reporting limit (MRL, i.e., <0.5 µg/L) in the bloom waters, but the lab culture contained approximately 30% extracellular MC-LR relative to the total.

Pre-oxidation exposures resulted in the release of intracellular MC from both natural bloom and lab-cultured cyanobacteria as summarized in Table 2 Two of the oxidants, monochloramine and potassium permanganate, saw limited release of intracellular MC. For all three waters, the highest dose ratio of monochloramine (CTs of 23–72 mg-min/L) resulted in a release less than 0.15 μg/L of extracellular MC and no significant total MC degradation. This corresponds with the low rate of monochloramine inactivation and reaction with cell membrane [15,28]. In past work, with lab-cultured cells the release of MC-LR by monochloramine required a CT value more than an order of magnitude greater (640 mg-min/L for 5×10^4 cells/mL) [13].

Table 2. CT, oxidant: DOC ratio and stagnation time resulting in the release of intracellular MCs.
* Select natural blooms did not release detectable concentrations of microcystins at this point.

	Cl_2	NH_2Cl	ClO_2	O_3	$KMnO_4$
Oxidant:DOC ratio (t ≤ 20 min)	0.5	No release	0.5 *	0.75/0.80	>2 *
CT_{lab} (mg-min/L)	11	23	12	0.72	117
CT_{USA} (mg-min/L)	15	60	50	3.0	486
CT_{CA} (mg-min/L)	21	72	58	4.1	391
Stagnation time *	≥2 h	≥8 h	≥20 min	≥8 h	≥2 h

Potassium permanganate did not affect the MC levels in the two bloom samples until the two highest doses of $KMnO_4$: DOC (CTs of 88–782 mg-min/L). Both blooms then saw a minor decrease in the total MCs with the CA bloom releasing an average of 0.35 μg/L as extracellular MC-LR. The lab-cultured *M. aeruginosa* water released intracellular MC at the two highest dose ratios, with the highest dose containing 3 μg/L of extracellular MC-LR. Prior to the release of intracellular MC, extracellular MC-LR was oxidized by $KMnO_4$ [29,30].

The impacts of chlorine, chlorine dioxide, and ozone on the total and extracellular MC in the lab-cultured and bloom-containing waters are illustrated in Figure 1A–C, respectively. During the application of ozone, no extracellular MC was detected in the two blooms and this was attributed to the rapid oxidation of MC by ozone (0.24–4.1×10^5 $M^{-1}s^{-1}$) and hydroxyl radical (1.1×10^{10} $M^{-1}s^{-1}$) prior to measurement [31]. Intracellular MC-YR in the USA bloom was removed to below the MRL at 0.5 Cl_2: DOC and 0.15 ClO_2: DOC, with no extracellular MC detected. Although chlorine (33 $M^{-1}s^{-1}$) and chlorine dioxide (1 $M^{-1}s^{-1}$) both have significantly lower reaction rates than ozone, a similar result was observed [29,32]. The CA bloom released extracellular MC-LR at 0.15 Cl_2: DOC and 0.25 ClO_2: DOC. Subsequent treatments with chlorine degraded the total MC from the CA bloom, but chlorine dioxide allowed both intra- and extracellular MC to remain. In contrast, lab-cultured *M. aeruginosa* released intracellular MC at the lowest dose ratios of 0.05 Cl_2: DOC or ClO_2: DOC (CT = 0.047 mg-min/L for chlorine and 0.060 mg-min/L for chlorine dioxide). This trend continued for both oxidants until the highest doses at which point the total MC was entirely extracellular. Ozonation of the *M. aeruginosa* degraded the extracellular MC until 0.5 O_3: DOC, at which point 1.3 μg/L was released.

To model the degradation of total MC (k_{total}), Equation (1) with CT was applied [12,14,16] with results shown in Table 1 and Table S3. The USA bloom, CA bloom, and the lab-cultured *M. aeruginosa* had ozone k_{total} rates of 143 $M^{-1}s^{-1}$, 56 $M^{-1}s^{-1}$ and 2664 $M^{-1}s^{-1}$, respectively. The bloom water rates were lower than those observed in a previous study of mixed species from a bloom (400–450 $M^{-1}s^{-1}$), likely due to the consumption of oxidant by the background organic matter in this work [20]. Under chlorination, the USA bloom's MC-YR decayed at a rate of 102 $M^{-1}s^{-1}$, which was greater than the CA bloom MC-LR rate at 82.9 $M^{-1}s^{-1}$. This behavior could be partly due to the reactivity difference between the congeners with MC-YR more rapidly degraded as compared to MC-LR [33,34]. The lab-cultured sample in this work had a rate of 136 $M^{-1}s^{-1}$, a value that was close to those generated in past work (range of 10–96 $M^{-1}s^{-1}$) [12,14]. Although the cell-damage rates were an order of magnitude lower than those observed in previous studies, pre-oxidation k_{total} decay rates were close to those from previous studies. This reflects that lower oxidant doses are required to release MC relative to the doses necessary to induce changes in chl-*a* and PC fluorescence or release DOC.

These results demonstrate that extracellular MC can be released from lab-cultured *M. aeruginosa* cells at low oxidant: DOC (0.05 Cl_2: DOC, 0.05 ClO_2: DOC, 0.25 O_3: DOC, and 2.7 $KMnO_4$: DOC). Although this result has not been previously reported for chlorine dioxide nor ozone, a recent study with chlorine saw similar behavior [17]. It is important to note that the natural bloom waters were substantially more resistant to oxidation as compared to cultured cells as a result of the presence of multiple species, cell-specific resilience, and the interference from background organic matter [16,20]. In the USA bloom, no intracellular MC was released; however, total MCs were completely removed at 0.5 Cl_2: DOC, 0.15 ClO_2: DOC, and 0.75 O_3: DOC. In the CA bloom, extracellular MC was observed at 0.15 Cl_2: DOC, 0.25 ClO_2: DOC, 0.8 O_3: DOC, and 2 $KMnO_4$: DOC, but complete removal did not occur.

2.3. Impact of Stagnation (Time Max = 96 or 168 h) on Pigment and DOC

After the initial 20 min oxidant exposure time, the remaining oxidant was quenched, and samples were subsequently held for up to 96 or 168 h. During the stagnation period, the release of DOC and cell damage (as shown via pigments concentration) was then evaluated. The dose ratios applied for chlorine, monochloramine, chlorine dioxide, ozone, and potassium permanganate were 0.15, 0.15, 0.15, 0.15, and 0.4/0.5, respectively. These dose ratios were selected as they produced minimal or no MC release during the initial oxidant exposure period.

Interestingly, the reduction in pigment during the oxidant exposure time followed an expected degradation path and pseudo-first order reaction kinetics were applied to model the reduction and interpret the results. However, the model method does not fit the pigment decay rates analysis after quenching and stagnation time, i.e., the majority of R^2 values are below 0.75. Therefore, the fraction of damaged cells after oxidation (time $\leq$ 20 min) were compared against the cells damaged during the stagnation period (Figure 2). All the ozone dose ratios produced a similar level of damage during the initial oxidation period and following stagnation. Following chlorine, potassium permanganate, and monochloramine treatment, the stagnation period more than doubled the fraction of cells classified as damaged relative to those affected during the oxidation period. In the CA bloom, after stagnation, the level of cell damage was the same for chlorine, monochloramine, chlorine dioxide, and potassium permanganate. The USA bloom and the lab-cultured *M. aeruginosa* saw that potassium permanganate, chlorine, and chlorine dioxide damaged a higher fraction of cells than ozone did after stagnation. This is in contrast with the general reactivity of these oxidants with cell membranes as well as their reactivity with organic matter functional groups [28,35]. It also does not agree with the pigment decay rates that were observed in this work after time $\leq$ 20 min contact. These results imply that oxidant-induced cell damage is strongly underestimated by the rates calculated using oxidant exposures that only measure cells immediately following oxidant exposure.

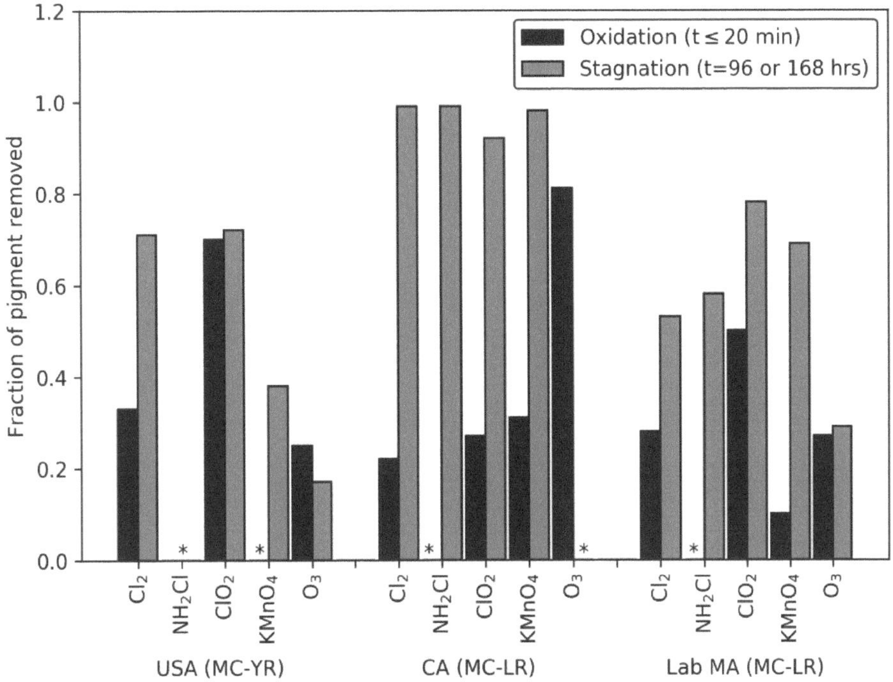

Figure 2. Fraction of pigment removed after 20 min exposure to oxidant: DOC ratios of 0.15. Cl_2: DOC, ClO_2: DOC, NH_2Cl: DOC, and O_3: DOC, and 0.5/0.4 $KMnO_4$: DOC, and fraction of pigment removed after oxidation at the same level, quenching and stagnation for 96 or 168 h. Columns with * are those in which the pigment concentrations did not change or increased.

A similar trend was observed for the DOC released following stagnation (Figure 3). The initial exposure period produced limited releases of DOC above 0.15 mg/L, but after stagnation every oxidant produced a release of DOC. In addition, the changes in DOC concentration did not follow the trend of oxidant reactivity. This ranged from 0.63–2.93 mg/L for the USA bloom, 0.6–1.1 mg/L for the CA bloom, and 0.16–0.3 mg/L for the lab-cultured *M. aeruginosa*. As with the release of DOC during the initial oxidation period, the intracellular DOC could contribute to the formation of disinfection byproducts and interfere with the degradation of MC.

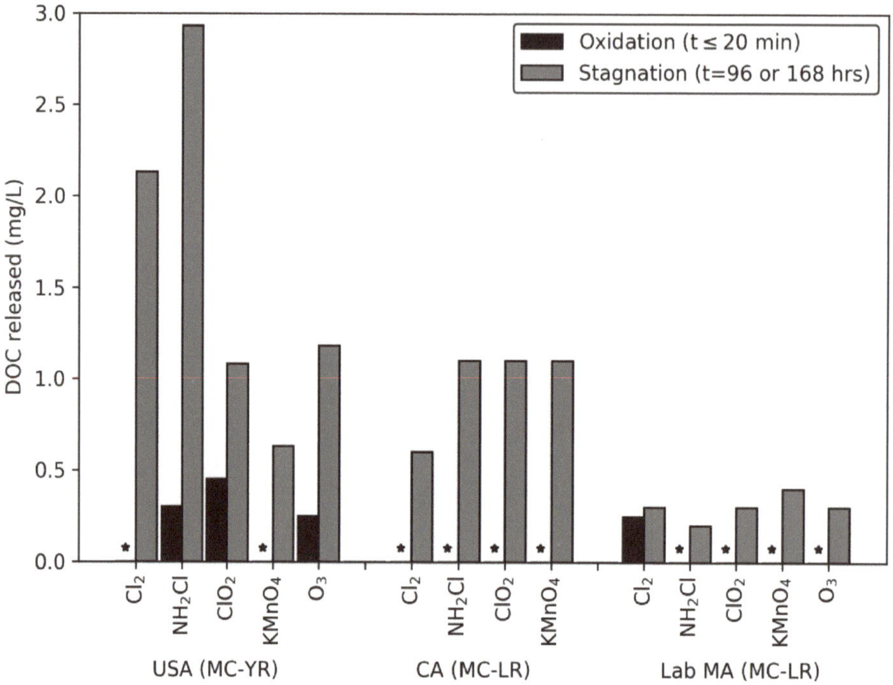

Figure 3. Release of DOC following exposure to oxidant: DOC ratios of 0.15 Cl_2: DOC, ClO_2: DOC, NH_2Cl: DOC, and O_3: DOC, and 0.5/0.4 $KMnO_4$: DOC with 20 min contact time. Stagnation samples were collected 96 or 168 h after quenching. Columns with * represent releases below the MRL of 0.15 mg/L.

2.4. *Impact of Stagnation (Time Max = 96 or 168 h) on Microcystin*

While many studies have evaluated the impact of oxidation on the release of MC, limited work has been done to evaluate the impact of stagnation post-quenching. This is a particularly important process to understand for the use of pre-oxidation, as the immediate release might not occur (results shown above) leading to a misevaluation of the risk associated with a given oxidation dose. In this work, one dose ratio was selected, and the extracellular and total MC were tracked after quenching for up to 96 or 168 h.

Across all oxidants, the naturally occurring bloom samples did not immediately release MC during stagnation (Figure 4). For chlorine, monochloramine, ozone, and potassium permanganate, the release of MC-YR in the USA sample occurred at 33 h, 24 h, 8 h, and 33 h, respectively. Except for potassium permanganate and ozone where 0.61 and 0.68 µg/L of intracellular MC remained, the final time points were below the MRL for both intra- and extracellular MC-YR. For the CA bloom, extracellular MC-LR was observed after 10 h for chlorine, 24 h for monochloramine, 10 h for chlorine dioxide, and 10 h for potassium permanganate. Despite the limited release of MC during the initial oxidant exposure, the cells were damaged to the point where after additional time, they released MC. In both bloom waters, a decrease in total MC was observed over time, despite the conversion of this MC from intracellular to extracellular. This decrease was attributed to biodegradation, which will be further discussed below.

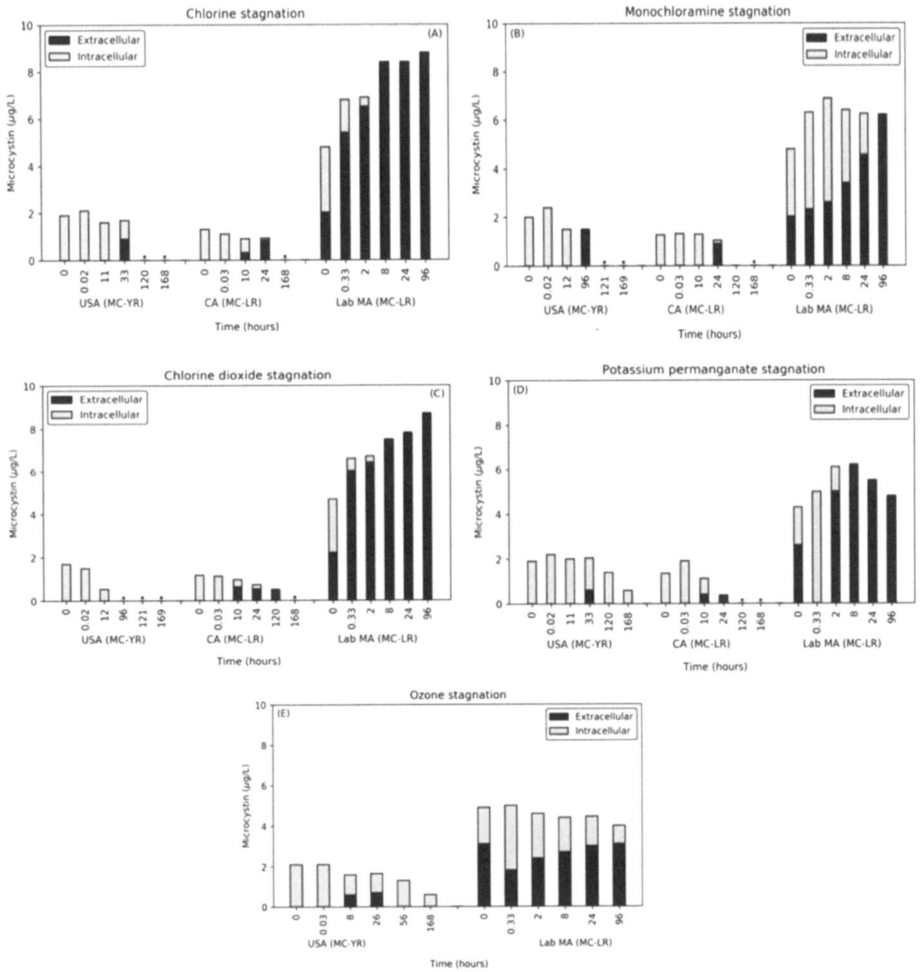

Figure 4. Effect of stagnation time on the presence of extracellular and intracellular MC after. oxidation and quenching with (**A**) chlorine, (**B**) monochloramine, (**C**) chlorine dioxide, (**D**) potassium permanganate, and (**E**) ozone. * Concentrations were below the MRL for both extracellular and intracellular MC.

The lab-cultured *M. aeruginosa* was significantly more susceptible to the release of intracellular MC during stagnation with extracellular MC-LR detected after < 0.6 hrs. In contrast to the natural bloom samples, the final time point for the lab-cultured cells contained almost entirely extracellular MC. At 96 h, chlorine, monochloramine, chlorine dioxide, and potassium permanganate stagnation contained 8.8 µg/L, 6.2 µg/L, 8.7 µg/L, and 4.8 µg/L, respectively. Ozone was the only oxidant in which a small concentration of intracellular MC remained during stagnation with 3.1 µg/L extracellular and 0.9 µg/L intracellular. As with the bloom waters, the ozone stagnation resulted in a degradation of the total MC present. Stagnation after chlorine and chlorine dioxide exposure saw continued increases in the extracellular MC from first to last time point. The total MC in the potassium permanganate stagnation sample increased until 8 h at which point the MC was entirely extracellular and subsequent decay was observed.

The delayed release of MC from damaged or lysed cells highlights the need to monitor MC throughout water treatment and solids-handing, not only following the pre-oxidation process. It also provides another potential source of MC release during treatment, which had previously been attributed to the accumulation of cells during treatment, e.g., in filter beds, in the sludge bed of sedimentation tanks, and in sludge thickeners [36–38]. Using the release of extracellular MC-LR from the lab-cultured *M. aeruginosa*, the rate at which this process occurs was calculated using Equation (1) and ranged from 1.61 to 4.12 × 10^{-6} s^{-1} (Table 3). These rates were generated from the release following chlorine, chlorine dioxide, ozone, and monochloramine, and did not follow a specific trend with oxidant efficacy. For the bloom samples, total MC concentration decreased during stagnation and the continuous release of extracellular MC was not observed. In both the bloom samples, regardless of the oxidant, similar total MC decay rates were observed, indicating that the process was likely not related to the oxidant applied. For the USA bloom, the rates ranged from 1.79 to 2.57 × 10^{-6} s^{-1} after chlorination, ozonation, potassium permanganate, and monochloramine exposure. The CA bloom had slightly higher rates, but the spread between oxidants was also low, with potassium permanganate, chlorine dioxide, monochloramine, and chlorine at 2.85 to 4.08 × 10^{-6} s^{-1}.

Table 3. Total MC biodegradation during stagnation ($k_{biodegradation}$) and the release rate for extracellular MC after stagnation ($k_{release}$) for the *M. aeruginosa* (lab MA), USA and CA blooms. Decay rates with R^2 values below 0.75 were not included. * Differentiates *$\mathbf{k_{release}}$ from $\mathbf{k_{biodegradation}}$.

Oxidant	Water	Stagnation (Time Max = 96 or 168 h)	
		$k_{biodegradation}$ (R^2)/*$k_{release}$ (s^{-1}) (R^2)	Half-Life (Days)
Chlorine	USA	2.57 × 10^{-6} (0.93)	3.12
	CA	3.37 × 10^{-6} (0.99)	2.38
	Lab MA	*1.61 × 10^{-6} (0.67)	-
Mono-chloramine	USA	2.44 × 10^{-6} (0.88)	3.29
	CA	4.08 × 10^{-6} (0.95)	1.97
	Lab MA	*4.12 × 10^{-6} (0.79)	-
Chlorine dioxide	USA	-	-
	CA	2.85 × 10^{-6} (0.90)	2.81
	Lab MA	*2.79 × 10^{-6} (0.83)	-
$KMnO_4$	USA	1.79 × 10^{-6} (0.83)	4.48
	CA	3.93 × 10^{-6} (0.81)	2.04
	Lab MA	-	-
Ozone	USA	1.90 × 10^{-6} (0.90)	4.22
	Lab MA	7.44 × 10^{-7} (0.76)/*2.50 × 10^{-6} (0.84)	10.8

The degradation in these waters was attributed to biological activity and the rates were translated to half-lives for comparison with past work. The extracellular MC-LR half-life ranged from 3.12–4.48 days in the USA bloom and 1.97–2.81 days in the CA bloom; these half-lives were similar to those observed in surface waters (1.22–7.66 days for MC-LR) [39] and lake waters (5.4 days for both MC-LR and MC-YR) [40]. A lag phase was observed, but there was not a discernible relationship between the lag phase and the oxidant applied. The similarity in behavior between the two blooms indicated that both waters had the required biomass and the enzymes present for biodegradation, which is indicative of a previous exposure to cyanobacteria producing MC in these waters [39,40]. With the exception of ozonation, the lab-cultured *M. aeruginosa* in Colorado River Water (CRW) did not exhibit the biodegradation behavior observed in the bloom waters. Following the application of ozone, total MC-LR degraded at a rate of 7.44 × 10^{-7} s^{-1}, which translated to a half-life of 10.8 days. This lower rate was indicative of the lower biomass present in the CRW and/or limited prior exposure to MC [40].

3. Conclusions

This study demonstrated differences between lab-cultured and natural-bloom cyanobacteria cell lysis rates, identified the CT values required for the complete intracellular release of microcystin, and identified the importance of stagnation following partial damage to cyanobacteria cells. Lab-cultured *M. aeruginosa* cells were found to be more susceptible to chemical oxidation than natural bloom cells. For lab-cultured *M. aeruginosa*, the level of extracellular MC increased after dose ratios of 0.05 Cl_2: DOC, 0.05 ClO_2: DOC, 0.25 O_3: DOC, and 2 $KMnO_4$: DOC. In contrast, the mixed species found in the two bloom waters required higher doses of 0.15 Cl_2: DOC, 0.15 ClO_2: DOC, 0.75/0.80 O_3: DOC, and 2/2.7 $KMnO_4$: DOC. Oxidant: DOC ratios and CT values that resulted in complete intracellular release were identified for each oxidant across the three water evaluated.

Following the oxidation phase and quenching, the effect of stagnation time was evaluated using partially damaged cells. Stagnation resulted in greater release of intracellular DOC and microcystin. At dose ratios of 0.15 for ozone, chlorine, chlorine dioxide, and monochloramine and a dose ratio of 0.4/0.53 for potassium permanganate, all three of the waters saw the release of extracellular MC. Increases in the extracellular fraction of MC started anywhere from 20 min after quenching for the lab-cultured sample up to 96 h after quenching for the USA bloom.

These data highlight the need to monitor the downstream time points for the potential release of extracellular toxin because partially damaged cells may be retained on filter surfaces or in sludge where continued MC release can threaten finished water quality and/or solids handling processes.

4. Materials and Methods

4.1. Cyanobacteria Culturing, Sampling, and Characterization

4.1.1. Laboratory-Cultured *Microcystis aeruginosa* Cells Transferred into Colorado River Water (CRW)

A microcystin-LR (MC-LR) producing cyanobacteria strain, unicellular *Microcystis aeruginosa* (LB 2385, UTEX Austin Culture Collection, Austin, TX, USA), was cultured as described in prior work [41]. Briefly, cells were cultured in Bold3N media for ~30 days and subsequently rinsed with 10 mM phosphate buffer (pH 7.5) after three centrifugations, respectively. A cell stock solution was prepared by suspending cells in ~20 mL of the phosphate buffer. Cell concentration of the stock was determined via optical density at 730 nm (OD730), which was previously correlated with cell counts using a digital flow cytometer [23]. Subsequently, cells were spiked into Colorado River water (CRW), which had DOC of 2.5 mg/L, alkalinity of 138 mg/L as $CaCO_3$, and pH of 8.0, to obtain 1.0×10^6 cells/mL. The background DOC of CRW (2.5 mg/L) was used to determine the applied oxidant doses.

4.1.2. United States (USA) Bloom: Grand Lake St. Marys

Grand Lake St. Marys in Celina, OH was sampled during October 2016, and the water was shipped in cubitainers on ice to Southern Nevada Water Authority (SNWA) for experiments. The USA bloom water had a DOC of 9.3 mg/L and pH of 7.9. The predominant cyanobacteria present was *Planktothrix agardhii/suspensa* (2.65×10^6 cells/mL), followed by *Planktolyngbya spp.* (3.72×10^5 cells/mL). The cyanobacteria were confirmed to produce microcystin-YR (MC-YR), which historically has been generated by *Planktothrix agardhii/suspensa* in the USA bloom [42]. Additional details and pictures of the bloom can be found in the Supporting Information (SI) Section S1.

4.1.3. Canadian (CA) Bloom: Lake Champlain

Anabaena spiroides (1.58×10^5 cells/mL), *Aphanothece clathrata brevis* (1.01×10^5 cells/mL), and *Microcystis aeruginosa* (4.03×10^4 cells/mL) complex naturally occurred on the Canadian side of Lake Champlain. Water was collected and transported to Polytechnique Montréal on ice by cubitainers. The Canadian bloom had a DOC of 6.1 mg/L and pH of 7.9.

4.2. Pre-Oxidation of Cyanobacteria Suspensions

4.2.1. Varied Oxidant: DOC Ratios

Cyanobacteria suspensions were placed in 1 L amber glass bottles for exposure to oxidants. Chlorine (Cl_2), chlorine dioxide (ClO_2), monochloramine (NH_2Cl), ozone (O_3), and potassium permanganate ($KMnO_4$) were assessed as pre-oxidants. Chlorine was obtained as a ~5% sodium hypochlorite (NaOCl) solution (Fisher Scientific). Chlorine dioxide was obtained as a ~3000 mg/L solution (CDG Environmental). $KMnO_4$ was obtained as ~900 mg/L solution (Ricca). Monochloramine was prepared using ammonium chloride, sodium hydroxide (NaOH), and NaOCl as described in [43]. Oxidant stock concentrations were measured to ensure accurate dosing.

Chlorine, chlorine dioxide, monochloramine, and ozone were spiked into suspensions with oxidant: DOC mass ratios of 0.05, 0.10, 0.15, 0.25, and 0.5. Ozone was also assessed at O_3: DOC of 0.75. The DOC for ratio calculations was based on the background DOC of the source water. Therefore, any released DOC following cyanobacteria cell lysis was not factored into these ratios. Potassium permanganate was spiked into suspensions with ratios of 0.1, 0.25, 0.4, 2, and 4 for the CA bloom and ratios of 0.13, 0.33, 0.5, 2.7, and 5.3 for the USA bloom and lab-cultured *M. aeruginosa*. Oxidants were allowed to react for a maximum of 20 min at room temperature (20 °C). Aliquots (10 mL) were taken from reactors throughout experiments for residual measurements; volume taken did not exceed 10% of total reactor volume. When no residual remained or 20 min had elapsed, sodium thiosulfate ($Na_2S_2O_3$) was added to reactors (targeting 100 mg/L in the reactor to ensure excess) to quench residual and/or to give all reactors the same treatment. Samples were then immediately taken for analysis of total and extracellular microcystins (MC), chl-*a* or phycocyanin (PC), and DOC. The middle oxidant: DOC ratio for each oxidant (except for ozone, which had an additional ratio assessed) had a duplicate reactor to assess reproducibility. Control reactors without chemical addition and with $Na_2S_2O_3$ addition were sampled as well; $Na_2S_2O_3$ did not have a measurable effect on MC, pigments, nor DOC.

4.2.2. Varied Stagnation Times

For select oxidants: DOC ratios (0.15 for Cl_2, ClO_2, NH_2Cl, and O_3; 0.4 for $KMnO_4$), discrete reactors with 500 mL of cell suspensions in 500 mL amber glass bottles were spiked with oxidants for sampling at different time points. Reactors were quenched with $Na_2S_2O_3$ at 20 min (or earlier when sampled for time points before 20 min). Sample times ranged from within 1 min of oxidant addition to either 65 or 168 h to look at the delayed release of MCs following partial pre-oxidation of cells. Samples were collected for total and extracellular MC, chl-*a* or PC, and DOC. Control reactors (no oxidant exposure; with and without $Na_2S_2O_3$) were also sampled at each time point.

4.3. Sample Analyses

4.3.1. Cyanobacteria Bloom Characterization and Water Quality Parameters

Optical density at 730 nm (OD730)—used for estimating cell density of laboratory-cultured *Microcystis aeruginosa* stock and suspensions—was measured using an ultraviolet-visible light spectrophotometer (Hach DR 5000). A correlation between OD730 and cell counts was established using a digital flow cytometer (FlowCAM, Fluid Imaging Technologies, Yarmouth, ME, USA) as described in previous work [23]. USA bloom samples—one as collected and one preserved with 1% Lugol's iodine—were shipped overnight to BSA Environmental Services (Beachwood, OH, USA) for cyanobacteria imaging, identification, and enumeration. CA bloom samples were also preserved with 1% Lugol's iodine for identification and enumeration at Polytechnique Montréal.

DOC was measured using Standard Method (SM) 5310 B and pH by SM 4500-H+ B. Chl-*a* was analyzed via SM 10,200 H (APHA, 2012). PC was measured with the Total Algae sensor on a YSI EXO2 Multiparameter Sonde (YSI, Yellow Springs, OH, USA). The sensor was blanked with deionized water and samples were measured by relative fluorescence units (RFU) as described in past work by [44].

4.3.2. Oxidant Residuals

Free and total chlorine residuals were measured by SM 4500-Cl G. Ozone residuals were measured using indigo trisulfonate via SM 4500-O3 [45]. Chlorine dioxide residuals were measured by 4500-ClO2 D. Permanganate residuals were measured by SM 4500-Cl G, with 1 mg/L free chlorine measurement equivalent to 0.891 mg/L $KMnO_4$ [45]. Permanganate samples were first filtered with 0.22 μm pore size polyvinylidene fluoride filters (Millex) to remove particulate manganese oxides which can interfere with residual measurement.

4.3.3. Microcystin

For extracellular MC, samples were immediately filtered using 0.45 μm pore size glass microfiber syringe filters (Whatman). To lyse all cells and quantify total MC, 10 mL samples were first frozen at −20 °C, thawed at 25 °C, and sonicated for 5 min prior to filtration. Samples were sonicated with a probe sonicator (Q500, QSonica, Newtown, CT, USA) outfitted with a $\frac{1}{4}$ inch microtip operated at 20 kHz with 200 μm amplitude/tip displacement requiring ~30 W power. Samples were kept in an ice bath to prevent overheating of the sample and probe. The sonication was pulsed, with 5 s on and 1 s off, until total sonication time reached 5 min. Freeze/thaw, sonication, and filtration of a control (20 μg/L MC-LR spiked into CRW) did not cause degradation of MC-LR.

MC concentrations were measured with liquid chromatography tandem mass spectrometry (LC-MS/MS) and enzyme-linked immunosorbent assay (ELISA). Samples were analyzed using EPA Method 546 for total MC via ELISA. LC-MS/MS samples were analyzed for eight microcystin congeners (MC-LA, MC-LF, MC-LR, MC-LW, MC-LY, MC-RR, MC-WR, and MC-YR) via liquid chromatography-tandem mass spectrometry (LC-MS/MS) as described in previous work [13] with method reporting limits (MRLs) for each congener at 0.5 μg/L.

4.3.4. Calculation of Oxidant Decay, MC Decay or Release Rates, and Cell Damage Rates

To analyze the data generated in this work, four different rate constants were calculated using Equation (1).

$$C = C_0 \times e^{-kR} \quad (1)$$

In Equation (1), C was the concentration [total or extracellular MC (μg/L), and chl-*a* (μg/L) or PC (RFU)] at a given time, C_0 was the starting concentration [total or extracellular MC (μg/L), and chl-*a* (μg/L) or PC (RFU) prior to oxidant exposure], and k was the oxidant decay rate [s^{-1} or $M^{-1}s^{-1}$]. Two of the rates calculated employed time as R ($k_{biodegradation}$ and $k_{release}$) and two used CT (k_{total} for total MC and k_{damage} to model cell damage) as R.

Supplementary Materials: The following are available online at http://www.mdpi.com/2072-6651/12/5/335/s1, Figures S1–S5: Oxidant decay in the USA bloom for the five different oxidant: DOC ratios applied in this work, Figures S6–S10: Oxidant decay in the lab-cultured *M. aeruginosa* in CRW for the five different oxidant: DOC ratios applied in this work, Figure S11: Photos of USA bloom cyanobacteria. Table S1: Oxidant decay rates for lab-cultured *M. aeruginosa* (lab MA) and USA bloom. Table S2: Oxidant exposure generated from the oxidant decay rates for lab-cultured *M. aeruginosa* (lab MA), USA and CA blooms. Table S3: Cell damage (k_{damage}) and total MC decay (k_{total}) decay after oxidation (t < 20 min) in the lab-cultured *M. aeruginosa*.

Author Contributions: The author contribution is as follows: conceptualization, all authors; methodology, all authors; formal analysis, all authors; investigation, all authors; resources, all authors; writing—original draft preparation, K.E.G., C.M.G.; writing—review and editing, all authors; visualization, K.E.G., C.M.G.; supervision, A.Z. and E.C.W.; project administration, A.Z. and E.C.W.; funding acquisition, A.Z., C.A., E.R., E.C.W. All authors have read and agreed to the published version of the manuscript.

Funding: This research was funded by The Water Research Foundation grant number [4692].

Acknowledgments: This work was funded by the Water Research Foundation (Project #4692). Additional funding provided by the Natural Sciences and Engineering Research Council of Canada (NSERC) and the Fonds de Recherche du Québec–Nature et technologies (FRQNT). The authors acknowledge the following people at SNWA for assistance with analytical and experimental work: Brett Vanderford, Rebecca Trenholm, Julia Lew,

Glen de Vera, Mary Murphy, James Park, Shandra Staker, Janie Holady, and Brittney Stipanov. The authors also thank Chris Mihalkovic from Hazen and Sawyer for help in acquiring water from the USA bloom.

Conflicts of Interest: The authors declare no conflict of interest.

References

1. USEPA Drinking Water Health Advisory for the Cyanobacterial Microcystin Toxins: 820R15100. Available online: https://www.epa.gov/sites/production/files/2017-06/documents/microcystins-report-2015.pdf (accessed on 18 November 2018).
2. USEPA Drinking Water Health Advisory for the Cyanobacterial Toxin Cylindrospermopsin: 820R15101. Available online: https://www.epa.gov/sites/production/files/2017-06/documents/cylindrospermopsin-report-2015.pdf (accessed on 18 November 2018).
3. Health Canada Guidelines for Canadian Drinking Water Quality Cyanobacterial Toxins. Available online: https://www.canada.ca/en/health-canada/services/publications/healthy-living/guidelines-canadian-drinking-water-quality-guideline-technical-document-cyanobacterial-toxins-document.html (accessed on 28 December 2018).
4. American Water Works Association (AWWA) Cyanotoxins in US Drinking Water: Occurrence, Case Studies and State Approaches to Regulation. Available online: https://www.awwa.org/Portals/0/AWWA/Government/201609_Cyanotoxin_Occurrence_States_Approach.pdf?ver=2018-12-13-101832-037 (accessed on 18 November 2018).
5. USEPA Recommendations for Public Water Systems to Manage Cyanotoxins in Drinking Water: 815R15100. Available online: https://www.epa.gov/sites/production/files/2017-06/documents/cyanotoxin-management-drinking-water.pdf (accessed on 18 November 2018).
6. USEPA. *Cyanobacteria and Cyanotoxins: Information for Drinking Water Systems*; EPA-810F11; USEPA: Washington, DC, USA, 2019.
7. American Water Works Association (AWWA); Water Research Foundation (WRF) A Water Utility Manager's Guide to Cyanotoxins. Available online: https://www.awwa.org/Portals/0/AWWA/Government/WaterUtilityManagersGuideToCyanotoxins.pdf?ver=2018-12-13-101839-130 (accessed on 28 November 2018).
8. Global Water Research Coalition (GWRC); Water Quality Research Australia (WQRA). *International Guidance Manual for the Management of Toxic Cyanobacteria*; Newcombe, G., Ed.; Alliance House: London, UK, 2009; ISBN 978-90-77622-21-6.
9. Newcombe, G. International Guidance Manual for the Management of Toxic Cyanobacteria. Available online: https://www.iwapublishing.com/books/9781780401355/international-guidance-manual-management-toxic-cyanobacteria (accessed on 27 April 2020).
10. Crittenden, J.C.; Trussell, R.R.; Hand, D.W.; Howe, K.J.; Tchobanoglous, G. *MWH's Water Treatment: Principles and Design: Third Edition*; John Wiley and Sons: Hoboken, NJ, USA, 2012; ISBN 9780470405390.
11. Oregon Department of Human Services: Public Health Division Fast Facts: Blue-green Alage Q&A for Water System Operators. Available online: https://www.oregon.gov/oha/PH/HEALTHYENVIRONMENTS/RECREATION/HARMFULALGAEBLOOMS/Pages/resources_for_samplers.aspx (accessed on 28 December 2018).
12. Zamyadi, A.; Fan, Y.; Daly, R.I.; Prévost, M. Chlorination of *Microcystis aeruginosa*: Toxin release and oxidation, cellular chlorine demand and disinfection by-products formation. *Water Res.* **2013**, *47*, 1080–1090. [CrossRef]
13. Wert, E.C.; Korak, J.A.; Trenholm, R.A.; Rosario-Ortiz, F.L. Effect of oxidant exposure on the release of intracellular microcystin, MIB, and geosmin from three cyanobacteria species. *Water Res.* **2014**, *52*, 251–259. [CrossRef] [PubMed]
14. Daly, R.I.; Ho, L.; Brookes, J.D. Effect of chlorination on *Microcystis aeruginosa* cell integrity and subsequent microcystin release and degradation. *Environ. Sci. Technol.* **2007**, *41*, 4447–4453. [CrossRef] [PubMed]
15. Ding, J.; Shi, H.; Timmons, T.; Adams, C. Release and removal of microcystins from *Microcystis* during oxidative-, physical-, and UV-Based disinfection. *J. Environ. Eng.* **2010**, *136*, 2–11. [CrossRef]
16. He, X.; Wert, E.C. Colonial cell disaggregation and intracellular microcystin release following chlorination of naturally occurring *Microcystis*. *Water Res.* **2016**, *101*. [CrossRef] [PubMed]

17. Zhang, H.; Dan, Y.; Adams, C.D.; Shi, H.; Ma, Y.; Eichholz, T. Effect of oxidant demand on the release and degradation of microcystin-LR from *Microcystis aeruginosa* during oxidation. *Chemosphere* **2017**, *181*, 562–568. [CrossRef]
18. Zamyadi, A.; Greenstein, K.E.; Glover, C.M.; Adams, C.; Rosenfeldt, E.; Wert, E.C. Impact of hydrogen peroxide and copper sulfate on the delayed release of microcystin. *Water* **2020**, *12*, 1105. [CrossRef]
19. Moradinejad, S.; Glover, C.M.; Mailly, J.; Seighalani, T.Z.; Peldszus, S.; Barbeau, B.; Dorner, S.; Prévost, M.; Zamyadi, A. Using advanced spectroscopy and organic matter characterization to evaluate the impact of oxidation on cyanobacteria. *Toxins* **2019**, *11*, 278. [CrossRef]
20. Zamyadi, A.; Coral, L.A.; Barbeau, B.; Dorner, S.; Lapolli, F.R.; Prévost, M. Fate of toxic cyanobacterial genera from natural bloom events during ozonation. *Water Res.* **2015**, *73*, 204–215. [CrossRef]
21. Coral, L.A.; Zamyadi, A.; Barbeau, B.; Bassetti, F.J.; Lapolli, F.R.; Prévost, M. Oxidation of *Microcystis aeruginosa* and *Anabaena flos-aquae* by ozone: Impacts on cell integrity and chlorination by-product formation. *Water Res.* **2013**, *47*, 2983–2994. [CrossRef]
22. Zamyadi, A.; Ho, L.; Newcombe, G.; Bustamante, H.; Prévost, M. Fate of toxic cyanobacterial cells and disinfection by-products formation after chlorination. *Water Res.* **2012**, *46*, 1524–1535. [CrossRef] [PubMed]
23. Wert, E.C.; Dong, M.M.; Rosario-Ortiz, F.L. Using digital flow cytometry to assess the degradation of three cyanobacteria species after oxidation processes. *Water Res.* **2013**, *47*, 3752–3761. [CrossRef] [PubMed]
24. Li, L.; Zhu, C.; Xie, C.; Shao, C.; Yu, S.; Zhao, L.; Gao, N. Kinetics and mechanism of *Pseudoanabaena* cell inactivation, 2-MIB release and degradation under exposure of ozone, chlorine and permanganate. *Water Res.* **2018**, *147*, 422–428. [CrossRef]
25. Zamyadi, A.; Choo, F.; Newcombe, G.; Stuetz, R.; Henderson, R.K. A review of monitoring technologies for real-time management of cyanobacteria: Recent advances and future direction. *TrAC Trends Anal. Chem.* **2016**, *85*, 83–96. [CrossRef]
26. Zhou, S.; Shao, Y.; Gao, N.; Li, L.; Deng, J.; Zhu, M.; Zhu, S. Effect of chlorine dioxide on cyanobacterial cell integrity, toxin degradation and disinfection by-product formation. *Sci. Total Environ.* **2014**, *482–483*, 208–213. [CrossRef] [PubMed]
27. Laszakovits, J.R.; MacKay, A.A. Removal of cyanotoxins by potassium permanganate: Incorporating competition from natural water constituents. *Water Res.* **2019**, *155*, 86–95. [CrossRef]
28. Ramseier, M.K.; von Gunten, U.; Freihofer, P.; Hammes, F. Kinetics of membrane damage to high (HNA) and low (LNA) nucleic acid bacterial clusters in drinking water by ozone, chlorine, chlorine dioxide, monochloramine, ferrate(VI), and permanganate. *Water Res.* **2011**, *45*, 1490–1500. [CrossRef]
29. Rodríguez, E.; Onstad, G.D.; Kull, T.P.J.; Metcalf, J.S.; Acero, J.L.; von Gunten, U. Oxidative elimination of cyanotoxins: Comparison of ozone, chlorine, chlorine dioxide and permanganate. *Water Res.* **2007**, *41*, 3381–3393. [CrossRef]
30. Fan, J.; Ho, L.; Hobson, P.; Daly, R.; Brookes, J. Application of various oxidants for cyanobacteria control and cyanotoxin removal in wastewater treatment. *J. Environ. Eng.* **2014**, *140*, 04014022. [CrossRef]
31. Onstad, G.D.; Strauch, S.; Meriluoto, J.; Codd, G.A.; Von Gunten, U. Selective oxidation of key functional groups in cyanotoxins during drinking water ozonation. *Environ. Sci. Technol.* **2007**, *41*, 4397–4404. [CrossRef]
32. Acero, J.L.; Rodriguez, E.; Meriluoto, J. Kinetics of reactions between chlorine and the cyanobacterial toxins microcystins. *Water Res.* **2005**, *39*, 1628–1638. [CrossRef] [PubMed]
33. He, X.; Stanford, B.D.; Adams, C.; Rosenfeldt, E.J.; Wert, E.C. Varied influence of microcystin structural difference on ELISA cross-reactivity and chlorination efficiency of congener mixtures. *Water Res.* **2017**, *126*, 515–523. [CrossRef] [PubMed]
34. Ho, L.; Onstad, G.; Von Gunten, U.; Rinck-Pfeiffer, S.; Craig, K.; Newcombe, G. Differences in the chlorine reactivity of four microcystin analogues. *Water Res.* **2006**, *40*, 1200–1209. [CrossRef] [PubMed]
35. Lee, Y.; von Gunten, U. Oxidative transformation of micropollutants during municipal wastewater treatment: Comparison of kinetic aspects of selective (chlorine, chlorine dioxide, ferrateVI, and ozone) and non-selective oxidants (hydroxyl radical). *Water Res.* **2010**, *44*, 555–566. [CrossRef]
36. Pestana, C.J.; Reeve, P.J.; Sawade, E.; Voldoire, C.F.; Newton, K.; Praptiwi, R.; Collingnon, L.; Dreyfus, J.; Hobson, P.; Gaget, V.; et al. Fate of cyanobacteria in drinking water treatment plant lagoon supernatant and sludge. *Sci. Total Environ.* **2016**, *565*, 1192–1200. [CrossRef]

37. Zamyadi, A.; MacLeod, S.L.; Fan, Y.; McQuaid, N.; Dorner, S.; Sauvé, S.; Prévost, M. Toxic cyanobacterial breakthrough and accumulation in a drinking water plant: A monitoring and treatment challenge. *Water Res.* **2012**, *46*, 1511–1523. [CrossRef]
38. Zamyadi, A.; Romanis, C.; Mills, T.; Neilan, B.; Choo, F.; Coral, L.A.; Gale, D.; Newcombe, G.; Crosbie, N.; Stuetz, R.; et al. Diagnosing water treatment critical control points for cyanobacterial removal: Exploring benefits of combined microscopy, next-generation sequencing, and cell integrity methods. *Water Res.* **2019**, *152*, 96–105. [CrossRef]
39. Chen, W.; Song, L.; Peng, L.; Wan, N.; Zhang, X.; Gan, N. Reduction in microcystin concentrations in large and shallow lakes: Water and sediment-interface contributions. *Water Res.* **2008**, *42*, 763–773. [CrossRef]
40. Maghsoudi, E.; Fortin, N.; Greer, C.; Duy, S.V.; Fayad, P.; Sauvé, S.; Prévost, M.; Dorner, S. Biodegradation of multiple microcystins and cylindrospermopsin in clarifier sludge and a drinking water source: Effects of particulate attached bacteria and phycocyanin. *Ecotoxicol. Environ. Saf.* **2015**, *120*, 409–417. [CrossRef]
41. Wert, E.C.; Rosario-Ortiz, F.L. Intracellular organic matter from cyanobacteria as a precursor for carbonaceous and nitrogenous disinfection byproducts. *Environ. Sci. Technol.* **2013**, *47*. [CrossRef]
42. Dumouchelle, D.H.; Stelzer, E.A. Chemical and biological quality of water in Grand Lake St. Marys, Ohio, 2011–12, with emphasis on cyanobacteria. *U.S. Geol. Surv. Sci. Investig. Rep.* **2014**, *5210*. [CrossRef]
43. Lyon, B.A.; Milsk, R.Y.; Deangelo, A.B.; Simmons, J.E.; Moyer, M.P.; Weinberg, H.S. Integrated chemical and toxicological investigation of UV-chlorine/chloramine drinking water treatment. *Environ. Sci. Technol.* **2014**, *48*, 6743–6753. [CrossRef] [PubMed]
44. Almuhtaram, H.; Cui, Y.; Zamyadi, A.; Hofmann, R. Cyanotoxins and cyanobacteria cell accumulations in drinking water treatment plants with a low risk of bloom formation at the source. *Toxins* **2018**, *10*, 430. [CrossRef] [PubMed]
45. APHA; AWWA; WEF. *Standard Methods for Examination of Water and Wastewater*; American Public Health Association: Washington, DC, USA, 2012.

toxins

MDPI

Article

Seaweed Essential Oils as a New Source of Bioactive Compounds for Cyanobacteria Growth Control: Innovative Ecological Biocontrol Approach

Soukaina El Amrani Zerrifi [1], Fatima El Khalloufi [2], Richard Mugani [1], Redouane El Mahdi [1], Ayoub Kasrati [3,4], Bouchra Soulaimani [4], Lillian Barros [5], Isabel C. F. R. Ferreira [5], Joana S. Amaral [5,6], Tiane Cristine Finimundy [5], Abdelaziz Abbad [4], Brahim Oudra [1], Alexandre Campos [7] and Vitor Vasconcelos [7,8,*]

1. Water, Biodiversity and Climate Change Laboratory, Phycology, Biotechnology and Environmental Toxicology Research Unit, Faculty of Sciences Semlalia Marrakech, Cadi Ayyad University, P.O. Box 2390, 40000 Marrakech, Morocco; soukainaelamranizerrifi@gmail.com (S.E.A.Z.); richardmugani@gmail.com (R.M.); redouane.elmahdii@gmail.com (R.E.M.); oudra@uca.ac.ma (B.O.)
2. Laboratory of Chemistry, Modeling and Environmental Sciences, Polydisciplinary Faculty of Khouribga, Sultan Moulay Slimane University of Beni Mellal, P.B. 145, 25000 Khouribga, Morocco; elkhalloufi.f@gmail.com
3. Department of Health and Agro-Industry Engineering, High School of Engineering and Innovation of Marrakesh (E2IM), Private University of Marrakesh (UPM), 42312 Marrakech, Morocco; ayoub.kasrati@gmail.com
4. Laboratory of Microbial Biotechnologies, Agrosciences and Environment, Faculty of Science Semlalia Marrakech, Cadi Ayyad University, P.O. Box 2390, 40000 Marrakech, Morocco; bouchrasoulaimanigebc@gmail.com (B.S.); abbad.abdelaziz@gmail.com (A.A.)
5. Centro de Investigação de Montanha (CIMO), Instituto Politécnico de Bragança, Campus de Santa Apolónia, 5300-253 Bragança, Portugal; lillian@ipb.pt (L.B.); iferreira@ipb.pt (I.C.F.R.F.); jamaral@ipb.pt (J.S.A.); tcfinimu@hotmail.com (T.C.F.)
6. REQUIMTE-LAQV, Faculdade de Farmácia, Universidade do Porto, 4050-313 Porto, Portugal
7. CIIMAR, Interdisciplinary Centre of Marine and Environmental Research, University of Porto, Terminal de Cruzeiros do Porto de Leixões, Av. General Norton de Matos, s/n, 4450-208 Matosinhos, Portugal; acampos@ciimar.up.pt
8. Departament of Biology, Faculty of Sciences, University of Porto, Rua do Campo Alegre, 4169-007 Porto, Portugal

* Correspondence: vmvascon@fc.up.pt; Tel.: +351-223401817

Received: 17 July 2020; Accepted: 14 August 2020; Published: 17 August 2020

Abstract: The application of natural compounds extracted from seaweeds is a promising eco-friendly alternative solution for harmful algae control in aquatic ecosystems. In the present study, the anti-cyanobacterial activity of three Moroccan marine macroalgae essential oils (EOs) was tested and evaluated on unicellular *Microcystis aeruginosa* cyanobacterium. Additionally, the possible anti-cyanobacterial response mechanisms were investigated by analyzing the antioxidant enzyme activities of *M. aeruginosa* cells. The results of EOs GC–MS analyses revealed a complex chemical composition, allowing the identification of 91 constituents. Palmitic acid, palmitoleic acid, and eicosapentaenoic acid were the most predominant compounds in *Cystoseira tamariscifolia, Sargassum muticum*, and *Ulva lactuca* EOs, respectively. The highest anti-cyanobacterial activity was recorded for *Cystoseira tamariscifolia* EO (ZI = 46.33 mm, MIC = 7.81 μg mL^{-1}, and MBC = 15.62 μg mL^{-1}). The growth, chlorophyll-*a* and protein content of the tested cyanobacteria were significantly reduced by *C. tamariscifolia* EO at both used concentrations (inhibition rate >67% during the 6 days test period in liquid media). Furthermore, oxidative stress caused by *C. tamariscifolia* EO on cyanobacterium cells showed an increase of the activities of superoxide dismutase (SOD) and catalase (CAT), and malondialdehyde (MDA) concentration was significantly elevated after 2 days of exposure. Overall, these experimental findings can open a promising new natural pathway based on the use of seaweed essential oils to the fight against potent toxic harmful cyanobacterial blooms (HCBs).

Keywords: anti-cyanobacterial activity; bio-control; seaweed essential oils; *Microcystis aeruginosa*

Key Contribution: *Microcystis* species are among the most important worldwide freshwater bloom-forming cyanobacteria. Seaweed essential oils were chosen for inhibitory experiment on *Microcystis aeruginosa*. The inhibition rate and oxidative stress caused by *C. tamariscifolia* EO on cyanobacterium cells were quite similar to that obtained by copper sulphate ($CuSO_4$).

---

1. Introduction

On the grounds of climate warming and increased nutrient inputs due to anthropogenic activities, harmful cyanobacterial blooms (HCBs) become a severe hazard for freshwater ecosystems [1–4]. Due to the critical economic and public health issues caused by HCBs, extensive research on this topic has been conducted aiming to disclose the detrimental effects of HCBs and mitigation strategies. Recent research has been focused on the strategies applied in HCBs control including chemical, physical, mechanical, and biological methods [5–7]. Cyanobacteria blooms are controlled with ultrasound, artificial mixing, and ultraviolet irradiation in the case of physical and mechanical strategies [8–11]. Chemical products such as photosensitizers (molecules that can be activated by light in order to generate reactive oxygen species which may damage cell structures), metals, and other chemical molecules are among the most commonly used chemical methods [12–14]. The introduction of grazers and competitors of cyanobacteria, such as zooplankton, microorganisms (viruses, pathogenic bacteria, or fungi), and macrophytes have been proposed as the most popular biomanipulation for the bio-control of toxic cyanobacteria [15–19]. However, the application of these strategies is not recommended because of their unforeseen ecological consequences, high costs, energy-intensive, and low efficiency [20]. In order to develop effective anti-HCBs agents that are more eco-friendly to the environment, scientists are looking for natural substances released by other aquatic organisms with activity against the growth of cyanobacteria. The marine environment is an excellent source of natural bioactive compounds with unique structures, different from those found in terrestrial natural compounds [21]. Among marine organisms, seaweeds produce active secondary metabolites with a wide range of biological activities, including antibacterial, antifungal, antioxidant, and anti-cyanobacterial compounds [7].

Owing to its specific geographical position, from the Mediterranean Sea to the North and the Atlantic Ocean to the West, Morocco holds a large bio-ecological diversity of seaweeds. This diversity was documented for instance by Chalabi et al. (2015) [22] who reported a particular richness of 489 species distributed at the Mediterranean Coast (381 species) and the Atlantic Coast (323 species). Several compounds with anti-cyanobacterial properties have been purified from the extracts of some seaweed species, such as palmitelaidic acid and 2,3 dihydroxypropyl ester extracted from the methanol extract of *Ulva prolifera* [6], and gossonorol and margaric acid purified from *Gracilaria lemaneiformis* ethanolic extract [23]. Currently and according to our knowledge, unlike the essential oils of plants and plant parts that have been evaluated for their potential anti-cyanobacterial properties, there is no scientific report describing the anti-prokaryotic activity of the essential oils from seaweeds [24–28]. In this respect, the present study aims to uncover for the first time the possible inhibitory effects of essential oils (EOs) extracted from three Moroccan seaweeds that are broadly known by their antimicrobial activities namely *Cystoseira tamariscifolia, Sargassum muticum,* and *Ulva lactuca,* on the growth of *Microcystis aeruginosa* a cyanobacteria species that commonly form HABs in Moroccan freshwaters. In addition, this study also provides first insights regarding the anti-cyanobacterial mechanism of EOs, analyzing the growth inhibition power through the following indicators: measurement of chlorophyll-*a*, protein contents, and activity of cellular stress response enzymes of the stated strain.

2. Results

2.1. Chemical Composition of Moroccan Seaweed EOs

The EO total content based on the dry weight of the seaweed materials are presented in Table 1. The highest total content was achieved with the EO extracted from the green macroalgae *U. lactuca* (0.187% ± 0.078%) followed by the brown seaweed *S. muticum* EO (0.106% ± 0.017%). While, the lowest total content was recorded by the brown seaweed *C. tamariscifolia* EO (0.062% ± 0.018%).

Table 1. Total content of studied seaweed essential oils.

Species	EO Total Content (%, *v/w*)
Ulva lactuca	0.19 ± 0.08
Sargassum muticum	0.11 ± 0.02
Cystoseira tamariscifolia	0.06 ± 0.02

The chemical composition of seaweed EOs was identified qualitatively and quantitatively by GC-MS analysis. The content, expressed in percentage, of the individual components of each seaweed, and retention indices, are summarized in Table 2. Forty constituents were determined in *C. tamariscifolia* EO, corresponding to 59.6% from the total compounds in this species. The major components in this EO were palmitic acid (7.7%) followed by dihydroactinidioide (6.57%), hexahydrofarnesyl acetone (5.1%), heptadecane (4.14%), and phytol (4.1%), while other compounds were present below 4%. In total, 41 compounds (corresponding to 45.7% from the total compounds in this species) were identified in the total *S. muticum* EO composition, the most predominant compounds were found to be palmitoleic acid (7.8%), dihydroactinidiolide (6.97%) and benzeneacetaldehyde (4.62%). Whereas, the EO extracted from *U. lactuca* revealed the presence of the highest number of compounds; 45 compounds (corresponding to 55% from the total compounds in this species), with dominance of eicosapentaenoic acid (8%), dihydroactinidioide (7.8%), and β-ionone (7.6%).

Table 2. Chemical composition of essential oils extracted from Moroccan seaweeds (%). Values in bold represent the major compounds present in each sample.

					Relative % [c]		
N°	Compound	RT (min)	LRI [a]	LRI [b]	Ct	Sm	Ul
1	(E)-2-Pentenal	6.27	750	744	-	-	0.086 ± 0.002
2	4-Methyl-2-pentanol	6.42	754	745	-	-	0.0063 ± 0.0002
3	Toluene	6.69	763	756	-	-	0.143 ± 0.004
4	Hexanal or *n*-Caproylaldehyde	7.92	800	801	0.196 ± 0.001	0.44 ± 0.02	0.23 ± 0.01
5	Furfural	9.24	828	828	0.19 ± 0.01	0.069 ± 0.002	-
6	4-Hexen-3-one	9.41	832	-	-	-	0.075 ± 0.001
7	3-Hexen-2-one	9.73	839	834 *	-	-	0.046 ± 0.003
8	(E)-2-Hexenal	10.17	848	846	0.085 ± 0.003	0.35 ± 0.01	0.347 ± 0.003
9	2-Furanmethanol	10.37	852	853 *	0.57± 0.01	-	-
10	1-Hexanol	11.11	868	863	-	0.063 ± 0.005	0.023 ± 0.001
11	4-Cyclopentene-1,3-dione	11.67	880	880 *	0.6 ± 0.01	0.02 ± 0	-
12	2-Heptanone	12.05	888	889	-	0.029 ± 0.001	0.35 ± 0.01
13	*cis*-4-Heptenal	12.47	897	893	-	0.115 ± 0.005	0.28 ± 0.01
14	*n*-Heptanal	12.57	900	901	0.29 ± 0.01	0.136 ± 0.003	0.293 ± 0.002
15	Acetylfuran	12.99	908	909	0.2 ± 0.005	0.176 ± 0.002	-
16	2-Cyclohexen-1-one	13.98	927	927 *	0.102 ± 0.002	-	-
17	α-Pinene	14.20	931	932	-	0.0151 ± 0.0001	-
18	Cyclohexen-2-one	14.22	931	-	-	-	0.07 ± 0.001
19	Hept-3-en-2-one	14.37	934	927	-	0.038 ± 0.004	0.052 ± 0.003
20	Benzaldehyde	15.00	956	952	0.31 ± 0.01	0.38 ± 0.01	0.569 ± 0.004
21	5-Methyl-furfural	15.71	960	957	0.737 ± 0.001	0.988 ± 0.02	3.39 ± 0.05
22	3,5,5-Trimethyl-2-hexene	16.47	975	-	-	-	0.192 ± 0.004
23	1-Octen-3-ol	16.63	978	974	0.103 ± 0.004	0.2036 ± 0.0001	-
24	2-methyl-3-Octanone	16.91	983	985 *	-	0.318 ± 0.004	-
25	6-Methyl-5-heptene-2-one	16.99	985	986 *	-	0.056 ± 0.004	-
26	3-Methyl-3-cyclohexen-1-one	17.00	985	-	-	-	0.69 ± 0.02
27	Octanal	17.81	1000	998	0.206 ± 0.003	-	-

Table 2. *Cont.*

N°	Compound	RT (min)	LRI [a]	LRI [b]	Relative % [c]		
					Ct	Sm	Ul
28	Pyrrole-2-carboxaldehyde	18.12	1006	1008 *	-	0.17 ± 0.01	-
29	(E,E)-2,4-Heptadienal	18.16	1007	1005	0.29 ± 0.01	0.09 ± 0.01	0.466 ± 0.003
30	4-Oxohex-2-enal	19.49	1033	-	1.43 ± 0.03	-	
31	2,2,6-Trimethyl-Cyclohexanone	19.56	1034	1036 *	-	-	0.83 ± 0.04
32	Benzeneacetaldehyde	19.93	1041	1036	1.9 ± 0.03	**4.62 ± 0.04**	0.822 ± 0.003
33	γ-Hexalactone	20.39	1050	1047	-	0.42 ± 0.01	1.14 ± 0.01
34	2,4,4-Trimethyl-2-cyclohexen-1-ol	20.42	1051	-	0.65 ± 0.01	-	-
35	(E)-2-Octenal	20.66	1055	1049	0.104 ± 0.001	0.149 ± 0.01	-
36	(R)-3,5,5-Trimethylcyclohex-3-en-1-ol	21.09	1063	-	-	-	0.187 ± 0.003
37	3-Methyl-benzaldehyde	21.27	1067	1064	-	-	0.26 ± 0.01
38	1-Octanol	21.39	1069	1063	0.23 ± 0.004	-	-
39	3,5-Octadien-2-one	22.60	1093	1093	-	-	0.63 ± 0.02
40	Phenylethyl Alcohol	23.49	1110	1115 *	-	1.14 ± 0.02	-
41	Isophorone	23.95	1120	1118	-	-	0.51 ± 0.01
42	4-Oxoisophorone	25.01	1141	1142 *	0.29 ± 0.01	0.813 ± 0.005	0.341 ± 0.01
43	Isomenthone	25.56	1152	1162 *	-	0.485 ± 0.003	-
44	2,6-Nonadienal, (E,Z)	25.61	1153	1150	-	-	0.35 ± 0.005
45	(E)-2-Nonenal	25.81	1157	1157	0.279 ± 0.005	-	-
46	1-Phenyl-1-propanone	26.19	1165	-	-	-	0.49 ± 0.01
47	2,2,6-Trimethyl-1,4-cyclohexanedione	26.33	1167	-	-	-	0.274 ± 0.003
48	2,4-Dimethyl-benzaldehyde	26.67	1175	1175 *	-	-	0.305 ± 0.01
49	1-(4-Methylphenyl)-ethanone	27.02	1181	1182	0.43 ± 0.01	-	-
50	*p*-Methylacetophenone	27.15	1184	1179	-	-	0.54 ± 0.03
51	Safranal	27.82	1198	1197	0.97 ± 0.01	-	1.8 ± 0.1
52	β-Cyclocitral	28.84	1219	1219	-	0.479 ± 0.001	0.614 ± 0.002
53	Ethylmethylmaleimide	29.67	1237	1234	1.562 ± 0.03	3.69 ± 0.03	0.59 ± 0.01
54	Pulegone	29.73	1238	1233	-	0.375 ± 0.003	-
55	2,6,6-Trimethyl-1-Cyclohexene-1-acetaldehyde	30.70	1259	1253 *	-	-	0.53 ± 0.01
56	2,3,6-Trimethyl-7-octen-3-ol	31.53	1277	-	-	1.846 ± 0.01	-
57	Indole	32.47	1297	1290	-	-	1.026 ± 0.003
58	Carvacrol	32.69	1302	1298	0.98 ± 0.02	-	-
59	γ-Amylbutyrolactone	35.29	1361	1362 *	-	1.03 ± 0.02	-
60	Capric acid	35.97	1376	-	1.15 ± 0.02	-	-
61	Fumaric acid, ethyl 2-methylallyl ester	36.27	1383	-	-	1.45 ± 0.03	-
62	β-Caryophyllene	37.91	1421	1417	-	0.02 ± 0.004	-
63	α-Ionone	38.22	1428	1428	3.23 ± 0.02	3.07 ± 0.03	1.1 ± 0.01
64	Nerylacetone	39.27	1454	1434	-	-	0.26 ± 0.01
65	β-Ionone	40.88	1492	1488	1.3 ± 0.03	-	**7.6 ± 0.2**
66	Dihydroactinidiolide	42.49	1536	1538 *	**6.577 ± 0.004**	**6.971 ± 0.003**	**7.8 ± 0.2**
67	Lauric acid	43.66	1562	1565	2.9 ± 0.1	0.49 ± 0.01	1.2 ± 0.1
68	Fumaric acid, ethyl 2-Methylallyl ester	44.46	1583	-	-	-	3.0 ± 0.1
69	Tridecanoic acid	47.61	1666	1662	-	-	0.1979 ± 0.0003
70	3-Keto-β-ionone	47.77	1670	1661 *	-	1.29 ± 0.02	-
71	4-(4-hydroxy-2,2,6-trimethyl-7-oxabicyclo [4.1.0]hept-1-yl)-3-Buten-2-one	48.40	1687	1690	-	1.2 ± 0.1	-
72	Heptadecane	48.77	1697	1700	**4.14 ± 0.04**	-	-
73	Pentadecanal	49.37	1711	1713	-	-	0.27 ± 0.01
74	Myristic acid	50.61	1769	1765 *	2.2 ± 0.1	2.16 ± 0.01	1.855 ± 0.001
75	Pentadecanoic acid	51.81	1820	1869	-	-	0.121 ± 0.004
76	Hexahydrofarnesyl acetone	52.07	1847	-	**5.1 ± 0.1**	-	-
77	2-Pentadecanone, 6,10,14-trimethyl	52.11	1843	1847	-	-	0.23 ± 0.01
78	Methyl 4,7,10,13-hexadecatetraenoate	52.68	1885	-	-	-	0.15 ± 0.01
79	Eicosane	52.80	1895	-	0.22 ± 0.02	-	-
80	Palmitoleic acid	53.41	1948	1953*	-	**7.8 ± 0.1**	-
81	Eicosapentaenoic acid	53.57	1962	-	-	-	**8.0 ± 0.2**
82	Palmitic acid	53.64	1968	1959	**7.7 ± 0.1**	0.73 ± 0.01	2.887 ± 0.02
83	Phytol	55.00	2113	2111 *	**4.1 ± 0.1**	0.38 ± 0.03	0.23 ± 0.01
84	Linolenic acid	55.36	2159	2134 *	-	-	1.2 ± 0.1
85	Eicosanal	55.83	2223	2224	0.7 ± 0.1	-	-
86	1-Hexacosanol	56.25	2283	2906	1.39 ± 0.05	-	-
87	Henicosanal	56.52	2325	2329	0.89 ± 0.04	-	-
88	Docosanal	57.15	2427	2434	1.38 ± 0.04	-	-
89	1-Docosanol	57.51	2488	2470	2.423 ± 0.004	-	-
90	Tricosanal	57.75	2529	2534	2.8 ± 0.1	-	-
91	Bis (2-ethylhexyl) phthalate	57.95	2562	2550 *	-	-	0.5 ± 0.1
	Total identified (%)				59.6 ± 0.1	45.7 ± 0.1	55 ± 1
	Not identified (%)				40.4 ± 0.1	54.3 ± 0.1	45 ± 1

[a] LRI, linear retention index determined on a DB-5 MS fused silica column relative to a series of *n*-alkanes (C8–C40). [b] Linear retention index reported in literature (Adams, 2017). [c] Relative % is given as mean ± SD, *n* = 3. * NIST Standard Reference Database 69: NIST Chemistry WebBook. Ct. *Cystoseira tamariscifolia*; Sm. *Sargassum muticum*; Ul. *Ulva lactuca*.

2.2. Screening of Anti-Cyanobacterial Activity

The potential anti-cyanobacterial properties of seaweed EOs was evaluated qualitatively using the disk diffusion methods. After 1 week of incubation, the inhibition zones were measured and the results are presented in Table 3 and Figure 1. The results show that the tested green macroalgae *U. lactuca* did not reveal any inhibitory activity against *M. aeruginosa*, while both brown algae showed algicidal activity against the tested cyanobacteria. The most relevant activity was observed with *C. tamariscifolia* EO with zones of inhibition greater than 46 mm. Notably, the growth inhibition of *C. tamariscifolia* EO was approximately similar to the positive control, copper sulphate ($CuSO_4$), that presented a growth inhibition diameter of 45.3 mm. Furthermore, *S. muticum* EO showed moderate activity (zone of inhibition was 32 mm). The activity of seaweed EOs against *M. aeruginosa* was determined quantitatively by means of broth microdilution technique. Calculation of minimal inhibitory concentrations (MIC) and minimal bactericidal concentrations (MBC) was performed after 1 week of incubation and the results are summarized in Table 4. $CuSO_4$ as positive control displayed a high potency (MIC = MBC = 3.12 μg mL^{-1}). The greatest effectiveness was achieved with the EO extracted from *C. tamariscifolia*, with MIC being equal to 7.81 μg mL^{-1} and the MBC equal to 15.62 μg mL^{-1}. Whereas, *S. muticum* EO showed a moderate potency with MIC and MBC values of 62.5 and 125 μg mL^{-1}, respectively.

Table 3. Inhibition-zone diameters, minimal inhibitory concentrations (MIC) and minimal bactericidal concentrations (MBC) of Moroccan seaweed essential oils (EOs).

Treatments	Inhibition Zone (mm)	MIC (μg mL^{-1})	MBC (μg mL^{-1})
C. tamariscifolia	46.3 ± 0.6 ***	7.81	15.62
S. muticum	32.3 ± 0.6 ***	62.5	125
U. lactuca	n.a	n.a	n.a
$CuSO_4$	45.3 ± 0.6 ***	3.12	3.12
DMSO	n.a	n.a	n.a

Each value representing mean ± SD of six replicates, *** $p < 0.001$ indicates significant differences compared with DMSO, n.a not active.

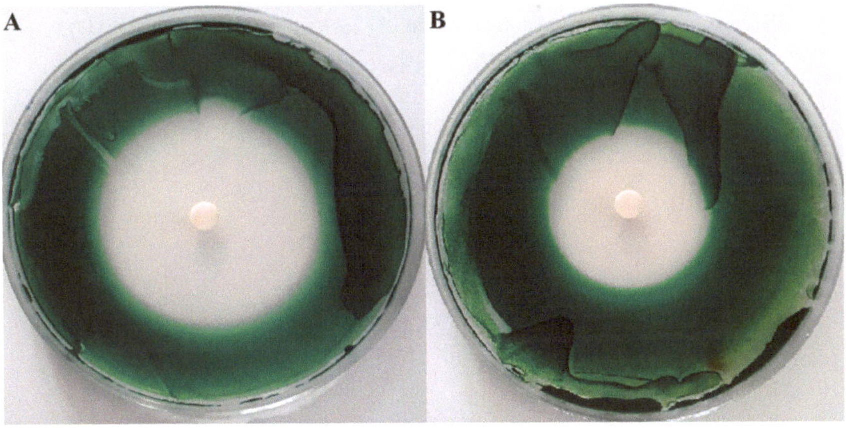

Figure 1. Anti-cyanobacterial activity of the active tested EOs against *M. aeruginosa* on solid media. (**A**) *C. tamariscifolia*; (**B**) *Sargassum muticum*.

Table 4. Inhibitory rate of *Cystoseira tamariscifolia* EO on *Microcystis aeruginosa*.

Treatments	Inhibition Rate (%)					
	Time (Days)					
	1	2	3	4	5	6
MIC	68.0 ± 0.4 ***	87.6 ± 0.4 ***	90.2 ± 0.5 ***	95.4 ± 0.1 ***	96.16 ± 0.08 ***	97.85 ± 0.05 ***
MBC	74 ± 1 ***	89.9 ± 0.2 ***	94.4 ± 0.3 ***	97.8 ± 0.1 ***	98.81 ± 0.07 ***	99.24 ± 0.07 ***
$CuSO_4$	71 ± 2 ***	88.9 ± 0.2 ***	94.12 ± 0.07 ***	97.54 ± 0.05 ***	98.5 ± 0.02 ***	98.87 ± 0.01 ***
DMSO	−0.3 ± 0.5	−0.08 ± 1.67	−0.15 ± 0.39	−1.0 ± 0.6	−1 ± 2	−1.1 ± 0.9

MIC: minimum inhibitory concentration of *C. tamariscifolia* EO (7.81 μg mL^{-1}) and MBC: minimum bactericidal concentration of *C. tamariscifolia* EO (15.62 μg mL^{-1}). Each value representing mean ± SD of three replicates, *** $p < 0.001$ indicate significant differences compared with DMSO.

2.3. Physiological Effects of C. tamariscifolia EO on M. aeruginosa

2.3.1. Inhibitory and Growth Rates of C. tamariscifolia EO on Tested Cyanobacteria

The physiological effects caused by the action of the essential oil on the cyanobacteria cells were accessed only for the seaweed EO that showed the highest activity on the qualitative assay (Table 3 and Figure 1). The algicidal effects of *C. tamariscifolia* EO at the MIC and MBC concentrations (7.81 and 15.68 μg mL^{-1}, respectively) against *M. aeruginosa* are shown as the inhibition and growth rates in Table 4 and Figure 2. The results indicate that *C. tamariscifolia* EO had a significant inhibitory effect on the tested cyanobacteria compared with the negative control (DMSO), which did not show any inhibitory effect. On the first day of the experience, *C. tamariscifolia* EO showed strong growth inhibition for both tested concentrations in a concentration-dependent way. The IR was 67.95% ± 0.38% and 73.93% ± 0.98% at the MIC (7.81 μg mL^{-1}) and MBC (15.68 μg mL^{-1}), respectively. Thereafter, the inhibition rates increased and set at more than 85% along the experiment, for both used concentrations. The maximum IR was recorded at the last day of treatment with IR of 99.24% ± 0.07% at the MBC concentration. The generation time of *M. aeruginosa* was 1.23/day with the growth rates of 0.75/day under standard culturing conditions and negative control treatment. Under treatment with $CuSO_4$, the growth rate and the generation time of the tested cyanobacteria decreased significantly with μ value of −0.18/day and generation time of −3.77/day. The *C. tamariscifolia* EO revealed a strong effect on the growth of *M. aeruginosa* at both tested concentration (the μ value was less than −0.08/day).

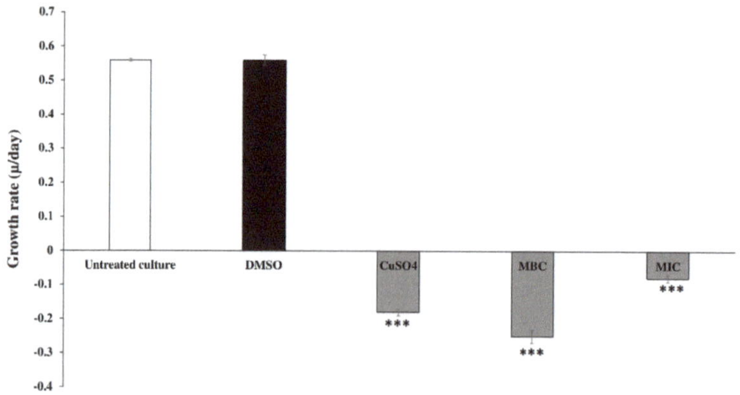

Figure 2. Effect of MIC and MBC, *C. tamariscifolia* EO on the growth rate of *M. aeruginosa*. MIC: minimum inhibitory concentration and MBC: minimum bactericidal concentration. Each value representing mean ± SD of three replicates. *** $p < 0.001$ indicate significant differences compared with the untreated culture.

2.3.2. Morphological Changes of *M. aeruginosa* Cells

The visual observation of the tested unicellular *M. aeruginosa* cultures under *C. tamariscifolia* EO at both used concentrations showed that after 3 days of exposure, a blue color appeared in the treated groups and became colorless after the fifth day of treatment. While, microscope observation at a magnification of ×40 showed that the unicellular cyanobacteria strain becomes colonial on the second day of treatment under stress with *C. tamariscifolia* EO. Moreover, similar morphological changes were also observed under treatment with the positive control ($CuSO_4$) (Figure 3).

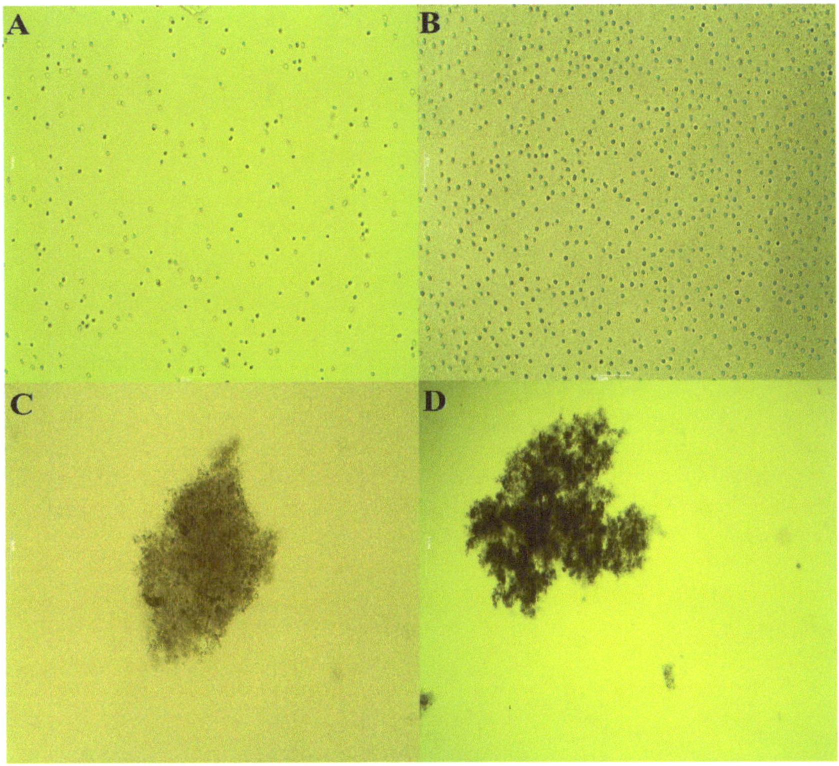

Figure 3. Micrographs of *M. aeruginosa* (Gr. × 40) (**A**) At the first day of treatment; (**B**) untreated culture at the end of treatment; (**C**) culture under treatment with *C. tamariscifolia* EO at second day of the experience; (**D**) culture under treatment with $CuSO_4$ at second day of the experience.

2.3.3. Effects of *C. tamariscifolia* EO on *M. aeruginosa* Chlorophyll-*a* and Protein Contents

To explore whether chlorophyll-*a* synthesis of the treated cells was inhibited by *C. tamariscifolia* EO, the chlorophyll-*a* content of *M. aeruginosa* cells after exposure to *C. tamariscifolia* EO are shown in Figure 4. At the first day of experience, the chlorophyll-*a* content was the same at all treatments (including control). With exposure time, the chlorophyll-*a* content increased significantly for the untreated culture and the negative control. While, *M. aeruginosa* chlorophyll-*a* content was significantly decreased by both used concentrations of *C. tamariscifolia* EO. Compared to $CuSO_4$ used as positive control, *C. tamariscifolia* EO recorded a similar effect on *M. aeruginosa* chlorophyll-*a* content during the 6 days of experiment.

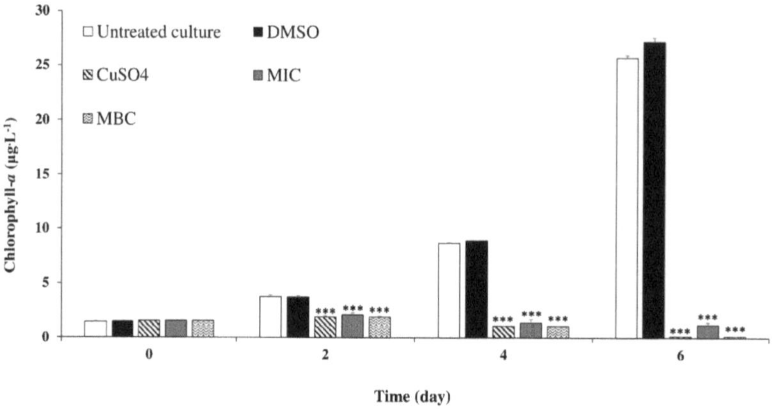

Figure 4. Effect of *C. tamariscifolia* EO on *M. aeruginosa* chlorophyll-*a* concentration. Results are presented as mean ± SD of three independent assays (*** indicates $p < 0.001$ relative to the untreated culture by ANOVA).

To investigate the effects of *C. tamariscifolia* EO on *M. aeruginosa* cells, the protein content of *M. aeruginosa* cells was determined after exposure to *C. tamariscifolia* EO, and the results are shown in Figure 5. Similar to chlorophyll-*a*, for the untreated culture and the negative control, the protein content increased significantly with exposure time. On day 2, the differences in protein content for the MIC and MBC concentrations of *C. tamariscifolia* EO were significantly different, compared with the negative control (untreated culture and DMSO). Similar results were recorded on days 4 and 6, *C. tamariscifolia* EO (at both used concentrations) and $CuSO_4$ (positive control) recorded a significant reducing effect on the protein content of *M. aeruginosa* cells. Furthermore, the protein contents were highly coherent with the cell density and the chlorophyll-*a* results.

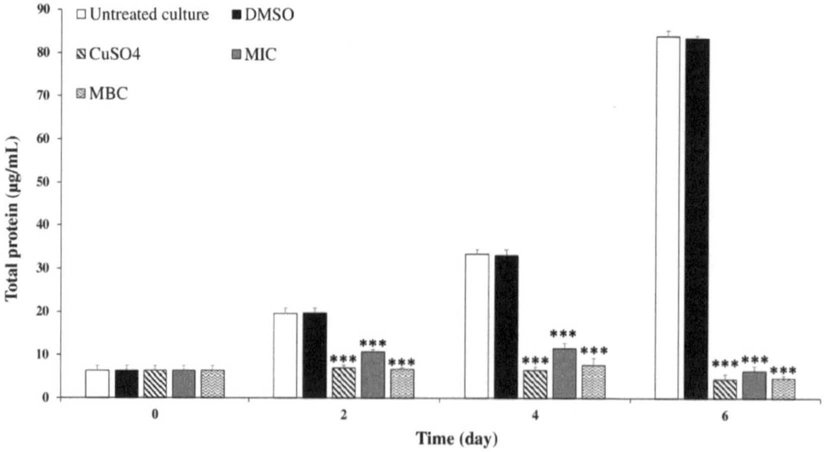

Figure 5. Effect of *C. tamariscifolia* EO on *M. aeruginosa* protein content. Results are presented as mean ± SD of three independent assays (*** indicates $p < 0.001$ relative to the untreated culture by ANOVA).

2.3.4. Effects of *C. tamariscifolia* EO Superoxide Dismutase (SOD) and Catalase (CAT) Activities and Malondialdehyde (MDA) Concentration in *M. aeruginosa* Cells

In order to determine whether the cellular oxidative defense system was activated, the superoxide dismutase (SOD) activity, as the first defense against Reactive Oxygen Species (ROS) among antioxidant systems, which catalyzes the superoxide anion into H_2O_2 and O_2, was investigated (Figure 6A). The results demonstrate that the differences in the SOD activities in cyanobacterial cells between control and treatment groups were significant and visible from the second day of treatment. From the second day of experiment, the SOD activity under the MBC concentration (15.62 μg mL^{-1}) of *C. tamariscifolia* EO treatment became higher than the positive control ($CuSO_4$) and reached the peak of 150.25 U/mg protein. Thereafter, the SOD activity began to decrease gradually in the following time in the control and treatment groups.

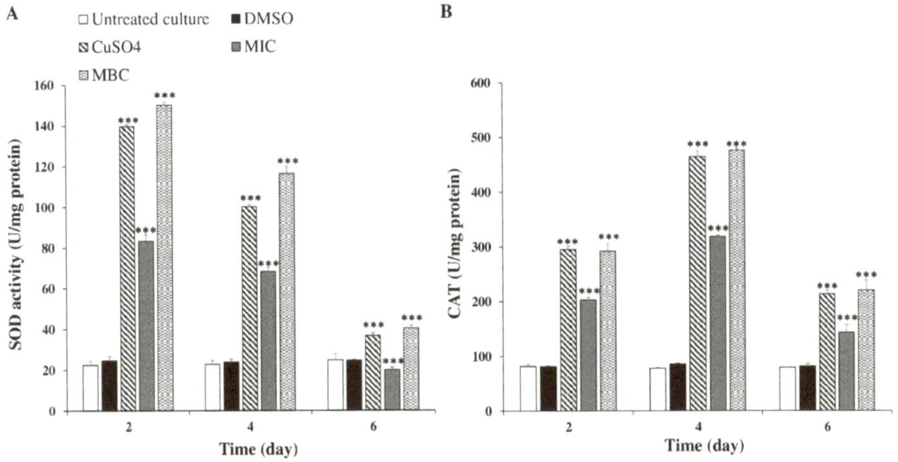

Figure 6. Superoxide dismutase (SOD) (**A**) and catalase (CAT) (**B**) activities in *M. aeruginosa* cells after treatment with *C. tamariscifolia* EO. Results are presented as mean ± SD of three independent assays (*** indicates $p < 0.001$ relative to the untreated culture by ANOVA).

The second defense against reactive oxygen species (ROS) is catalase (CAT), which can convert H_2O_2 into H_2O and directly eliminates H_2O_2 in the peroxisome. As shown in Figure 6B the CAT activity in *M. aeruginosa* cells exposed to *C. tamariscifolia* EO exhibited a significant increase, while the CAT activity in the untreated culture and the negative control remained unchanged over time. The differences between treated and control groups were apparent from the second day of treatment. The CAT activity at 15.62 μg mL^{-1} of *C. tamariscifolia* EO (MBC concentration) was higher than that at 7.81 μg mL^{-1} of *C. tamariscifolia* EO (MIC concentration); however, the differences between them were not significant. After 4 days of exposure, the activity of CAT greatly increased and reached a maximum value with the exposure to 15.62 μg mL^{-1} *C. tamariscifolia* EO. At the sixth day of treatment, the CAT activity with *C. tamariscifolia* EO decreased but was still significantly different compared with the untreated culture.

Malondialdehyde (MDA, the final product of lipid peroxidation) was used as an indicator of lipid peroxidation, as indicated in Figure 7, the MDA content increased significantly at both tested concentrations of *C. tamariscifolia* EO from the second day of treatment. While, the MDA level in the negative control groups remained unchanged over time. After 48 h of exposure, 15.62 and 7.81 μg mL^{-1} *C. tamariscifolia* EO induced an increase in MDA compared with the control. The MDA concentrations increased with the increased concentration of *C. tamariscifolia* EO. The maximal MDA value was

94 µmol/L at day 4 of treatment with 15.62 µg mL^{-1} *C. tamariscifolia* EO, which was 4.95 times higher than that in the negative control groups.

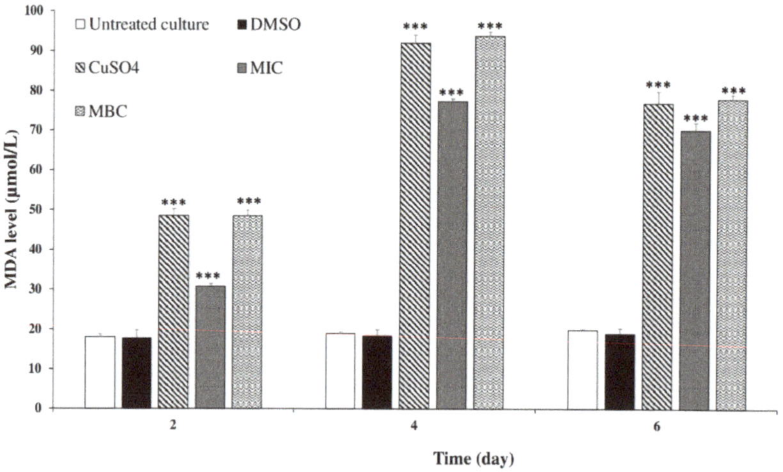

Figure 7. Malondialdehyde (MDA) concentration in *M. aeruginosa* cells after treatment with *C. tamariscifolia* EO. Results are presented as mean ± SD of three independent assays (*** indicates $p < 0.001$ relative to the untreated culture by ANOVA).

3. Discussion

Seaweeds are one of the most primitive and dominant organisms in aquatic ecosystems. They could provide an eco-friendly approach for HCBs control due to their ability to produce a large range of bioactive compounds [7,29,30]. Therefore, we proceeded to assess the anti-cyanobacterial activity of three Moroccan seaweed EOs. To the best of our knowledge, the present study constitutes the first attempt to extract and characterize Moroccan seaweed EOs. The total content percentage of Moroccan seaweed EOs was lower compared to that reported previously by Patra et al. (2017a, 2015) and Patra and Baek (2016a, 2016b) [31–34] who found that the total content of seaweeds collected from the Korean coast was usually higher than 0.26%. This difference in seaweed total content percentage could be due to the geographical locations, the species used, harvesting time, and the used extraction method. An important richness and variability of compounds was observed after seaweed EOs chemical analysis. Among the total of 91 compounds identified in the three selected algal EOs, 14 compounds showed to belong to the terpenes group. In general, terpenoids are compounds that have been associated with several bioactive properties, including antimicrobial activity [35].

The most abundant constituents in the two brown seaweeds *C. tamariscifolia* and *S. muticum* EOs were fatty acids, namely palmitic acid and palmitoleic acid, respectively. Previous studies on EOs from other species of seaweeds have shown the presence of hexadecenoic acid (palmitoleic acid). In particular, Patra et al. (2017a) [36] reported the presence of this unsaturated fatty acid as one of the main compounds (22.39%) of the brown edible seaweed *Undaria pinnatifida* EO collected from the Korean coast. A high content of palmitic acid (9.2% and 16.57%) was also found by Patra et al. (2017b, 2015a) [31,37] on *Porphyra tenera* and *Laminaria japonica* EOs, respectively. Previously, the EO composition of seaweed species of *Cystoseira* genus, other than *C. tamariscifolia*, has been described. Ozdemir et al. (2006) [38] reported that the brown seaweed *C. barbata* EO (*Cystoseira* genus) consists of several compounds different from those we have determined in *C. tamariscifolia* EO, such as docosane (7.61%), tetratriacontane (7.47%), eicosane (5.05%), tricosane (4.43%), hexadecane (4.16%), and heptadecane (1.35%) as major compounds. Additionally, 1-chloro-2,2-diethoxyethane (21.5%), 2,3-butanediol (6.5%), chloroacetic acid (3.7%), and 1,1-dichloro-2,2-diethoxyethane (2%) were identified in the volatile compounds composition of *C. crinite* (*Cystoseira* genus) collected from the eastern Mediterranean [39]. Nevertheless, it should be noticed that research on the chemical composition of seaweed EOs is still very scarce. Regarding the chemical composition of the green macroalgae *U. lactuca* EO, the polyunsaturated fatty acid eicosapentaenoic acid (8%) was the most dominant constituent. The presence of other polyunsaturated fatty acids, such as cis- and trans-5,8,11,14-eicosatetraenoic acids were also detected in the *Dictyopteris polypodioides* EO collected from the Algerian coast [40].

Among the identified compounds, some terpenes such as β-caryophyllene (0.02%) and α-pinene (0.015%) were only found in minor amounts, while others such as dihydroactinidiolide (6.6–7.8%), β-Ionone (not detected to 7.6%), and phytol (0.23–4.1%) were present in higher amounts. Terpenes such as safranal and others have been recently reported in the essential oil extracted from the brown algae *D. polypodioides* [40]. Similarly, the presence of the terpenic compounds dihydroactinidiolide, β-Ionone and phytol has been previously described in marine algae [41]. Besides terpenes, several compounds belonging to different groups such as alcohols, aldehydes, and ketones were present in minor amounts. The presence of such compounds is in good agreement with Gressler et al. (2009) [41], who mentioned the ability of marine algae to produce a wide range of metabolites including hydrocarbons, terpenes, fatty acids, esters, alcohols, aldehydes, and ketones.

The qualitative screening using the paper disk diffusion method in solid medium demonstrated that the green macroalgae *U. lactuca* did not show any inhibitory activity against the tested Gram-negative bacteria *M. aeruginosa*. These results are in good agreement with those found by Zerrifi et al. (2019) [30] who investigated the anti-cyanobacterial activity of *U. lactuca* methanolic extract collected from the Moroccan coast. Their results showed that the *U. lactuca* extract also did not have an effect against *M. aeruginosa*. Similar results were obtained by Salvador et al. (2007) [42] who reported that the seaweeds of the genus *Ulva* did not show antibacterial activity against any assayed Gram-negative bacteria. On the contrary, Mishra (2018) [43] observed that methanol, butanol, and ethyl acetate extract of *U. lactuca* display moderate activity against *Pseudomonas aeruginosa*. Additionally, Begum et al. (2018) [44] reported that the methanolic extract of *U. reticulate* (*Ulva* genus) revealed the maximum inhibitory activity against Gram-negative bacteria *P. aeruginosa*. Furthermore, we found that *S. muticum* EO showed a strong activity against *M. aeruginosa* (32.33 mm). Our results are in agreement with those found by Kumaresan et al. (2018) [45] who investigated the antimicrobial activity of *S. wightii* (*Sargassum* genus) aqueous extract and showed that this extract had an important antibacterial activity against Gram-negative bacteria *Escherichia coli* with inhibition zone of 13 mm. Sujatha et al. (2019) [46] observed that *S. swartzii* ethanolic extract exhibited a high antibacterial activity against Gram-negative bacteria. *S. muticum* extract was active against *Enterobacter aerogenus, Proteus vulgaris, Salmonella typhymurium,* and *Salmonella paratyphi* [47]. In what concerns the activity against *M. aeruginosa*, Zerrifi et al. (2019) [30] evaluated the methanolic extract of *S. muticum*, which did not reveal any inhibiting capacity, most probably due to the different chemical composition between the methanolic extract and that obtained by hydrodistillation. As mentioned, in the present study the highest activity was recorded for *C. tamariscifolia* EO (46.33 mm). However, this finding is in disagreement with those found by Farid et al. (2009) [48]. These authors investigated the antibacterial activities of *C. tamariscifolia* collected from Morocco, and their results suggest that *C. tamariscifolia* dichloromethane/methanol extract did not show antimicrobial activity against the Gram-negative bacteria assayed (*E. coli*). Salvador et al. (2007) [42] observed that *C. tamariscifolia* extract did not inhibit the growth of any tested Gram-negative bacteria. On the other hand, Ozdemir et al. (2006) [38] reported that the volatile oil of the genus *Cystoseira* recorded a moderate effect on the tested Gram-negative bacteria (7 mm against *E. coli* and *S. typhimurium*). Contrarily, Zerrifi et al. (2019) [30] observed that *C. tamariscifolia* extract conferred an important activity against Gram-negative bacteria *M. aeruginosa* with inhibition zone equal to 13.33 mm. In another report, Chiheb et al. (2009) [49] reported that different *Cystoseira* species show a potent antibacterial activity against tested Gram-negative bacteria *Salmonella typhi, E. coli, P. aeruginosa*, and *Klebsiella* sp. Ainane et al. (2014) [50] found that *C. tamariscifolia* extract produce interesting zones of inhibition against both tested Gram-negative bacteria, *Enterobacter cloacae* and *Klebsiella'pneumoniae* (inhibition diameter between 10 and 15 mm). Likewise, a recent study showed the high antibacterial activity against Gram-negative bacteria of another species of the *Cystoseira* genus (*C. mediterranea*) [51].

The quantitative screening using the broth microdilution technique confirmed the results of the qualitative assay since *C. tamariscifolia* EO achieved the greatest effectiveness against *M. aeruginosa* (MIC = 7.81 μg mL^{-1} and MBC = 15.62 μg mL^{-1}). These results are in accordance with those reported earlier by Wang et al. (2015, 2014) and Zerrifi et al. (2020) [27,28,52] who found that other EOs also showed a high activity against the toxic cyanobacteria *M. aeruginosa*. Our results in the liquid medium revealed that *C. tamariscifolia* EO recorded a significant anti-cyanobacterial activity against the toxic cyanobacteria *M. aeruginosa* with a percentage inhibition of more than 67%. Several studies were conducted to investigate growth inhibition by EOs extracted from many aquatic and terrestrial plants and solvent extract of seaweeds on *M. aeruginosa*. Wang et al. (2014) [52] reviewed the anti-cyanobacterial activity of two emergent plant EOs (*Typha latifolia* and *Arundo donax*) on *M. aeruginosa*. The authors reported inhibition rates of more than 40% at 50.0 mg L^{-1} of both tested EOs. The *Rosmarinus officinalis* EO recorded significant growth inhibition against *M. aeruginosa* [25]. Moreover, Wang et al. (2015) [27] showed that the growth of *M. aeruginosa* was strongly inhibited by *Vallisneria spinulosa* EO at 50.0 mg L^{-1} with an inhibition rate equal to 41.7%. Xian et al. (2006) [53] found that *Ceratophyllum demersum* EO composed of fatty compounds, terpenoids, and phenolic compounds, recorded a high inhibitory activity on *M. aeruginosa* growth. Furthermore, Zerrifi et al. (2019) [30] tested the effect of *C. tamariscifolia* methanolic extract on the growth of *M. aeruginosa*. Their results show that the reached inhibition rates were higher than 49% after the first day of treatment at 0.6 mg L^{-1}. The morphological changes observed in *M. aeruginosa* culture in this study are quite similar to those observed by Harada et al. (2009), Huang et al. (2002), and Zerrifi et al. (2020) [28,54,55]. The chlorophyll-*a* and protein content that reflect *M. aeruginosa* growth, was decreased after *C. tamariscifolia* EO treatment. This decrease could be related to malfunctions of normal physiological metabolism in cyanobacterial cells (e.g., disruption of Photosystem I and destruction of Photosystem II) [56,57]. The mentioned changes can be related to the chemical composition of *C. tamariscifolia* EO, namely to the presence, although in low amounts, of several terpenoids, including oxygenated compounds such as alcohols and aldehydes. Besides, the possibility of synergisms, between the more abundant terpenoids, such as dihydroactinidiolide (6.57%) and hexahydrofarnesyl acetone (5.1%) with compounds present in minor amounts, should also be considered. Finally, all the three studied seaweeds presented a very complex composition, with a large abundance of compounds, several of which were not possible to be identified by the used technique (GC-MS). Some of those unidentified compounds can possibly explain the different activity observed for the three samples.

The activation of SOD and CAT activities were responsible for the protection of *M. aeruginosa* cells against oxidative exposure (eliminate ROS or reduce damaging effects). Our finding of SOD and CAT activities in *M. aeruginosa* were also observed in response to glyphosate treatment [58]. Meng et al. (2015) [59] observed a significant increase in the SOD activity of *M. aeruginosa* exposed to different concentrations of *Ailanthus altissima* extract. Similar results are also found from other treatments, such as rice straw aqueous extract [60], heptanoic acid and benzoic acid [61], fenoxaprop-*p*-ethyl [62], 17b-estradiol [63], and juglone (5-hydroxy-1,4-naphthoquinone) [64] on *M. aeruginosa*. The last product of lipid peroxidation is MDA, which is an indicator of oxidative stress [65]. The increase of MDA concentration is in agreement with Zhang et al. (2017) [66] who detected a significant increase in MDA levels of *M. aeruginosa* exposed to 5 and 10 mg L^{-1} of glufosinate, comparing to control, 0.5, and 1 mg L^{-1}, indicates the occurrence of damage to the lipid membranes. The treatment of *M. aeruginosa* cells with pyrogallol (polyphenol) caused lipid peroxidation and altered MDA levels [67]. Contrariwise, Xie et al. (2019) [68] found that the MDA concentrations on *M. aeruginosa* cells showed no apparent change under napropamide and acetochlor treatments.

4. Conclusions

After screening of EOs extracted from Moroccan seaweeds for their anti-cyanobacterial activity, our results revealed that marine macroalgae EOs are potential producers of anti-cyanobacteria compounds. Consequently, they should be subject to a comprehensive study as natural sources of bioactive substances. Accordingly, to better understand the potential effects and the mechanisms of action of the studied EOs on *M. aeruginosa*, the search of the active EOs major components effects on *M. aeruginosa* will be the next step of our research. Moreover, further research will need to be conducted using other seaweeds and/or phytoplankton species in macrocosms and natural field conditions, studying the toxicity, nature, and stability of the compounds and their potentially synergistic interactions in the aquatic ecosystem.

5. Materials and Methods

5.1. Seaweed Material Sampling and Extraction of Essential Oils (EOs)

Three seaweed species were selected for the study: *C. tamariscifolia* (Phaephyceae, Sargassaceae), *S. muticum* (Phaephyceae, Sargassaceae), and *U. lactuca* (Ulvophyceae, Ulvaceae). These macro-algae were harvested from two Moroccan coastal regions (Table 5).

Table 5. Date of harvesting and location of the Moroccan seaweeds studied.

Species	Species Code	Harvesting Place	Date of Harvesting	Latitude/Longitude
C. tamariscifolia	Ct	Souiria Laqdima	February 2019	N 32°03′04.6″/ W 9°20′30.2″
S. muticum	Sm	El jadida	April 2019	N 3°15′45.9″/ W 8°30′03.4″
U. lactuca	Ul	El jadida	March 2019	N 3°15′45.9″/ W 8°30′03.4″

The samples were rinsed with seawater and distilled water to remove debris. After identification of each species according to their morphological and histological features [69], seaweed materials were dried in the shade at room temperature (≈25 °C) and subjected to hydro-distillation, using a Clevenger-type apparatus for 3 h until total recovery of oil. The EOs obtained were dried over anhydrous sodium sulfate and stored at 4 °C in the dark.

5.2. Gas Chromatography/Mass Spectrometry (GC/MS) Analyses

The seaweed essential oils were analyzed by GC-MS following a protocol previously described by Falcão et al. (2018) [70]. Analyses were performed in a GC-2010 Plus (Shimadzu, Kioto, Japan) gas chromatography system equipped with a AOC-20iPlus (Shimadzu, Kioto, Japan) automatic injector, a SH-RXi-5ms column (30 m × 0.25 mm × 0.25 μm; Shimadzu, Kioto, Japan), and a mass spectrometry detector, operated using an injector temperature of 260 °C and the following oven temperature profiles: an isothermal hold at 40 °C for 4 min, an increase of 3 °C/min to 175 °C, followed by an increase of 15 °C/min to 300 °C and an isothermal hold for 10 min. The transfer line temperature was set at 280 °C and the ion source at 220 °C; the carrier gas, helium, was adjusted to a linear velocity of 30 cm/s; the ionization energy was 70 eV, the scan range was set at 35–500 u, with a scan time of 0.3 s. A quantity of 1 μL of each sample diluted in n-hexanewas injected using the split injection mode at 1:10. The identification of the essential oil components was carried out by comparison of the obtained spectra with those from the NIST17 mass spectral library and by determining the linear retention index (LRI) based on the retention times of an n-alkanes mixture (C8–C40, Supelco, Darmstadt, Germany). When possible, comparisons were also performed with commercial standard compounds and with published data. Compounds were quantified as relative percentage of total volatiles using relative peak area values obtained from total ion current (TIC).

5.3. Screening for Anti-Cyanobacterial Activity

5.3.1. Cyanobacteria Strain

In this study, the cyanobacteria strain *M. aeruginosa* was sampled from the eutrophic reservoir Lalla Takerkoust (31°21′36″ N; 8°7′48″ W), Morocco, in bloom period (October 2017) and then the strain was isolated, separated into single cells, and maintained in culture in BG11 medium under a controlled culture chamber endowed with the following conditions: temperature of 26 ± 2 °C, light intensity of 63 μmol m^{-2} s^{-1}, and a light/dark cycle of 15 h/9 h [30].

5.3.2. Disc Diffusion Method

In vitro anti-cyanobacterial activity of seaweed EOs of each of the three algae was evaluated using the agar diffusion method [71]. The suspension of tested *M. aeruginosa*, containing about 10^8 cells/mL using a Malassez counting cell, was spread on BG11 medium with 4% of agarose. Subsequently, 10 μL of each EO and $CuSO_4$, prepared at a concentration of 50 μg mL^{-1} in ultrapure water, as positive control was dropped on sterile filter paper discs, 6 mm in diameter (Whatman no. 1, Little Chalfont, UK) and placed on the agar surface. Before incubation in the culture chamber under the described condition, all treated plates were stored in a refrigerator at 4 °C for more than 4 h to prevent the cyanobacteria growth and allow the diffusion of the bioactive substances contained in the EOs into the medium. Each experiment was repeated six times to statistically confirm the results.

5.3.3. Determination of the Minimum Inhibitory Concentration (MIC) and Minimum Bactericidal Concentration (MBC)

The determination of the MIC values of the EOs that showed activity in the disc diffusion assay, was carried out in a 96-well microplate using the microdilution assay according to the NCCLS guidelines M7-A4 [72]. The MIC values represent the lowest EO concentration that prevents the cyanobacteria growth. Succinctly, 200 μL of tested cyanobacteria culture with density of 3 × 10^6 cells/mL (exponential growth phase) was added to each microplate well. The EOs were dissolved in DMSO (1%) and added to the tested culture to obtain final concentrations from 4000 to 1.953 μg mL^{-1}. Subsequently, the prepared microplates were incubated for 5 days under the described controlled conditions in the culture chamber. In order to determine the MBC values, which represent the lowest EOs concentration that induces 100% cell death of incubated cyanobacteria, 100 μL of each wells without visible cyanobacteria growth was spread on BG11 and incubated for 5 days in the culture chamber.

5.4. Determination of Cyanobacteria Growth Rates

The effects of the most bioactive EO on *M. aeruginosa* strain, namely the EO of *C. tamariscifolia*, were accessed by measuring the inhibition and growth rates estimation. The growth test was conducted, in triplicate, under the determined MIC and MBC concentration of the most bioactive EO and $CuSO_4$ (positive control) in Erlenmeyer flasks (150 mL) containing 9 mL of cyanobacteria inoculum and 71 mL of BG11 medium. The initial density of the tested cyanobacteria culture was adjusted by addition of BG11 medium and counting cells until a value of 2 × 10^6 cells/mL (the exponential growth phase). DMSO was employed as negative control. Whereas, another untreated cyanobacteria culture was used for the performance of all calculations necessary for the results treatment. The inhibition (IR) and growth rates were estimated by cells counting using a hemocytometer under a microscope every 24 h [73] and calculated using the following Equations (1) and (2), respectively:

$$IR(\%) = (((Nc - Nt))/(Nc) \times 100) \quad (1)$$

where, Nc and Nt represent the cell concentrations (cells/mL) in the control and treatment samples, respectively [74].

$$\mu = (\ln Ne - \ln Nb)/\Delta t \quad (2)$$

In which μ is the average growth rate; Ne and Nb (cells/mL) are the cell densities on the last day and the first day of the experiment, respectively, and Δt denotes the duration of the experiment.

5.5. Biochemical Parameters in M. aeruginosa

5.5.1. Determination of Chlorophyll-*a* and Total Protein Contents

Chlorophyll-*a* (Chl-*a*) concentration was measured in triplicate and calculated following the method previously described by Lichtenthaler and Wellburn (1983) [75]. Shortly, 5 mL of the culture sample was centrifuged at 4000× *g* for 15 min to collect algal cells. Cells were then re-suspended with boiling ethanol (95%). The three replicas were incubated at 4 °C for 48 h. Subsequently, another centrifugation for 5 min at 3400× *g* was performed to eliminate the pellet. The supernatant optical density (OD) was read at different wavelengths absorbance (649 and 665 nm). Chlorophyll-*a* concentration was calculated by the following Equation (3).

$$[\text{Chl-}a] = 13.95 \times DO665 - 6.88 \times DO649 \quad (3)$$

The enzyme extracts were prepared according to Li et al. (2016) protocol [76]. Briefly, *M. aeruginosa* cells were collected by centrifugation of each culture (5 mL) at 4000× *g* for 25 min. The pellet was re-suspended in 0.1 M phosphate buffer (pH 6.5) containing 1% (*w*/*v*) polyvinylpyrrolidone (PVP). Then the cells were disrupted and homogenized by an ultrasonic cell pulverizer for 5 min in an ice bath. The homogenate was then centrifuged 10,000× *g* at 4 °C for 10 min. The supernatant was used for total protein measurement and antioxidant enzyme activity assays. The total protein content was determined by the application of Bradford (1976) method [77]. Briefly, 100 μL of the enzyme extract was added to 2 mL of Bradford's reagent and incubated at room temperature in obscurity for 20 min. Furthermore, a mixture of the assay buffer (100 μL) and the Bradford's reagent (2 mL) was used as a blank. The absorbance was read at 595 nm and the protein content was calculated from a calibration curve of Bovine Serum Albumin (BSA).

5.5.2. Activity of Antioxidant Response Enzymes, CAT and SOD

The SOD activity was assayed in triplicate according to Beauchamp and Fridovich (1971) method [78]. The reaction mixture contained 0.8 mL PBS solution (50 mM, pH 7.8), 0.3 mL methionine solution (130 mM), 0.3 mL Na_2EDTA solution (100 μM), 0.3 mL riboflavin solution (20 μM), 0.3 mL nitroblue tetrazolium (NBT) solution (750 μM), and 1 mL enzyme extract for a total volume of 3 mL. As SOD has the ability to inhibit the photochemical reduction of NBT, this assay utilized negative controls (silver paper wrapped around the test tube to mimic fully dark condition without any photochemical reduction of NBT), positive controls (deficiency of SOD activity in light with full photochemical reduction of NBT), and treatment groups (in light with SOD inhibition on photochemical reduction of NBT). The absorbencies of all experimental tubes were measured at 560 nm after a 20 min irradiance of 40–60 mmol photons m^{-2} s^{-1}. One unit of SOD activity was defined as the amount of enzyme that inhibited 50% of photochemical reduction of NBT. CAT activity was assayed in triplicate by absorbance decrease being proportional to the breakdown rate of hydrogen peroxide (H_2O_2) at 240 nm according to the method of Rao et al. (1996) [79]. The reaction mixture contained 1 mL H_2O_2, 1.9 mL H_2O, and 1 mL crude enzyme. Samples were incubated for 2 min at 37 °C and the absorbance of the sample was monitored for 5 min at 240 nm using a Varian Cary® 50 UV-Vis Spectrophotometer (Agilent Technologies, Santa Clara, CA, USA).

5.5.3. Determination of MDA Content

The lipid peroxidation level was reflected by changes of malondialdehyde (MDA) content, which was determined in triplicate, according to Du et al. (2017) [62]. Samples were collected every 2 days and centrifuged at 4000× *g* for 20 min. The cell pellets were homogenized with 2 mL of 10% (*w*/*v*) trichloroacetic acid (TCA) and centrifuged at 12,000× *g* for 10 min at 4 °C. After centrifugation, 2 mL of the supernatant was mixed with 2 mL of 0.6% thiobarbituric acid (in 10% TCA) and heated in boiling water for 15 min. The reaction was stopped by transferring the reaction tubes into an ice bath. Following cooling, the samples were then centrifuged at 12,000× *g* for 10 min. The absorbance of the supernatant was measured at 532, 600, and 450 nm, taking a mixture of 2 mL ultrapure water and 2 mL 0.6% TBA as reference. The MDA level (μmol/L) was calculated according to Equation (4):

$$\text{MDA} = 6.45 \times \text{OD532} - \text{OD600} - 0.56 \times \text{OD450} \quad (4)$$

5.6. Statistical Analysis

The experiments were done in six replicates in solid medium ($n = 6$) and three replicates in liquid medium ($n = 3$) with each independent assay. Statistical analysis between experimental groups and the control were performed by applying a one-way and two-way ANOVA analysis. Post hoc differences between group means was carried out with the Tukey test using Sigma Plot software (sigmaplot 12.5 for windows; Systat Software Inc., San Jose, CA, USA) for Windows. Values of $p < 0.001$ were considered statistically significant.

Author Contributions: Conceptualization, S.E.A.Z., B.O. and V.V.; Funding acquisition, B.O., A.C. and V.V.; Investigation, S.E.A.Z., L.B., I.C.F.R.F., J.S.A., T.C.F., B.O., A.C. and V.V.; Methodology, S.E.A.Z., F.E.K., R.M., R.E.M., A.K., B.S., L.B., I.C.F.R.F., J.S.A., T.C.F., A.A., B.O., A.C. and V.V.; Supervision, B.O.; Writing—original draft, S.E.A.Z.; Writing—review and editing, S.E.A.Z., L.B., I.C.F.R.F., J.S.A., T.C.F., B.O., A.C. and V.V. All authors have read and agreed to the published version of the manuscript.

Funding: This project received funding from the European Union's Horizon 2020 research and innovation programme under the Marie Skłodowska-Curie grant agreement No 823860; Foundation for Science and Technology (FCT, Portugal) for financial support through national funds FCT/MCTES to UIDB/04423/2020, UIDP/04423/2020 and UIDB/00690/2020 (CIMO), and also FCT, P.I., through the institutional scientific employment program-contract for L. Barros contract.

Conflicts of Interest: No potential conflict of interest was reported by the authors.

References

1. Chapra, S.C.; Boehlert, B.; Fant, C.; Bierman, V.J.; Henderson, J.; Mills, D.; Mas, D.M.L.; Rennels, L.; Jantarasami, L.; Martinich, J.; et al. Climate change impacts on harmful algal blooms in u.s. freshwaters: A screening-level assessment. *Environ. Sci. Technol.* **2017**, *51*, 8933–8943. [CrossRef] [PubMed]
2. O'Neil, J.M.; Davis, T.W.; Burford, M.A.; Gobler, C.J. The rise of harmful cyanobacteria blooms: The potential roles of eutrophication and climate change. *Harmful Algae* **2012**, *14*, 313–334. [CrossRef]
3. Rigosi, A.; Carey, C.C.; Ibelings, B.W.; Brookes, J.D. The interaction between climate warming and eutrophication to promote cyanobacteria is dependent on trophic state and varies among taxa. *Limnol. Oceanogr.* **2014**, *59*, 99–144. [CrossRef]
4. Redouane, E.M.; Zerrifi, S.E.A.; El Khalloufi, F.; Oufdou, K.; Oudra, B.; Lahrouni, M.; Campos, A.; Vasconcelos, V. Mode of action and faith of microcystins in the complex soil-plant ecosystems. *Chemosphere* **2019**. [CrossRef]
5. Mohamed, Z.A.; Hashem, M.; Alamri, S.A. Growth inhibition of the cyanobacterium *Microcystis aeruginosa* and degradation of its microcystin toxins by the fungus *Trichoderma citrinoviride*. *Toxicon* **2014**, *86*, 51–58. [CrossRef]
6. Sun, Y.; Wang, H.; Guo, G.; Pu, Y.; Yan, B.; Wang, C. Isolation, purification, and identification of antialgal substances in green alga *Ulva prolifera* for antialgal activity against the common harmful red tide microalgae. *Environ. Sci. Pollut. Res.* **2016**, *23*, 1449–1459. [CrossRef]

7. Zerrifi, S.E.A.; El Khalloufi, F.; Oudra, B.; Vasconcelos, V. Seaweed bioactive compounds against pathogens and microalgae: Potential uses on pharmacology and harmful algae bloom control. *Mar. Drugs* **2018**, *16*, 55. [CrossRef] [PubMed]
8. Marzbali, M.H.; Mir, A.A.; Pazoki, M.; Pourjamshidian, R.; Tabeshnia, M. Removal of direct yellow 12 from aqueous solution by adsorption onto *spirulina* algae as a high-efficiency adsorbent. *J. Environ. Chem. Eng.* **2017**, *5*, 1946–1956. [CrossRef]
9. Park, J.; Church, J.; Son, Y.; Kim, K.; Lee Hyoung, W. Recent advances in ultrasonic treatment: Challenges and field applications for controlling harmful algal blooms (HABs). *Ultrason. Sonochem.* **2017**, *38*, 326–334. [CrossRef] [PubMed]
10. Pathak, J.; Singh, P.R.; Häder, D.P.; Sinha, R.P. UV-induced DNA damage and repair: A cyanobacterial perspective. *Plant Gene* **2019**, *19*, 100194. [CrossRef]
11. Visser, P.M.; Ibelings, B.W.; Bormans, M.; Huisman, J. Artificial mixing to control cyanobacterial blooms: A review. *Aquat. Ecol.* **2016**, *50*, 423–441. [CrossRef]
12. Huh, J.; Ahn, J.-W. A Perspective of chemical treatment for cyanobacteria control toward sustainable freshwater development. *Environ. Eng. Res.* **2017**, *22*, 1–11. [CrossRef]
13. Nagai, T.; Aya, K.; Yoda, I. Environmental toxicology comparative toxicity of 20 herbicides to 5 periphytic algae and the relationship with mode of action. *Environ. Toxicol.* **2016**, *35*, 368–375. [CrossRef] [PubMed]
14. Pohl, J.; Saltsman, I.; Mahammed, A.; Gross, Z.; Roder, B. Inhibition of green algae growth by corrole-based photosensitizers. *J. Appl. Microbiol.* **2015**, *118*, 305–312. [CrossRef]
15. Coloma, S.E.; Dienstbier, A.; Bamford, D.H.; Sivonen, K.; Roine, E.; Hiltunen, T. Newly isolated *Nodularia phage* influences cyanobacterial community dynamics. *Environ. Microbiol.* **2017**, *17*, 273–286. [CrossRef]
16. Gerphagnon, M.; Macarthur, D.J.; Latour, D.; Gachon, C.M.M.; Ogtrop, F.; Van Gleason, F.H.; Sime-ngando, T. Microbial players involved in the decline of filamentous and colonial cyanobacterial blooms with a focus on fungal parasitism. *Environ. Microbiol.* **2015**, *17*, 2573–2587. [CrossRef]
17. Liu, Q.; Sun, B.; Huo, Y.; Liu, M.; Shi, J.; Jiang, T.; Zhang, Q.; Tang, C.; Bi, H.; He, P. Nutrient bioextraction and microalgae growth inhibition using submerged macrophyte *Myriophyllum spicatum* in a low salinity area of East China Sea. *Mar. Pollut. Bull.* **2018**, *127*, 67–72. [CrossRef]
18. Montemezzani, V.; Duggan, I.C.; Hogg, I.D.; Craggs, R.J. Screening of potential zooplankton control technologies for wastewater treatment High Rate Algal Ponds. *Algal Res.* **2017**, *22*, 1–13. [CrossRef]
19. Wichelen, J.; Van Vanormelingen, P.; Codd, G.A.; Vyverman, W. The common bloom-forming cyanobacterium *Microcystis* is prone to a wide array of microbial antagonists. *Harmful Algae* **2016**, *55*, 97–111. [CrossRef]
20. Huisman, J.; Codd, G.A.; Paerl, H.W.; Ibelings, B.W.; Verspagen, J.M.H.; Visser, P.M. Cyanobacterial blooms. *Nat. Rev. Microbiol.* **2018**, *16*, 471–483. [CrossRef]
21. Schwartz, N.; Dobretsov, S.; Rohde, S.; Schupp, P.J. Comparison of antifouling properties of native and invasive *Sargassum* (Fucales, Phaeophyceae) species. *Eur. J. Phycol.* **2017**, *52*, 116–131. [CrossRef]
22. Chalabi, A.; Semroud, R.; Grimes, S. Plan d'action Stratégique Pour La Conservation de La Diversité Biologique en Région Méditerranéenne. 2015. Available online: http://www.abhatoo.net.ma/maalama-textuelle/developpement-economique-et-social/developpement-economique/planification/planification-de-l-environnement/plan-d-action-strategique-pour-la-conservation-de-la-diversite-biologique-en-region-mediterraneenne-rapport-national-maroc (accessed on 17 August 2020).
23. Sun, Y.; Meng, K.; Su, Z.; Guo, G.; Pu, Y.; Wang, C. Isolation and purification of antialgal compounds from the red alga *Gracilaria lemaneiformis* for activity against common harmful red tide microalgae. *Environ. Sci. Pollut. Res.* **2017**, *24*, 4964–4972. [CrossRef] [PubMed]
24. Barani, M.; Yousefzadi, M.; Moezi, M. Essential oils, new source of algicidal compounds. *J. Appl. Phycol.* **2015**, *27*, 267–273. [CrossRef]
25. Najem, A.M.; Abed, I.J.; Al-haidari, A.M.D. Evaluation the activity of *Rosemary* (*Rosmarinus officinalis* L.) essential oil against some cyanobacteria. *Iraqi J. Biotechnol.* **2016**, *15*, 97–102.
26. Najem, A.M.; Abed, I.J. Potential use of rosemary (*Rosmarinus officinalis* L.) essential oil as Anti- bacterial and anti-algal. *J. Pharm. Biol. Sci.* **2018**. [CrossRef]
27. Wang, H.; Liang, F.; Zhang, L. Composition and anti-cyanobacterial activity of essential oils from six different submerged macrophytes. *Pol. J. Environ. Stud.* **2015**, *24*, 333–338. [CrossRef]

28. Zerrifi, E.A.S.; Kasrati, A.; Redouane, E.; Tazart, Z.; El khaloufi, F.; Abbad, A.; Oudra, B.; Campos, A.; Vasconcelos, V. Essential oils from Moroccan plants as promising ecofriendly tools to control toxic cyanobacteria blooms. *Ind. Crops Prod.* **2020**, *143*, 111922. [CrossRef]
29. Zerrif, S.E.A.; El Ghazi, N.; Douma, M.; El Khalloufi, F.; Oudra, B. Potential uses of seaweed bioactive compounds forharmful microalgae blooms control: Algicidal effects and algal growth inhibition of *Phormidium* sp (freshwater toxic cyanobacteria). *Smetox J.* **2018**, *1*, 59–62.
30. Zerrifi, S.E.A.; Tazart, Z.; El Khalloufi, F.; Oudra, B.; Campos, A.; Vasconcelos, V. Potential control of toxic cyanobacteria blooms with Moroccan seaweed extracts. *Environ. Sci. Pollut. Res.* **2019**, 1–11. [CrossRef]
31. Patra, J.K.; Lee, S.-W.; Kwon, Y.-S.; Park, J.G.; Baek, K.-H. Chemical characterization and antioxidant potential of volatile oil from an edible seaweed *Porphyra tenera* (Kjellman, 1897). *Chem. Cent. J.* **2017**, *11*, 34. [CrossRef]
32. Patra, J.K.; Das, G.; Baek, K. Antibacterial mechanism of the action of *Enteromorpha linza* L. essential oil against Escherichia coli and Salmonella Typhimurium. *Bot. Stud.* **2015**. [CrossRef]
33. Patra, J.K.; Baek, K. Anti-listerial activity of four seaweed essential oils against *Listeria monocytogenes*. *Jundishapur J. Microbiol.* **2016**, *9*. [CrossRef] [PubMed]
34. Patra, J.K.; Baek, K. Antibacterial activity and action mechanism of the essential oil from *Enteromorpha linza* L. against foodborne pathogenic bacteria. *Molecules* **2016**, *21*, 388. [CrossRef] [PubMed]
35. Gaysinski, M.; Ortalo-Magné, A.P.; Thomas, O.; Culioli, G. Extraction, purification, and NMR analysis of terpenes from brown algae. In *Natural Products from Marine Algae: Methods and Protocols*; Humana Press: New York City, NY, USA, 2015; pp. 207–223. ISBN 9781493926848.
36. Patra, J.K.; Lee, S.; Park, J.G.; Baek, K. Antioxidant and antibacterial properties of essential oil extracted from an edible seaweed *Undaria pinnatifida*. *J. Food Biochem.* **2017**, *41*, e12278. [CrossRef]
37. Patra, J.K.; Das, G.; Baek, K. Chemical composition and antioxidant and antibacterial activities of an essential oil extracted from an edible seaweed, *Laminaria japonica* L. *Molecules* **2015**, *20*, 12093–12113. [CrossRef]
38. Ozdemir, G.; Horzum, Z.; Sukatar, A.; Karabay-yavasoglu, N.U. Antimicrobial activities of volatile components and various extracts of *Dictyopteris membranaceae* and *Cystoseira barbata* from the coast of Izmir, Turkey. *Pharm. Biol.* **2006**, *44*, 183–188. [CrossRef]
39. Kamenarska, Z.; Yalçin, F.N.; Ersöz, T.; Çaliş, I.; Stefanov, K.; Popov, S. Chemical composition of *Cystoseira crinita* Bory from the Eastern Mediterranean. *Z. Nat.* **2002**, *57*, 584–590. [CrossRef]
40. Riad, N.; Zahi, M.R.; Trovato, E.; Bouzidi, N.; Daghbouche, Y.; Utczás, M.; Mondello, L.; El Hattab, M. Chemical screening and antibacterial activity of essential oil and volatile fraction of *Dictyopteris polypodioides*. *Microchem. J.* **2020**, *152*, 104415. [CrossRef]
41. Gressler, V.; Colepicolo, P.; Pinto, E. Useful strategies for algal volatile analysis. *Curr. Anal. Chem.* **2009**, *5*, 271–292. [CrossRef]
42. Salvador, N.; Garreta, A.G.; Lavelli, L.; Ribera, M.A. Antimicrobial activity of Iberian macroalgae. *Sci. Mar.* **2007**, *71*, 101–113. [CrossRef]
43. Mishra, A.K. *Sargassum*, *Gracilaria* and *Ulva* exhibit positive antimicrobial activity against human pathogens. *OALib* **2018**, *5*, 1–11. [CrossRef]
44. Begum, S.; Nyandoro, S.; Buriyo, A.; Makangara, J.; Munissi1, J.; Duffy, S.; Avery, V.; Erdelyi, M. Bioactivities of extracts, debromolaurinterol and fucosterol from macroalgae species. *Tanzan. J. Sci.* **2018**, *44*, 104–116.
45. Kumaresan, M.; Vijai Anand, K.; Govindaraju, K.; Tamilselvan, S.; Ganesh Kumar, V. Seaweed *Sargassum wightii* mediated preparation of zirconia (ZrO_2) nanoparticles and their antibacterial activity against gram positive and gram negative bacteria. *Microb. Pathog.* **2018**, *124*, 311–315. [CrossRef] [PubMed]
46. Sujatha, R.; Siva, D.; Mohideen, P.N.A. Screening of phytochemical profile and antibacterial activity of various solvent extracts of marine algae *Sargassum swartzii*. *World Sci. News* **2019**, *115*, 27–40.
47. Moorthi, P.V.; Balasubramanian, C. Antimicrobial properties of marine seaweed, *Sargassum muticum* against human pathogens. *J. Coast. Life Med.* **2015**, 1–5. [CrossRef]
48. Farid, Y.; Etahiri, S.; Assobhei, O. Activité antimicrobienne des algues marines de la lagune d'Oualidia (Maroc): Criblage et optimisation de la période de la récolte. *Appl. Biosci.* **2009**, *24*, 1543–1552.
49. Chiheb, I.; Hassane, R.; Martinez-Lopez, J.; Dominguez Seglar, J.F.; Gomez Vidal, J.A.; Bouziane, H.; Mohamed, K. Screening of antibacterial activity in marine green and brown macroalgae from the coast of Morocco. *Afr. J. Biotechnol.* **2009**, *8*, 1258–1262. [CrossRef]

50. Ainane, T.; Abourriche, A.; Kabbaj, M.; Elkouali, M.; Bennamara, A.; Charrouf, M.; Talbi, M.; Lemrani, M. Biological activities of extracts from seaweed *Cystoseira tamariscifolia*: Antibacterial activity, antileishmanial activity and cytotoxicity. *J. Chem. Pharm. Res.* **2014**, *6*, 607–611.
51. Saleh, B.; Al-Hallab, L.; Al-Mariri, A. Seaweed extracts effectiveness against selected Gram-negative bacterial isolates. *Biol. Sci. PJSIR* **2019**, *62*, 101–110.
52. Wang, H.; Liang, F.; Qiao, N.; Dong, J.; Zhang, L.; Guo, Y. Chemical composition of volatile oil from two emergent plants and their algae inhibition activity. *Pol. J. Environ. Stud.* **2014**, *23*, 2371–2374.
53. Xian, Q.; Chen, H.; Zou, H.; Yin, D. Allelopathic activity of volatile substance from submerged macrophytes on *Microcystin aeruginosa*. *Acta Ecol. Sin.* **2006**, *26*, 3549–3554. [CrossRef]
54. Harada, K.; Ozaki, K.; Tsuzuki, S. Blue color formation of cyanobacteria with β-cyclocitral. *J. Chem. Ecol.* **2009**, *35*, 1295–1301. [CrossRef] [PubMed]
55. Huang, J.J.; Kolodnyy, N.H.; Redfeam, J.T.; Allen, M.M. The acid stress response of the cyanobacterium *Synochocystis* sp. strain PCC 6308. *Arch. Microbiol.* **2002**, *177*, 486–493. [CrossRef]
56. Leu, E.; Krieger-liszkay, A.; Goussias, C.; Gross, E.M. Polyphenolic allelochemicals from the aquatic angiosperm. *Society* **2002**, *130*, 2011–2018. [CrossRef]
57. Ni, L.; Acharya, K.; Hao, X.; Li, S. Isolation and identification of an anti-algal compound from *Artemisia annua* and mechanisms of inhibitory effect on algae. *Chemosphere* **2012**, *88*, 1051–1057. [CrossRef]
58. Wu, L.; Qiu, Z.; Zhou, Y.; Du, Y.; Liu, C.; Ye, J.; Hu, X. Physiological effects of the herbicide glyphosate on the cyanobacterium *Microcystis aeruginosa*. *Aquat. Toxicol.* **2016**, *178*, 72–79. [CrossRef]
59. Meng, P.; Pei, H.; Hu, W.; Liu, Z.; Li, X.; Xu, H. Allelopathic effects of *Ailanthus altissima* extracts on *Microcystis aeruginosa* growth, physiological changes and microcystins release. *Chemosphere* **2015**, *141*, 219–226. [CrossRef]
60. Hua, Q.; Liu, Y.; Yan, Z.; Zeng, G.; Liu, S.; Wang, W. Allelopathic effect of the rice straw aqueous extract on the growth of *Microcystis aeruginosa*. *Ecotoxicol. Environ. Saf.* **2018**, *148*, 953–959. [CrossRef]
61. Chai, T.; Zhu, H.D.; Yan, H.Z.; Zhao, D.; Liu, X.Y.; Fu, H.Y. Allelopathic effects of two organic acids on *Microcystis aeruginosa*. *Earth Environ. Sci.* **2018**, *146*. [CrossRef]
62. Du, Y.; Ye, J.; Wu, L.; Yang, C.; Wang, L.; Hu, X. Physiological effects and toxin release in *Microcystis aeruginosa* and *Microcystis viridis* exposed to herbicide fenoxaprop-p-ethyl. *Environ. Sci. Pollut. Res.* **2017**, *24*, 7752–7763. [CrossRef]
63. Bai, L.; Cao, C.; Wang, C.; Zhang, H.; Deng, J.; Jiang, H. Response of bloom-forming cyanobacterium *Microcystis aeruginosa* to 17B-estradiol at different nitrogen levels. *Chemosphere* **2019**, *219*, 174–182. [CrossRef] [PubMed]
64. Hou, X.; Huang, J.; Tang, J.; Wang, N.; Zhang, L.; Gu, L.; Sun, Y.; Yang, Z.; Huang, Y. Allelopathic inhibition of juglone (5-hydroxy-1,4-naphthoquinone) on the growth and physiological performance in *Microcystis aeruginosa*. *J. Environ. Manag.* **2019**, *232*, 382–386. [CrossRef] [PubMed]
65. Wang, F.; Liu, D.; Qu, H.; Chen, L.; Zhou, Z.; Wang, P. A full evaluation for the enantiomeric impacts of lactofen and its metabolites on aquatic macrophyte Lemna minor. *Water Res.* **2016**, *101*, 55–63. [CrossRef] [PubMed]
66. Zhang, Q.; Song, Q.; Wang, C.; Zhou, C.; Lu, C.; Zhao, M. Effects of glufosinate on the growth of and microcystin production by *Microcystis aeruginosa* at environmentally relevant concentrations. *Sci. Total Environ.* **2017**, *575*, 513–518. [CrossRef] [PubMed]
67. Wang, J.; Liu, Q.; Feng, J.; Lv, J.; Xie, S. Effect of high-doses pyrogallol on oxidative damage, transcriptional responses and microcystins synthesis in *Microcystis aeruginosa* TY001 (Cyanobacteria). *Ecotoxicol. Environ. Saf.* **2016**, *134*, 273–279. [CrossRef] [PubMed]
68. Xie, J.; Zhao, L.; Liu, K.; Liu, W. Enantiomeric environmental behavior, oxidative stress and toxin release of harmful cyanobacteria *Microcystis aeruginosa* in response to napropamide and acetochlor. *Environ. Pollut.* **2019**, *246*, 728–733. [CrossRef]
69. Gayral, P. Les algues de la côte Atlantique marocaine. *Bull. Société Des Sci. Nat. Phys. Maroc.* **1958**, *42*, 1–34.
70. Falcão, S.; Bacém, I.; Igrejas, G.; Rodrigues, P.J.; Vilas-Boas, M.; Amaral, J.S. Chemical composition and antimicrobial activity of hydrodistilled oil from *Juniper berries*. *Ind. Crops Prod.* **2018**, *124*, 878–884. [CrossRef]
71. NCCLS. *Performance Standards for Antimicrobial Disk Susceptibility Test*, 6th ed.; NCCLS: Annapolis Junction, MD, USA, 1997.

72. NCCLS. *Methods for Dilution Antimicrobial Susceptibility Tests for Bacteria That Grow Aerobically*, 4th ed.; NCCLS: Annapolis Junction, MD, USA, 1997.
73. Sbiyyaa, B.; Loudiki, M.; Oudra, B. Capacité de stockage intracellulaire de l'azote et du phosphore chez *Microcystis aeruginosa* Kütz. t *Synechocystis* sp.: Cyanobactéries toxiques occasionnant des blooms dans la région de Marrakech (Maroc). *Int. J. Limnol. Ann. Limnol.* **1998**, *34*, 247–257. [CrossRef]
74. Xu, H.; Paerl, H.W.; Qin, B.Q.; Zhu, G.W.; Gao, G. Nitrogen and phosphorus inputs control phytoplankton growth in eutrophic Lake Taihu, China. *Limnol. Oceanogr.* **2010**, *55*, 420–432. [CrossRef]
75. Lichtenthaler, H.; Wellburn, A. Determinations of total carotenoids and Chlorophylls b of leaf extracts in different solvents. *Biochem. Soc. Trans.* **1983**, *11*, 591–592. [CrossRef]
76. Li, J.; Liu, Y.; Zhang, P.; Zeng, G.; Cai, X.; Liu, S.; Yin, Y.; Hu, X.; Hu, X.; Tan, X. Growth inhibition and oxidative damage of *Microcystis aeruginosa* induced by crude extract of *Sagittaria trifolia* tubers. *J. Environ. Sci.* **2016**, *43*, 40–47. [CrossRef] [PubMed]
77. Bradford, M.M. A rapid and sensitive method for the quantitation of microgram quantities of protein utilizing the principle of proteindye binding. *Anal. Biochem.* **1976**, *72*, 248–254. [CrossRef]
78. Beauchamp, C.; Fridovich, I. Superoxide Dismutase: Improved assays and an assay applicable to acrylamide gels. *Anal. Biochem.* **1971**, *44*, 276–287. [CrossRef]
79. Rao, M.V.; Paliyath, G.; Ormrod, D.P. Ultraviolet-B- and ozone-induced biochemical changes in antioxidant enzymes of Arabidopsis thaliana. *Plant. Physiol.* **1996**, *110*, 125–136. [CrossRef] [PubMed]

toxins

MDPI

Article

Large-Scale Green Liver System for Sustainable Purification of Aquacultural Wastewater: Construction and Case Study in a Semiarid Area of Brazil (Itacuruba, Pernambuco) Using the Naturally Occurring Cyanotoxin Microcystin as Efficiency Indicator

Maranda Esterhuizen [1,2,3,*] and **Stephan Pflugmacher** [1,2,3,4]

[1] Ecosystems and Environmental Research Programme, Faculty of Biological and Environmental Sciences, University of Helsinki, Niemenkatu 73, 15140 Lahti, Finland; Stephan.PflugmacherLima@umanitoba.ca

[2] Helsinki Institute of Sustainability Science (HELSUS), University of Helsinki, Fabianinkatu 33, 00014 Helsinki, Finland

[3] Korea Institute of Science and Technology Europe (KIST), Joint Laboratory of Applied Ecotoxicology, Campus 7.1, 66123 Saarbrücken, Germany

[4] Clayton H. Riddell Faculty of Environment, Earth, and Resources, University of Manitoba, Wallace Bldg, 125 Dysart Rd, Winnipeg, MB R3T 2N2, Canada

* Correspondence: maranda.esterhuizen@helsinki.fi; Tel.: +358-503-188-337

Received: 29 September 2020; Accepted: 29 October 2020; Published: 30 October 2020

check for updates

Abstract: The aquaculture industry in Brazil has grown immensely resulting in the production of inefficiently discarded wastewater, which causes adverse effects on the aquatic ecosystem. The efficient treatment of aquaculture wastewater is vital in reaching a sustainable and ecological way of fish farming. Bioremediation in the form of the Green Liver System employing macrophytes was considered as wastewater treatment for a tilapia farm, COOPVALE, in Itacuruba, Brazil, based on previously demonstrated success. A large-scale system was constructed, and the macrophytes *Azolla caroliniana*, *Egeria densa*, *Myriophyllum aquaticum*, and *Eichhornia crassipes* were selected for phytoremediation. As cyanobacterial blooms persisted in the eutrophic wastewater, two microcystin congeners (MC-LR and -RR) were used as indicator contaminants for system efficiency and monitored by liquid-chromatography–tandem-mass-spectrometry. Two trial studies were conducted to decide on the final macrophyte selection and layout of the Green Liver System. In the first trial, 58% MC-LR and 66% MC-RR were removed and up to 32% MC-LR and 100% MC-RR were removed in the second trial. Additional risks that were overcome included animals grazing on the macrophytes and tilapia were spilling over from the hatchery. The implementation of the Green Liver System significantly contributed to the bioremediation of contaminants from the fish farm.

Keywords: phytoremediation; cyanobacterial toxins; microcystin degradation; water treatment; ecosystem services

Key Contribution: A large-scale Green Liver System was constructed and optimized for the successful remediation of wastewater from a tilapia farm in Itacuruba, Pernambuco, Brazil.

1. Introduction

Brazil, in particular the Pantanal and the Amazon areas, is well known for its majestic landscapes, which are linked to vast water resources. Brazil holds approximately 10% of the global freshwater

water supply stored in nearly 30,000 reservoirs covering a surface area of approximately 50,000 km^2. Most of these reservoirs have been built for energy production, irrigation, and drought mitigation purposes. These seemingly infinite water sources have attracted a large variety of commercial fish farming. Most of them are small-scale production units; however, they summarize to about 100,000 aquacultural units occupying approximately 80,000 ha [1]. Besides fish, other aquatic organisms, such as shrimp (mostly *Litopenaeus vannamei*), crayfish (*Procambarus clarkii*), bivalves (*Mytilidae* and *Oyster*), as well as the bullfrog (*Rana catesbeiana*) are reared on a commercial scale. The fish species typically used are native to, e.g., the Parana, Sao Francisco, and the Amazon; however, Brazil has a long-term history for introducing alien fish species for aquacultural purposes as well. For example, the Nile tilapia (*Oreochromis niloticus*), introduced to Brazil in the 1950s, has become one of the most important fish species used commercially [2]. With the introduction of sex-reversal technology using hormones, the small volume/high-density cage technology (SVHD), rich nutrient food, and fish antibiotics, the farming of Nile tilapia has become more economically feasible with production yields reaching approximately 133,000 metric tons by 2009 [1].

Aquaculture enterprises typically employ land-based pond systems for fish hatching and rearing [3], resulting in wastewaters released into nearby reservoirs. The wastewaters would not only include hormones and antibiotics commonly used in aquaculture [4,5] but also would be enriched with nutrients leading to the eutrophication of these waterbodies and subsequently, the domination of cyanobacterial blooms [6]. Cyanobacteria do not only cause aesthetic issues in surface water but also produce a range of toxins, the most commonly detected being the microcystins (MCs) [7]. The hepatotoxic MCs can occur in various isoforms, differentiated only by variations of two amino acids in fixed positions; e.g., MC-LR contains lysine and arginine, and MC-RR contains two arginine residues (Figure 1).

In semiarid and arid regions of Brazil, the water from these reservoirs is not only used for irrigation, but also as a drinking water resource. The water quality of these reservoirs, therefore directly affects human health in the region. Due to the still-growing aquacultural industry in Brazil, along with the known associated environmental impacts [8]; it has become increasingly necessary to address the development of technologies that will ensure the sustained high quality of freshwater reservoirs in the wake of the aquaculture boom and ensure successful implementation.

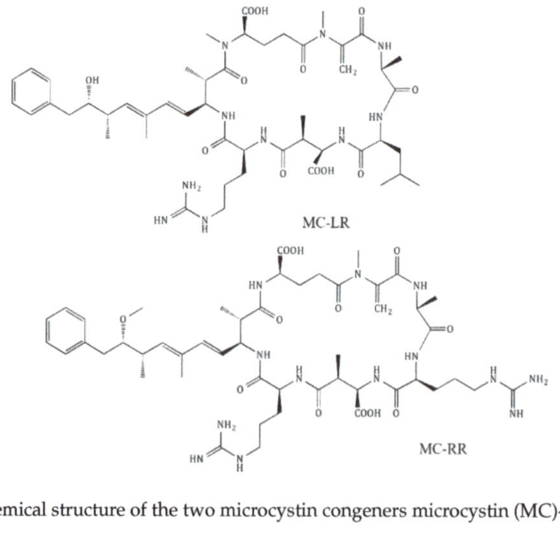

Figure 1. Chemical structure of the two microcystin congeners microcystin (MC)-LR and -RR.

Acknowledging freshwater as the limited resource that it is, it becomes apparent that tools are needed to purify water sustainably and cost-effectively to make it affordable for small unit enterprises. One option is the use of the Green Liver System®, which in contrast to the wetland system, is an entirely artificial system that utilizes the phytoremediation potential of aquatic plants, which take up and biotransform contaminants from water [9,10]. Plants and animals share many similarities regarding the biotransformation of xenobiotics, except in stage three, where animals secrete the biotransformed products and plants sequester them in cell wall fractions, the apoplast, or vacuoles [11,12]. Phytoremediation has shown promising results with respect to eutrophication, xenobiotics, and cyanobacterial toxin removal capabilities in the laboratory [10,13–18] as well as in a small pilot plant in Hefei (Anhui region, China) [19].

As with all water treatment options, several risks factors need to be considered before the Green Liver Systems can be implemented and continued sustainable operation assured. First, the main contaminants concerned have to be quantified to establish their concentrations in the wastewater. Here, the first step towards a customized Green Liver System takes place and plants are chosen according to the specific contaminants present. The selection of the individual species and combination of macrophytes is based on extensive laboratory research, which has established their remediation capabilities for the relevant contaminants and to avoid allelopathic effects between the different plant species [9].

After setting up the system, three major risk groups can disturb the running system; climate, nature, and humans. For the climate, factors such as heavy rain or drought might be a risk for the Green Liver System. Heavy rain can flush the system, causing the macrophytes to accumulate near the discharge points, causing them to die-off. Drought could cause decreasing water levels in combination with increasing water temperature leading to the death of the macrophytes as well. Besides climatic factors, animals, typically such as donkeys and wild goats in Brazil, could invade the system and feed on the fresh aquatic plants. Furthermore, as this system is designed to treat the wastewater of a tilapia farm, it is highly likely that juvenile fish might enter the system via the inflow. Tilapia fish are plant feeders and might significantly reduce the biomass of the aquatic macrophytes in the system, which would lead to a decrease in the efficiency of the system itself.

The present study aimed to construct a Green Liver System suitable for the treatment of wastewater from COOPVALE, a tilapia farm opened in 1999. Before the system had been constructed, water from the tilapia farm was directly released into the nearby Itaparica reservoir (now known as the Luiz Gonzaga Dam). The presented research tested the upscaling of a lab-scale [10] and small pilot-scale system [19] to a large-scale system in terms of flow rate and a retention time of three days to yield satisfactory water remediation. Due to the eutrophication resulting from the fish farm waste, cyanobacterial blooms occurred in the wastewater from COOPVALE. The two most commonly detected MC congeners, MC-LR and -RR, were selected as system efficiency indicator contaminants and were monitored at the fish farm hatchery, as well as the system inlet and outlet during two trials to investigate the remediation efficiency of the Green Liver System. The risks associated with the implementation of such a system was evaluated, and mitigation attempts were tested in the two trial studies.

2. Results and Discussion

The immense and still growing aquacultural industry in Brazil has created the need for sustainable wastewater purification to remove nutrients, toxins, and veterinary substances (such as hormones and antibiotics). The excessive application of fish food ad libitum caused the reservoir, as a sink for the wastewater, to become eutrophic [3] causing cyanobacterial blooms [6]. As the financial means of the Brazilian fish farmers in most cases are limited, it is necessary to use low-cost strategies to suit the financial situation of the farmers. Thus, treating aquacultural wastewater using aquatic plants, in particular, using the Green Liver System and employing native macrophytes, seems like the most feasible approach [9,20].

2.1. First Trial

Shortly after construction and planting, during the four-week macrophyte acclimation period, the large-scale Green Liver System was invaded by wild, formerly domestic, goats, which fed on the aquatic macrophytes. A goat-proof fence was constructed around the system to keep them out.

The aquatic fern *Azolla caroliniana*, known for its remediation of heavy metals [21], nitrate from water [22,23], and effluents from fish farms [23,24], was selected for the first compartment. In laboratory studies, *A. caroliniana* remediated up to 41% phosphate and 30% nitrate [24]. In the Green Liver System®, this free-floating fern was used to cover the water surface of the first compartment to limit sunlight penetration and thus the development of cyanobacterial blooms. The sunlight intensity in Brazil during summer (October to February), however, was too high and did not support the growth of *A. caroliniana*. After three weeks, the compartments housing *A. caroliniana* turned red and started dying. *A. caroliniana* requires low light intensities for growth, needing only 25–50% of sunlight (approximately 3200 lux) [25,26]. Optimal water temperature for the growth is 30 °C, whereas the growth rate was reduced significantly by water temperatures above 35 °C [27]. *A. caroliniana* can withstand a pH range between 3.5 and 10 and needs a relative humidity between 85–90% [27]. As the climate conditions during the first trial were unusually hot, with day temperatures between 28 and 35 °C, the water temperature in the system was between 29 °C ± 3 °C and pH 7.3 ± 0.1. Hence, this species was replaced in the second trial by the free-floating *Eichhornia crassipes*, which has been used previously for the effluent treatment of aquacultural wastewater, especially Nile tilapia [28]. The macrophytes in the other compartments flourished and thrived. Cyanobacteria, which were spilling over from the bloom in the hatchery water, did not thrive in the Green Liver System, probably as the macrophytes removed excess nutrients such as N and Ps and due to the inadequate conditions for cyanobacterial growth, such as shading from surface floating macrophytes.

2.2. Second Trial and Final Layout

E. crassipes, which replaced *A. caroliniana* in the first compartment for the second trial, is also a free-floating plant and thus with a similar capability for providing shade and thus cooling the water. The macrophyte is an invasive species and thus using it for phytoremediation serves as a benefit. *E. crassipes* has the enormous advantage of changing water parameters, most notably reducing the water temperature [29]. According to Greco and de Freitas [30], this macrophyte grows best at high temperatures, as experienced in Brazil. *E. crassipes* has previously been tested for the phytoremediation of urban wastewater [31], cyanide [32], and mercury [33] and has been assessed as useful and appropriate for such purposes. However, using *E. crassipes* in a Green Liver System can also have adverse effects, as this plant might reduce the oxygen level in the surface region of the water column and consequently provide the ideal environment for the development of mosquitos and snails, acting as vectors for diseases [34]. The nutrient uptake from water by *E. crassipes* is also controversial because plants seem to be able to recycle the nutrients in decaying leaves, which are still attached to the mother plant [35–37].

As the *E. crassipes* plants started to flower, fishnets were successfully installed close to the water surface to keep them in their compartments and to avoid spreading into the other compartments due to wind-triggered movement.

Since the wastewater comes from hatchery ponds, juvenile tilapia were flushed into the system. As they started growing and feeding on the macrophytes, the biomass within the system significantly reduced. Hence, compartments 2 to 5 had to be restocked with *Egeria densa*. To avoid the infiltration of fish via the hatchery pond systems, fish barriers were installed between the ponds and the inflow of the Green Liver System. Furthermore, a small overflow and a fish escape channel were built to avoid dead fish accumulating in front of the barrier and to allow fish to escape directly into the reservoir. To reduce the amount of tilapia already growing in the Green Liver System, a top-down approach was chosen and some predator fish were introduced into the system, specifically 30 Piranhas (*Pygocentrus nattereri*) and 10 Tucunare (common name is butterfly peacock bass; *Cichla ocellaris*). All the aforementioned preventative measures were successful in keeping each of the plants in their respective compartments and eliminating fish flowing down from the fish farm.

For Green Liver Systems, plant species with a high growth rate and the ability to produce high biomass are more efficient in the treatment of aquacultural wastewater. The *E. crassipes* and *E. densa* selected for this Green Liver System are especially capable of excessive growth. Green Liver Systems, as with most artificial systems, are easy to handle, customizable for the specific needs of stakeholders, and cost-effective. The Green Liver Systems is a suitable and promising way for aquaculture farmers to clean their wastewater before discharging them into freshwater bodies or reservoirs nearby. However, more ideas have to be developed on how to proceed with the contaminated plants. In general, when plants are saturated (up-take threshold reached), they have to be harvested to prevent die-off, which is accompanied by the possible re-release of the contaminants and their metabolites. Hence, the possibility to use the harvested plant material as fertilizer or animal feed has to be excluded.

2.3. Contaminant Monitoring

In general, free-floating plants should limit bloom the formation of potentially introduced cyanobacteria in the hatchery ponds as sunlight, and thus photosynthesis would be limited. Inhibited photosynthesis, in turn, will lead to the decay of the cyanobacteria and the release of any toxins, making it easier for the macrophytes in the subsequent compartments to remediate the toxins. Hence, it was expected that the submerged macrophytes, *E. densa* and *Myriophyllum aquaticum* stocked in compartment 2 to 6, would do the primary uptake of the cyanobacterial toxins. Laboratory experiments have proven the ability of both species to take up cyanotoxins [19].

Furthermore, *E. densa* and *M. aquaticum* are well known to remove heavy metals from wastewater [38,39]. Additionally, some reports exist on the removal capacity of *M. aquaticum* [40] and *E. densa* [17] with respect to the fish antibiotic oxytetracycline, which is used routinely in fish farms to keep the fish healthy despite the high fish density in the hatchery ponds as well as in the SVHD in the reservoir.

The first sampling results showed that the cyanotoxin burden in the hatchery water was 38.3 ± 2.6 ng·L^{-1} for MC-LR and 12.9 ± 7.3 ng·L^{-1} MC-RR. These concentrations were reduced by 58% for MC-LR and 66% for MC-RR in the first trail and by 32% MC-LR and 100% MC-RR in the second trial (Figure 2). MC-RR is ten times more toxic than MC-LR [41], making the results from the second trial more significant as 100% of the MC-RR is removed.

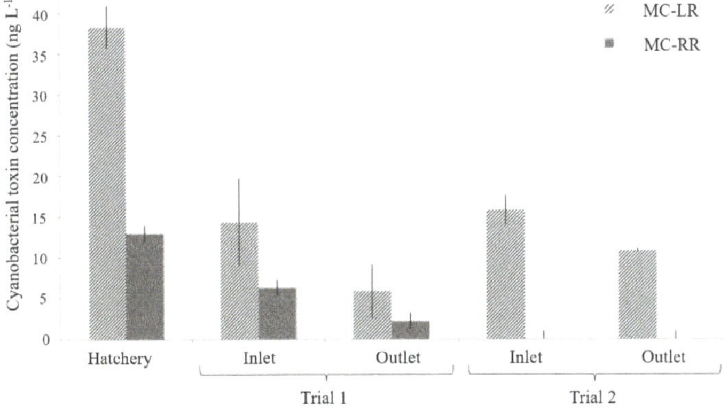

Figure 2. The concentration of the cyanobacterial toxins MC-LR and MC-RR in the hatchery and at the Green Liver System at the inlet (wastewater) and outlet (remediated water) during the first and second trials. Data present the average cyanobacterial toxin concentration (n = 7) ± standard deviation.

According to these analytical measurements of the cyanobacterial toxins, the use of a Green Liver System is an adequate and promising solution to remove contaminants from aquaculture wastewater before it is released into freshwater reservoirs. With the present system, it has been evidenced that it is an effective and low-cost way to treat fish farm wastewater. By choosing different plant species in a Green Liver System, the efficiency in exactly removing the targeted pollutants is very high. The utilization of the system in Brazil seems very practical due to the warm climate in this region; the year-round growth of the macrophytes guarantee the efficiency of the treatment [42].

In general, promising candidates from the group of macrophytes or other photosynthetic active aquatic species (ferns, moss, macroalgae) are tested in laboratory systems before being stocked in the outdoor system. This will ensure that the performance of a single plant species is known before use. In this sense, macrophytes represent the tools for the remediation of according contaminants. The selection of appropriate species has to be customized according to the contaminants in the wastewater and the requirements for the water quality to be achieved. The most crucial factor for a Green Liver System working in a semiarid and arid region seems to be the water/wastewater inflow. To ensure the adequate treatment of the wastewater, the inflow has to be relatively constant and should never be paused. Therefore, the Green Liver System can be expected to function well and sustain its performance as long as the hydrology and the growth of the macrophytes are maintained and detritus is removed from time to time.

The economic outlook is an essential factor for the sustainable cleaning-up of aquacultural wastewater as most Brazilian farmers are not able to invest vast sums of money for this issue. The required capital investment should be as small as possible, but these costs should consider the overall design and required size of the system, the construction and installation cost, the operation and maintenance and the sampling and analysis to check the performance of the system.

3. Conclusions

The constructed large-scale Green Liver System was not completely able to remove the MC congeners tested but could substantially diminish the MC contaminant concentrations resulting from aquacultural farms, ecologically and economically, as well as eliminate visible cyanobacterial blooms resulting from the land-based hatchery ponds before entering the Itaparica reservoir. This nature-based treatment is a first step in the right directions as untreated waste is not directly released into the environment. Other emerging contaminants in the wastewater, such as antibiotics (oxytetracycline), hormones (methyl-testosterone) and other cyanobacterial toxins will be evaluated in the future.

4. Materials and Methods

4.1. Customized Planning

A questionnaire was developed to determine whether a Green Liver System was suitable to fit the needs of according stakeholders (Figure A1). The questionnaire excluded specific contaminant groups since they were discussed at a later stage during the customized development of a Green Liver System®. In this questionnaire, the more fundamental questions were asked in a simple YES or NO manner. After the first visit to COOPVALE in 2012, the area for the Green Liver System construction was defined. Since the system had to be as cost-effective as possible, no electrical pumps were used to control the water flow. Therefore, the system was planned below the hatchery pond system of the tilapia company using the natural declination to transport the wastewater from the ponds to the Green Liver System.

The system had a final size of 100 m × 25 m × 2 m, summating to a total volume of 5000 m^3 (5 mL) (Figure 3). The system was divided into six compartments by curved brick stone barriers to control the water flow. Each of these barriers was 0.75 m wide to allow easy access for water sampling and management measures. The barriers, in total 15 m long, were constructed using commercially available bricks and water repellent cement. The tips of the barriers (2.5 m) were curved to minimize the accumulation of debris within the system, which further reduces the need for regular maintenance work while reducing the water flow velocity. To seal the system base, commercially available clay mineral was used at a thickness of 0.15 m. Regulated discharge from the hatchery ponds was realized through a 25 m long-channel using the natural decline from the land-based hatchery ponds, which are approximately 5 m higher than the Green Liver System. Water discharge was regulated at the hatchery ponds, to avoid the flooding of the Green Liver System.

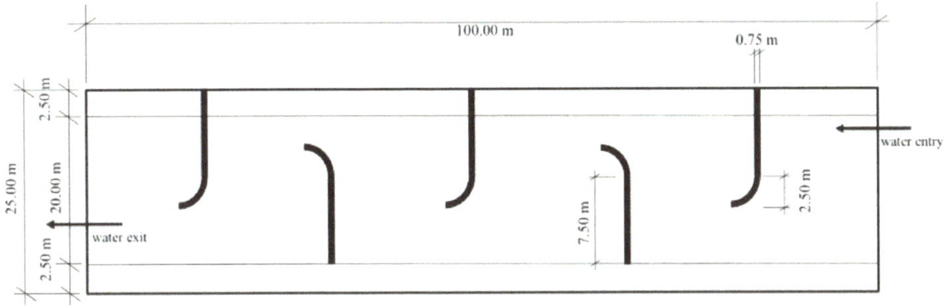

Figure 3. Customized construction plan for the Green Liver System on the premises of COOPVALE (Itacuruba, Brazil).

4.2. Flow Calculations and Upscaling

Using the volumetric flow rate Equation (1), the required water flow rate into the system could be calculated based on the three-day retention time required by the macrophytes to remediate the pollutants as determined in the laboratory [9,10]:

$$Q = \frac{V}{Rt} \quad (1)$$

where the flow rate (Q) in $L \cdot h^{-1}$ is equal to the bed volume (V) in L, divided by the residence time (Rt) in hours. Considering the system would operate 20 h a day, leaving four hours for maintenance or any other related works, in order for it to produce the desired volume in three days the flow rate needed in the system according to Equation (1) would be:

$$Q = \frac{(5,000,000 \text{ L})}{(20 \text{ h} \times 3 \text{ days})} = 8\,3333 \text{ L/h} \quad (2)$$

4.3. Aquatic Macrophytes Used for the Green Liver System

The Green Liver System was stocked with aquatic plants found in the same area as the Itaparica reservoir, namely the macrophytes *Egeria densa* (syn. *Anacharis densa* (Planch.) Vict., *Elodea densa* (Planch.) Casp.), *Myriophyllum aquaticum* (Vell.) Verdc., and *Eichhornia crassipes* (Mart.) Solms-Laubach as well as the aquatic pteridophyte *Azolla caroliniana* (Wild.), which were previously found in laboratory studies to efficiently remediate cyanobacterial toxins as well other chemicals associated with aquaculture [10,17]. The macrophytes were allowed to acclimate for four weeks prior to the first hatchery wastewater being allowed in the system for remediation.

4.4. First Trial

For the first trial, *A. caroliniana*, *E. densa*, and *M. aquaticum* were used in the system. Compartment 1 and 2 were stocked with *A. caroliniana* covering half of the water surface to encourage further plant growth within the system. Compartments 3 and 4 were stocked with *M. aquaticum* using 1 kg·FW of plant material per m^2, and in compartments 5 and 6, and *E. densa* was added at a density of 0.5 kg·FW of plant material per m^2. The system was allowed to run for seven days before samples were taken for the first trial to track MC remediation.

4.5. Second Trial and Final Layout

For the second trial, in the first compartment, *A. caroliniana* was replaced by *E. crassipes*. In an attempt to avoid *E. crassipes* from spreading into the other compartments due to wind-triggered movement, fishnets were installed close to the water surface to retain the plants in their assigned compartments.

To avoid the infiltration of fish via the hatchery pond systems, fish barriers were installed between the ponds and the inflow of the Green Liver System. The fish traps were made of concrete and had two chambers filled with loose gravel stones with three different sizes (very fine, fine, and coarse) as filter material. Furthermore, a small overflow and fish escape channel were built to avoid dead fish accumulating in front of the barrier and to allow fish to escape directly into the reservoir. To reduce the amount of tilapia already growing in the Green Liver System, a top-down approach was chosen and some predator fish were introduced into the system, i.e., 30 Piranhas (*Pygocentrus nattereri*) and 10 Tucunare, more commonly known as butterfly peacock bass (*Cichla ocellaris*). The newly modified system and macrophytes were again allowed to acclimate for three weeks before the hatchery water was introduced into the system, and samples for MC congener monitoring were taken after seven days. For the final working system, macrophytes that showed any signs of chlorosis of necrosis were replaced immediately to avoid the re-release of pollutants back into the water.

4.6. Microcystin Congener Monitoring

Water samples for the analysis of the MC congeners MC-LR and MC-RR were taken ($n = 7$) at the fish hatchery, the Green Liver System inlet and outlet, respectively, during the first and second trials. The determination and quantification of the MC congeners, MC-LR and -RR, were performed by liquid chromatography–tandem mass spectrometry (LC–MS/MS) on an Alliance 2695 UHPLC coupled to a Micromass Quattro Micro^{TM} (Waters Corp., Herts, UK). The matrix separation was achieved on a Kinetex^{TM} C18 reverse-phase column (2.1 × 50 mm, 2.6 µm pore size, Phenomenex, Torrance, CA, USA). Milli-Q water containing 0.1% trifluoroacetic acid (TFA) and 5% acetonitrile (ACN) served as mobile phase A and ACN containing 0.1% TFA served as mobile phase B. The flow rate was maintained at 0.2 mL·min^{-1} and an injection volume of 20 µL per sample was used. The separation of the congeners was achieved using a linear gradient with the mobile phases in the following program: 0 min 65% A; 3.75–7 min 35% A and 7.8–12 min 100% A. The column oven temperature was 40 °C. Elution peaks for the MC congeners were observed at 7.1 min for MC-RR, and 7.44 min for MC-LR. The parent compound and its fragment ions, respectively, were scanned at the following mass-to-charge ratio (m/z): MC-LR 995.5 → 135.1 and MC-RR 519.9 → 135.3. ESI+ conditions for all MCs were set as follows: capillary voltage of 3 kV, source temperature of 120 °C, desolvation temperature of 500 °C and a cone gas flow rate of 100 L·h^{-1}. For MC-LR, the collision energy was 65, cone voltage was 60 V, and for MC-RR the collision energy was 35, and the cone voltage was 20 V. Desolvation gas flow rate was 1000 L·h^{-1} [43]. Calibrations were linear ($R^2 = 0.999$) between 5 and 500 µg L^{-1}. The limit of detection (LOD) was set at 1 µg·L^{-1} (signal to noise S/N > 3) and limit of quantification at 5 µg·L^{-1} (S/N > 5) for both congeners.

Author Contributions: Conceptualization, S.P.; methodology, S.P. and M.E.; formal analysis, S.P. and M.E.; data curation, S.P. and M.E.; writing—original draft preparation; writing—review and editing, M.E.; supervision, S.P.; project administration, S.P.; funding acquisition, S.P. All authors have read and agreed to the published version of the manuscript.

Funding: This work was funded by BMBF, FONA and Sustainable Land Use Management within the project INNOVATE and the Joint Laboratory of Applied Ecotoxicology, KIST Europe. Open access funding provided by University of Helsinki.

Acknowledgments: The authors thank the staff and students at the Technical University of Berlin for assistance with this research as well as COOPVALE for the permission and help in building the system on their premises. Open access funding provided by University of Helsinki. We wish to acknowledge Tobias Block for taking the aerial photograph.

Conflicts of Interest: The authors declare no conflict of interest.

Appendix A

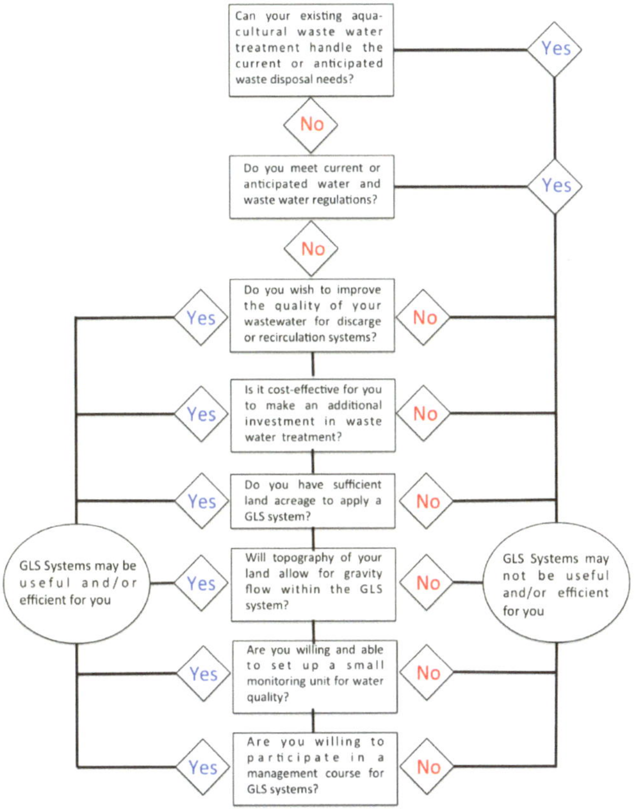

Figure A1. Questionnaire to figure out the needs of the stakeholder and the suitability of a Green Liver System (GLS) for their needs.

References

1. Kubitza, F. An overview of tilapia aquaculture in Brazil. In *New Dimensions on Farmed Tilapia, Proceedings of the 6th International Symposium on Tilapia in Aquaculture, Manila, Philippines, 12–16 September 2004*; Bolivar, R.B., Mair, G.C., Fitzsimmons, K., Eds.; NRAES: Ithaca, NY, USA, 2004; Available online: http://ag.arizona.edu/azaqua/ista/ista6/ista6web/pdf/709.pdf (accessed on 12 February 2019).
2. Pincinato, R.B.M.; Asche, F. The development of Brazilian aquaculture: Introduced and native species. *Aquac. Econ. Manag.* **2016**, *20*, 312–323. [CrossRef]
3. Henry-Silva, G.G.; Camargo, A.F.M. Impacto das atividades de aquicultura e sistemas de tratamento de efluentes com macrófitas aquáticas: Relato de caso. *Bol. Inst. Pesca* **2008**, *34*, 163–173.
4. Rose, P.E.; Pedersen, J.A. Fate of oxytetracycline in streams receiving aquaculture discharges: Model simulations. *Environ. Toxicol. Chem.* **2005**, *24*, 40–50. [CrossRef] [PubMed]

5. Hoga, C.A.; Almeida, F.L.; Reyes, F.G.L. A review on the use of hormones in fish farming: Analytical methods to determine their residues. *CyTA Food* **2018**, *16*, 679–691. [CrossRef]
6. Scholz, S.N.; Esterhuizen-Londt, M.; Pflugmacher, S. Rise of toxic cyanobacterial blooms in temperate freshwater lakes: Causes, correlations and possible countermeasures. *Toxicol. Environ. Chem.* **2017**, *99*, 543–577. [CrossRef]
7. Omidi, A.; Esterhuizen-Londt, M.; Pflugmacher, S. Still challenging: The ecological function of the cyanobacterial toxin microcystin—What we know so far. *Toxin Rev.* **2018**, *37*, 87–105. [CrossRef]
8. Mancuso, M. Effects of fish farming on marine environment. *J. FisheriesSciences.com* **2015**, *9*, 89–90.
9. Pflugmacher, S. Green Liver and Green Liver System—A sustainable way for future water purification. *Aperito J. Aquat. Mar. Ecosyst.* **2015**, *1*. [CrossRef]
10. Pflugmacher, S.; Kühn, S.; Lee, S.; Choi, J.; Baik, S.; Kwon, K.; Contardo-Jara, V. Green Liver Systems® for Water Purification: Using the phytoremediation potential of aquatic macrophytes for the removal of different cyanobacterial toxins from water. *Am. J. Plant Sci.* **2015**, *6*, 1607–1618. [CrossRef]
11. Sandermann, H. Plant metabolism of xenobiotics. *Trends Biochem. Sci.* **1992**, *17*, 82–84. [CrossRef]
12. Sandermann, H. Higher plant metabolism of xenobiotics: The "Green Liver" Concept. *Pharmacogenetics* **1994**, *4*, 225–241. [CrossRef]
13. DeBusk, T.A.; Williams, C.; Ryther, J.H. Removal of nitrogen and phosphorus from wastewater in a water hyacinth-based treatment system. *J. Environ. Qual.* **1983**, *12*, 257–262. [CrossRef]
14. Boyd, C. Guidelines for aquaculture effluent management at the farm-level. *Aquaculture* **2003**, *226*, 101–112. [CrossRef]
15. Baccarin, A.E.; Camargo, A.F.M. Characterization and evaluation of the impact of feed management on the effluents of Nile tilapia (*Oreochromis niloticus*) culture. *Braz. Arch. Biol. Technol.* **2005**, *48*, 81–90. [CrossRef]
16. Esterhuizen-Londt, M.; Pflugmacher, S.; Downing, T.G. β-N-Methylamino-L-alanine (BMAA) uptake by the aquatic macrophyte *Ceratophyllum demersum*. *Ecotoxicol. Environ. Saf.* **2011**, *74*, 74–77. [CrossRef]
17. Vilvert, E.; Contardo-Jara, V.; Esterhuizen-Londt, M.; Pflugmacher, S. The effect of oxytetracycline on physiological and enzymatic defense responses in aquatic plant species *Egeria densa*, *Azolla caroliniana*, and *Taxiphyllum barbieri*. *Toxicol. Environ. Chem.* **2017**, *99*, 104–116. [CrossRef]
18. Contardo-Jara, V.; Kuehn, S.; Pflugmacher, S. Single and combined exposure to MC-LR and BMAA confirm suitability of *Aegagropila linnaei* for use in Green Liver Systems®—A case study with cyanobacterial toxins. *Aquat. Toxicol.* **2015**, *165*, 101–108. [CrossRef]
19. Nimptsch, J.; Wiegand, C.; Pflugmacher, S. Cyanobacterial toxin elimination via bioaccumulation of MC-LR in aquatic macrophytes: An application of the "Green Liver Concept". *Environ. Sci. Technol.* **2008**, *42*, 8552–8557. [CrossRef]
20. Redding, T.; Todd, S.; Midlen, A. The treatment of aquaculture wastewaters—A botanical approach. *J. Environ. Manag.* **1997**, *50*, 283–299. [CrossRef]
21. Bennicelli, R.; Stepniewska, Z.; Banach, A.; Szajnocha, K.; Ostrowski, J. The ability of *Azolla caroliniana* to remove heavy metals (Hg(II), Cr(III), Cr (IV)) from municipal waste water. *Chemosphere* **2004**, *55*, 141–146. [CrossRef]
22. Singh, P.K.; Singh, D.P.; Singh, R.P. Growth, acetylene reduction activity, nitrate uptake and nitrate reductase activity of *Azolla caroliniana* and *Azolla pinnata* at varying nitrate levels. *Biochem. Physiol. Pflanz.* **1992**, *188*, 121–127. [CrossRef]
23. Forni, C.; Chen, J.; Tancioni, L.; Caiola, M.G. Evaluation of the fern Azolla for growth, nitrogen and phosphorus removal from wastewater. *Water Res.* **2001**, *35*, 1592–1598. [CrossRef]
24. Toledo, J.J.; Penha, J. Performance of *Azolla caroliniana* Wild. and *Salvinia auriculata* Aubl. on fish farming effluent. *Braz. J. Biol.* **2011**, *71*, 37–45. [CrossRef] [PubMed]
25. Laurinavichene, T.V.; Yakunin, A.F.; Gototov, I.N. Effect of temperature and photoperiod duration on growth and nitrogen fixation in Azolla. *Fiziol. Rast. (Moscow)* **1990**, *37*, 457–461.
26. Liu, X.; Chen, M.; Bian, Z.; Liu, C. Studies on urine treatment by biological purification using *Azolla* and UV photocatalytic oxidation. *Adv. Space Res.* **2008**, *41*, 783–786. [CrossRef]
27. Watanabe, I.; Berja, N.S. The growth of four species of Azolla as affected by temperature. *Aquat. Bot.* **1983**, *15*, 175–185. [CrossRef]
28. Henry-Silva, G.G.; Camargo, A.F.M. Efficiency of aquatic macrophytes to treat Nile Tilapia pond effluents. *Sci. Agric. (Piracicaba Braz.)* **2006**, *63*, 433–438. [CrossRef]

29. Rai, D.N.; Datta Mushi, J. The influence of thick floating vegetation (Water hyacinth: *Eichhornia crassipes*) on the physicochemical environment of a freshwater wetland. *Hydrobiologia* **1979**, *62*, 65–69. [CrossRef]
30. Greco, M.K.B.; Freitas, J.R. On two methods to estimate the reproduction of *Eicchornia crassipes* in the eutrophic Pampulha reservoir (MG/Brazil). *Braz. J. Biol.* **2002**, *62*, 463–471. [CrossRef]
31. Zimmles, Y.; Kirzhner, F.; Malkovskaja, A. Application of *Eichhornia crassipes* and *Pistia stratiotes* for treatment of urban sewage in Israel. *J. Environ. Manag.* **2006**, *81*, 420–428. [CrossRef]
32. Ebel, M.; Evangelou, M.W.H.; Schaeffer, A. Cyanide phytoremediation by water hyacinths (*Eichhornia crassipes*). *Chemosphere* **2006**, *66*, 816–823. [CrossRef]
33. Caldelas, C.; Santiago, I.T.; Araus, J.L.; Bort, J.; Febrero, A. Physiological response of *Eichhornia crassipes* (Mart.) Solms to the combined exposure to excess nutrients and Hg. *Braz. J. Plant Physiol.* **2009**, *21*, 1–12. [CrossRef]
34. NAS—National Academy of Sciences. *Making Aquatic Weeds Useful: Some Perspectives for Developing Countries*, 4th ed.; NAS: Washington, DC, USA, 1981; 174p.
35. Reddy, K.R.; Agami, M.; Tucker, J.C. Influence of nitrogen supply rates on growth and nutrient storage by water hyacinth (*Eichhornia crassipes*) plants. *Aquat. Bot.* **1989**, *36*, 33–43. [CrossRef]
36. Reddy, K.R.; Agami, M.; Tucker, J.C. Influence of phosphorus on growth and nutrient storage by water hyacinth (*Eichhornia crassipes*) plants. *Aquat. Bot.* **1990**, *37*, 355–365. [CrossRef]
37. Center, T.D.; Van, T.K. Alteration of water hyacinth (*Eichhornia crassipes* (Mart) Solms) leaf dynamics and phytochemistry by insect damage and plant density. *Aquat. Bot.* **1989**, *35*, 181–195. [CrossRef]
38. Lesage, E.; Mundia, C.; Rousseau, D.P.L. Sorption of Co, Cu, Ni and Zn from industrial effluents by the submerged aquatic macrophyte *Myriophyllum spicatum*. *Ecol. Eng.* **2007**, *30*, 320–325. [CrossRef]
39. Bakar, A.F.A.; Yusoff, I.; Fatt, N.T.; Othman, F.; Ashraf, M.A. Arsenic, zinc and aluminium removal from gold mine wastewater effluent and accumulation by submerged aquatic plants (*Caboma piauhyensis*, *Egeria densa* and *Hydrilla verticillata*). *BioMed Res. Int.* **2013**, 890803. [CrossRef]
40. Gujarathi, N.P.; Haney, B.J.; Linden, J.C. Phytoremediation potential of *Myriophyllum aquaticum* and *Pistia stratiotes* to modify antibiotic growth promoters, tetracycline and oxytetracycline in aqueous wastewater systems. *Int. J. Phytoremediat.* **2005**, *7*, 99–112. [CrossRef] [PubMed]
41. Bouaïcha, N.; Miles, C.O.; Beach, D.G.; Labidi, Z.; Djabri, A.; Benayache, N.Y.; Nguyen-Quang, T. Structural Diversity, Characterization and Toxicology of Microcystins. *Toxins* **2019**, *11*, 714. [CrossRef]
42. Garfí, M.; Pedescoll, A.; Bécares, E.; Hijosa-Valsero, M.; Sidrach-Cardona, R.; García, J. Effect of climatic conditions, season and wastewater quality on contaminant removal efficiency of two experimental constructed wetlands in different regions of Spain. *Sci. Total Environ.* **2012**, *437*, 61–67. [CrossRef]
43. Wu, L.; Pflugmacher, S.; Yang, A.; Esterhuizen-Londt, M. Photocatalytic degradation of microcystin-LR by modified high-energy {001} titanium dioxide: Kinetics and mechanism study of HF8. *SDRP J. Earth Sci. Environ. Stud.* **2018**, *3*, 408–416.

MDPI
St. Alban-Anlage 66
4052 Basel
Switzerland
Tel. +41 61 683 77 34
Fax +41 61 302 89 18
www.mdpi.com

Toxins Editorial Office
E-mail: toxins@mdpi.com
www.mdpi.com/journal/toxins

www.ingramcontent.com/pod-product-compliance
Lightning Source LLC
LaVergne TN
LVHW072331090526
838202LV00019B/2398
* 9 7 8 3 0 3 6 5 2 1 0 8 4 *